THE RUST FUNGI
of Cereals, Grasses and Bamboos

THE RUST FUNGI
of Cereals, Grasses and Bamboos

by GEORGE BAKER CUMMINS
Professor of Botany, Purdue University

Illustrations by the author

SPRINGER-VERLAG   NEW YORK · HEIDELBERG · BERLIN

1971

©1971 by Springer-Verlag New York Inc.
Library of Congress Catalog Card Number 75-147257

Printed in the United States of America

ISBN 0-387-05336-0 Springer-Verlag New York · Heidelberg · Berlin
ISBN 3-540-05336-0 Springer-Verlag Berlin · Heidelberg · New York

To the Memory

of

JOSEPH CHARLES ARTHUR
1850--1942

Preface

In the preparation of this descriptive manual of the rust fungi of the grasses of the world the principal goal was to produce a system by which these important pathogens might be recognized on the basis of their moropphology, without dependence on the identity of the host plant. This is an Utopian goal and, being Utopian, has doubtless not been attained. But it is better to have tried and partially failed than not to have tried at all.

The first attempt to revise the classification on a new basis utilized the rust fungi of the tribe Andropogoneae. A "Group System" was initiated (Uredineana 4:5-89. 1953) based on the uredinial stage. The attempt was satisfactory at the time, but was not adaptable when all grass rust fungi were considered. Consequently, an expanded system was employed when I attempted a summarization of all grass rust fungi. The expanded scheme (Plant Disease Reporter Supplement 237:1-52. 1956) of 9 Groups proved to be a most helpful organizational system and is used here (see explanations, p. xi) in <u>Puccinia</u>, <u>Uromyces</u>, and <u>Uredo</u>. The system is useful and does aggregate generally similar species, rather than segregating them as in a host-based arrangement. The characters used, i.e. presence or absence of paraphyses, arrangement of germ pores, and echinulate or verrucose spore surface, are subject to minimal intergradations. There are other morphological criteria that might be used to aggregate the species. Hopefully, he who attempts a successor to this manual may find that the system used here is useful as a point of departure to something better.

The listing of species of hosts is limited largely to the more poorly known fungi. A complete list of all host species of <u>Puccinia graminis</u> or <u>Puccinia recondita</u>, even if I were competent to provide it, seems to me to be relatively unimportant and, at best, of only transient and largely regional significance. The listing of host or fungus species by all countries or subdivisions of countries is not considered to be coincident with the purpose of this manual. Regional lists, and these are numerous, provide such information satisfactorily.

The concept of specific limits employed is conservative. This has the advantage of aggregating generally similar fungi into species; it has the disadvantage of grouping fungi which may later prove separable. In the interim one tends to lose sight of the fact that cited synonyms may not be as well understood as the reduction to synonymy implies. No one can guess the moropphology of the aecial stages of the approximately 65% of species whose aecia are unknown. About the only consistent factor in the concept, including mine, of species is inconsistency. For example, under <u>Puccinia graminis</u>, 18 perfect state names are listed in synonymy, for <u>Puccinia hordei</u>, 26, and for <u>Puccinia recondita</u>, 51. Yet <u>Puccinia graminis</u> has as great a range of variability as either of the others. The multiplication of "species" based on aecial host-telial host combinations accounts for most of the binomials under <u>Puccinia recondita</u>, and seems to be a particularly pernicious practice. <u>Puccinia graminis</u> probably has been saved from a similar fate because it can claim only <u>Berberis</u> as a haplont host.

The nomenclature of the grasses is in accord with present useage, insofar as I could determine. Synonyms are cited only when they involve the hosts of type specimens. Generally, one can only assume that the identity of the host is correct because the authenticity of the identification is seldom indicated. But there are notable exceptions, e.g. the Holway Collections were nearly always identified or verified by specialists. Misidentification of hosts is not uncommon and usually, because of the penchant to use the identity of the host plant as a "Character" of the fungus, this leads to a superfluous binomial. As examples, <u>Puccinia anthistiriae</u> Barcl. is <u>Puccinia graminis</u> Pers. and the grass an <u>Agropyron</u> and not an <u>Anthistiria</u>; the host of <u>Puccinia melanocephala</u> H. Syd. & P. Syd. is an <u>Erianthus</u> not an <u>Arundinaria</u>, and this results in reduction to synonymy of <u>Puccinia erianthi</u> Padwick and Khan; <u>Puccinia amianthina</u> H. Syd. & P. Syd. is <u>Puccinia deformata</u> Berk. & Curt. on <u>Olyra</u> and not <u>Bambusa</u>; and <u>Puccinia ischaemi</u> Diet. is <u>Puccinia zoysiae</u> Diet. on a <u>Zoysia</u> and not an <u>Ischaemum</u>. The collections of rust fungi on the bamboos are notorious for host notations that lack conviction or merely state "on a bamboo." Recent Japanese publications provide extensive lists of hosts according to current nomenclature of the Bambusoideae.

So many individuals and institutions have contributed to my studies that I shall not list all of them. But the assistance is hereby acknowledged and the source can be assumed in most cases from the standard herbarium abbreviation cited after the type. But because of the scope of the cooperation of some individuals and institutions it is my pleasure to acknowledge, by name: F.C. Deighton and the late G.R. Bisby, the Commonwealth Mycological Institute (IMI), Kew, Juan C. Lindquist, LaPlata (LPS), Naohide Hiratsuka, Tokyo, and the National Herbarium (PRE), Pretoria. For some 25 years, I have had free access to the National Fungus Collections (BPI), Beltsville, much of the time as Collaborator, and to the excellent John A. Stevenson Collection of books and reprints deposited there. Mr. Stevenson gave inestimable help with nomenclatural problems and constant encouragement, occasionally when more sorely needed than he knew, to my efforts. In recent years, C.R. Benjamin has been equally cooperative. During the past 15 years, the National Science Foundation has provided significant financial assistance to permit extensive field studies in the western United States and the northern half of Mexico. These grants have been acknowledged in the pertinent journal papers. Locally, Purdue University, through the Agricultural Experiment Station and the Department of Botany and Plant Pathology have provided salary, research space, equipment and financial support, and, equally important, congenial and helpful colleagues. Last, but by no means least, is my debt to the late J.C. Arthur. The nature of the debt is difficult to define in its entirety. An obvious part was the rare privilege of assisting, from A to Z, in the preparation of his Manual of the Rusts in the United States and

Canada. This was a real "cram session" in Uredinology for a
mere graduate student and an early and thorough introduction
to the trials and tribulations of writing a book. The greater
debt is less easily delimited but is in the mystic realm of per-
sonal influence. Perhaps inspiration is the appropriate word.

Purdue University   George B. Cummins
Lafayette, Indiana
August 1970

Scope

This descriptive manual provides a classification of the rust fungi (Uredinales) of the grasses of the world. The term "grasses" as used here includes the Gramineae, both cultivated and native. The bamboos are treated as a subfamily of the Gramineae, although they are sometimes accorded family rank.

The Keys to Fungus Species by Genera of Grasses

Following Fischer (Manual of the North American Smut Fungi, 1953) I have adopted the very useful system of keying the species of rust fungi by the genera of grasses parasitized. This is done solely, and somewhat reluctantly, in the interest of utility because the emphasis in this manual is on the fungus and not its hosts.

The Group System

The system used to group and key the species of Puccinia, Uromyces, and to group the species of the form genus Uredo is as follows:

Group I: Uredinia with paraphyses; urediniospores echinulate; germ pores equatorial or rarely basal.

Group II: Uredinia with paraphyses; urediniospores echinulate; germ pores scattered.

Group III: Uredinia with paraphyses; urediniospores verrucose; germ pores equatorial. No species known.

Group IV: Uredinia with paraphyses; urediniospores verrucose; germ pores scattered. One species of Uredo.

Group V: Uredinia without paraphyses; urediniospores echinulate; germ pores equatorial or rarely basal.

Group VI: Uredinia without paraphyses, urediniospores echinulate; germ pores scattered.

Group VII: Uredinia without paraphyses, urediniospores verrucose; germ pores equatorial.

Group VIII: Uredinia without paraphyses; urediniospores verrucose; germ pores scattered.

Group IX: Uredinia not produced (opsis-forms), or unknown; species of uncertain affinities.

Illustrations

The figures of teliospores were either traced from photomicrographs or drawn using a camera lucida, in either case from type specimens except those listed under Sources of Illustrations (p.xiii). Urediniospores usually were drawn from type specimens but the exceptions are not listed. The originals were drawn at a magnification of 800 diameters and reduced in reproduction to 640 diameters, except for Figure 1 which is reduced to 320 diameters. References are given to publication of photomicrographs of teliospores of type specimens.

Measurements

Spore sizes are mostly given as (30-)35-45(-48) x

(17-)20-23(-25)μ. The figures 35-45 x 20-23 would describe
the majority of the population, or what I consider to be the
typical size. Figures enclosed in parentheses are outside
of the typical size but not uncommon. Unusual measurement,
or what might be considered to be rare or freak sizes, are
not given.

Citation of Types
   Type specimens are cited only for the accepted name although
type specimens of one kind or another have been seen for
most synonyms. The repository of the cited types is given by
standard abbreviations. If no type was seen, this is stated.

Aecial Descriptions and Hosts
   Most descriptions of aecial stages are adapted from other
sources. Only minimal original study was devoted to the aecia
and their hosts are indexed only by genera and families.

Proof of Life Cycles
   Only the first experimental proof of life cycles is cited.
There seemed to be no reason to repeat the more complete
references cited by Arthur (Manual of the Rusts in United States
and Canada, 1934) and Gäumann (Die Rostpilze Mitteleuropas,
1959).

With the exception of the figures listed below, teliospores, but not always urediniospores, were taken from holotype, isotype, lectotype, or neotype specimens.

Figure Number	Source
3	Cheo 2904 on Andropogon sp.; China
6	Wiehe 134 on Pennisetum polystachyon; Nyasaland
15	Cummins 62-124 on Setaria macrostachya; Mexico
19	Sydow Ured. 1263 on Bambusae; Japan
25	Sydow Fungi exot. exs. 214 on Microstegium nudum; Japan
28	Deighton 2334 on Loudetia arundinacea; Sierra Leone
29	Clemens on Hemarthria uncinata; Australia
37	Hiratsuka Oct. 1925 on Miscanthus sinensis; Japan
38	Kern Sept. 1937 on Andropogon scoparius U. S. A.
40	Deighton 152 on Imperata cylindrica; Sierra Leone
50	Reliq. Holway. 118 on Olyra micrantha; Brazil
53	Hansford 2988 on Cymbopogon martinii; Uganda
56	Yasuda Mar. 1920 on Phyllostachys aurea; Japan
66	Sydow Ured. 1314 on Sasa tesselata; Japan
67	Hara, Dec. 1912 on Sasa purpurascens; Japan
68	Hara Apr. 1912 on Sasa purpurascens; Japan
71	Mayor July 1918 on Festuca altissima; Switzerland
72	Padwick 722 on Brachypodium sylvaticum; India
87	Type of Puccinia chloridis-incompletae; India
93	Hansford 2200 on Hyparrhenia pilgeriana; Uganda
94	Kellerman 6074 on Sorghum vulgare; Guatemala
97	Jaap Fungi sel.exs. 40 on Phragmites communis; Germany
104	Cummins 62-339 on Aristida arizonica; U. S. A.
110	Stakman 143 on Zea mays; Peru
118	Thuemen Mycoth. Univ. 1337 on Sesleria coerulea; Austria
119	Mains 3772 on Olyra latifolia; British Honduras
127	Nattrass 438 on Imperata cylindrica; Cyprus
132	Morimoto Sept. 1954 on Lophatherum gracile; Japan
141	Type of P. penniseti on Pennisetum typhoides; Tanganyika
143	From Korbonskaia Fig. 1
148	Reliq. Holw. 35 on Piptochaetium stipoides; Chile
168	Arndt Feb. 1935 on Arundinaria tecta; U. S. A.
169	Sydow Ured. 1313 on Pleioblastus simoni; Japan
170	Sydow Ured. 1172 on Sporobolus cryptandrus; U. S. A.
171	Jaap Fungi sel. exs. 138 on Molinia coerulea; Germany
173	Sydow Ured. 75 on Phragmites communis; Germany
175	Sydow Ured. 1270 on Phragmites communis; Switzerland
177	Sydow Ured. 1617 on Phragmites gigantea; Persia
181	Urban Aug. 1960 on Festuca ovina; Czechoslovakia
204	Sydow Ured. 436 on Phalaris arundinacea; England
209	Tobinaga on Agropyron ciliare; Japan
211	Vestergren Microm. rar. sel. 1383 on Koeleria cristata; Switzerland

CONTENTS

Key to the Genera of Rust Fungi

1. Teliospores sessile (2)
1. Teliospores pedicellate (4)
2. Teliospores irregularly arranged in
                        subepidermal crusts........2. Phakopsora
2. Teliospores catenulate (3)
3. Teliospores in subepidermal crusts............3. Physopella
3. Teliospores in erumpent, flabellate crusts.....1. Dasturella
4. Teliospores with 3 germ pores per cell......4. Stereostratum
4. Teliospores with 1 germ pore per cell (5)
5. Teliospores 1-celled only........................6. Uromyces
5. Teliospores with 2 or more cells; 1-celled
                spores may be intermixed..........5. Puccinia

Key to Species by Genera of Hosts
                (Uredo excluded)

Achnatherum see Stipa

Aegilops (Festucoideae:Triticeae)
1. Telia exposed; urediniospore pores
                        equatorial........98. Puccinia graminis
1. Telia covered; urediniospore pores
                        scattered (2)
2. Uredinia in chlorotic streaks.......58. Puccinia striiformis
2. Uredinia not in chlorotic streaks (3)
3. Urediniospore wall yellowish...........186. Puccinia hordei
3. Urediniospore wall brownish.........187. Puccinia recondita

Aegopogon (Eragrostoideae:Lappagineae)
1. Teliospores 2-celled..............196. Puccinia aegopogonis
1. Teliospores 1-celled...............53. Uromyces aegopogonis

Aeluropus (Festucoideae:Festuceae)
1. Teliospores 2-celled..............264. Puccinia aeluropodis
1. Teliospores 1-celled......58. Uromyces aeluropodis-repentis

Agropyron (Festucoideae:Triticeae)
1. Teliospores 1-celled................24. Uromyces fragilipes
1. Teliospore more than 1-celled (2)
2. Teliospores 2-celled (4)
2. Teliopsores mostly 3-or 4-celled (3)
3. Uredinia aparaphysate...........163. Puccinia agropyricola
3. Uredinia with paraphyses.............57. Puccinia naumovii
4. Teliospores with apical digitations...54. Puccinia coronata
4. Teliospores without digitations (5)

1

```
5.  Uredinia in chlorotic streaks.......58. Puccinia striiformis
5.  Uredinia without such streaks (6)
6.  Uredinia with paraphyses (7)
6.  Uredinia aparaphysate (8)
7.  Paraphyses clavate-capitate.........63. Puccinia brachypodii
7.  Paraphyses capitate.................59. Puccinia montanensis
8.  Telia covered.......................187. Puccinia recondita
8.  Telia exposed (9)
9.  Urediniospore pores equatorial.........98. Puccinia graminis
9.  Urediniospore pores scattered (10)
10. Teliospores striate..............168. Puccinia pattersoniana
10. Teliospores smooth...........189. Puccinia agropyri-ciliaris
```

Agrostis (Festucoideae:Aveneae)

```
1.  Teliospores 1-celled (2)
1.  Teliospores 2-celled (3)
2.  Telia firmly covered, spores
                        attached..............32. Uromyces dactylidis
2.  Telia loosely covered, spores
                        loose................24. Uromyces fragilipes
3.  Teliospores with apical digitations (4)
3.  Teliospores without digitations (5)
4.  Uredinia paraphysate...................54. Puccinia coronata
4.  Uredinia aparaphysate............161. Puccinia praegracilis
5.  Uredinia in chlorotic streaks.......58. Puccinia striiformis
5.  Uredinia not in such streaks (6)
6.  Uredinia paraphysate (7)
6.  Uredinia aparaphysate (8)
7.  Paraphyses thick-walled.............63. Puccinia brachypodii
7.  Paraphyses thin-walled, capitate........60. Puccinia pygmaea
8.  Urediniospore pores scattered (9)
8.  Urediniospore pores equatorial........98. Puccinia graminis
9.  Telia covered (10)
9.  Telia exposed (11)
10. Urediniospore wall nearly colorless.....185. Puccinia poarum
10. Urediniospore wall brownish..........187. Puccinia recondita
11. Teliospore wall pale golden............239. Puccinia moyanoi
11. Teliospore wall chestnut-
                        brown.........195. Puccinia agrostidicola
```

Aira (Festucoideae:Aveneae)

```
1.  Uredinia in chlorotic streaks.......58. Puccinia striiformis
1.  Uredinia not in such streaks...........98. Puccinia graminis
```

Alopecurus (Festucoideae:Phalarideae)

```
1.  Teliospores 1-celled; telia covered..32. Uromyces dactylidis
1.  Teliospores 2-celled (2)
2.  Telia exposed                        98. Puccinia graminis
2.  Telia covered (3)
3.  Teliospores with apical digitations....54. Puccinia coronata
3.  Teliospore without digitations (4)
4.  Uredinia in chlorotic streaks.......58. Puccinia striiformis
4.  Uredinia not in such streaks (5)
5.  Uredinia with thick-walled
                        paraphyses............63. Puccinia brachypodii
5.  Uredinia aparaphysate...............187. Puccinia recondita
```

2

<u>Ammophila</u> (Festucoideae:Aveneae)
1. Telia exposed; germ pores equatorial...98. <u>Puccinia graminis</u>
1. Telia covered; germ pores scattered (2)
2. Teliospores with apical digitations....54. <u>Puccinia coronata</u>
2. Teliospores without digitations (3)
3. Uredinia paraphysate....................60. <u>Puccinia pygmaea</u>
3. Uredinia aparaphysate (4)
4. Teliospores 2-celled.................174. <u>Puccinia ammophilae</u>
4. Teliospores mostly 3-celled..............164. <u>Puccinia elymi</u>

<u>Amphibromus</u> (Festucoideae:Aveneae)
1. Telia exposed; urediniospore pores
                      equatorial.........98. <u>Puccinia graminis</u>

<u>Amphilophis</u> see <u>Bothriochloa</u>

<u>Andropogon</u> (Andropogonoideae:Andropogoneae)
1. Uredinia with paraphyses (2)
1. Uredinia aparaphysate (8)
2. Urediniospore pores equatorial (3)
2. Urediniospore pores scattered (7)
3. Teliospore pedicels less than 25µ long (4)
3. Teliospore pedicels exceeding 25µ (5)
4. Teliospores mostly 28-35µ long.......14. <u>Puccinia microspora</u>
4. Teliospores mostly 36-50µ long.......21. <u>Puccinia posadensis</u>
5. Teliospore pedicels 75µ or less
                      long.....37. <u>Puccinia nakanishikii</u>
5. Teliospore pedicels 100µ or more long (6)
6. Teliospores mostly 35-41µ long.........50. <u>Puccinia duthiae</u>
6. Teliospores mostly 40-56µ
                      long........46. <u>Puccinia andropogonicola</u>
7. Teliospores pedicellate,
             2-celled...............71. <u>Puccinia eritraeensis</u>
7. Teliospores sessile, in crusts......2. <u>Phakopsora incompleta</u>
8. Teliospores 1-celled (9)
8. Teliospores 2-celled (10)
9. Urediniospores echinulate..............38. <u>Uromyces clignyi</u>
9. Urediniospores verrucose...........59. <u>Uromyces andropogonis</u>
10. Urediniospore pores equatorial (11)
10. Urediniospore pores scattered (14)
11. Urediniospore verrucose.............255. <u>Puccinia ellisiana</u>
11. Urediniospores echinulate (12)
12. Urediniospore wall 5-8µ apically........104. <u>Puccinia eucomi</u>
12. Urediniospore wall uniform (13)
13. Urediniospore pores 2.............124. <u>Puccinia erianthicola</u>
13. Urediniospore pores mostly 4..........122. <u>Puccinia tripsaci</u>
14. Urediniospore wall brown..........230. <u>Puccinia andropogonis</u>
14. Urediniospore wall colorless (15)
15. Urediniospore lumen strongly
             stellate..........197. <u>Puccinia versicolor</u>
15. Urediniospore lumen weakly or not
             stellate............200. <u>Puccinia agrophila</u>

<u>Aneurolepidium</u> see <u>Elymus</u>

<u>Anthistiria</u> see <u>Themeda</u>

Anthephora (Panicoideae:Anthephoreae)
1. Uredinia paraphysate, spores
                      echinulate.........4. Puccinia chaseana
1. Uredinia aparaphysate, spores
                      verrucose.......248. Puccinia anthephorae

Anthoxanthum (Festucoideae:Phalarideae)
1. Telia exposed; germ pores equatorial...98. Puccinia graminis
1. Telia covered; germ pores scattered (2)
2. Teliospores with apical digitations....54. Puccinia coronata
2. Teliospores without digitations (3)
3. Uredinia aparaphysate...............187. Puccinia recondita
3. Uredinia paraphysate................63. Puccinia brachypodii

Apera (Festucoideae:Aveneae)
1. Telia exposed; germ pores equatorial...98. Puccinia graminis
1. Telia covered; germ pores scattered (2)
2. Teliospores with apical digitations....54. Puccinia coronata
2. Teliospores without digitations.....63. Puccinia brachypodii

Apluda (Andropogonoideae:Andropogoneae)
1. Teliospores 2-celled...................32. Puccinia apludae
1. Teliospores 1-celled (2)
2. Urediniospore pores equatorial......15. Uromyces schoenanthi
2. Urediniospore pores scattered..........61. Uromyces inayati

Arctagrostis (Festucoideae:Festuceae)
1. Telia covered, uredinia paraphysate.63. Puccinia brachypodii

Aristida (Eragrostoideae:Aristideae)
1. Teliospores 1-celled (7)
1. Teliospores 2-celled (2)
2. Uredinia paraphysate (3)
2. Uredinia aparaphysate (5)
3. Urediniospore pores equatorial.........47. Puccinia sonorica
3. Urediniospore pores scattered
4. Teliospores globoid or nearly so........76. Puccinia eylesii
4. Teliospores ellipsoid or broadly so.......83. Puccinia unica
5. Urediniospores echinulate..............98. Puccinia graminis
5. Urediniospores verrucose (6)
6. Urediniospores pores equatorial......263. Puccinia aristidae
6. Urediniospore pores scattered...........275. Puccinia tarri
7. Urediniospores echinulate..............4. Uromyces aristidae
7. Urediniospores verrucose.............57. Uromyces seditiosus

Arrhenatherum (Festucoideae:Aveneae)
1. Telia with apical digitations..........54. Puccinia coronata
1. Telia without such digitations (2)
2. Telia exposed; germ pores equatorial...98. Puccinia graminis
2. Telia covered; germ pores scattered (3)
3. Uredinia in chlorotic streaks.......58. Puccinia striiformis
3. Uredinia not in such streaks (4)
4. Uredinia with thick-walled
                      paraphyses............63. Puccinia brachypodii
4. Uredinia aparaphysate (5)
5. Urediniospore wall pale yellowish.......186. Puccinia hordei
5. Urediniospore wall brownish.........187. Puccinia recondita

<u>Arthraxon</u> (Andropogonoideae:Andropogoneae)
1. Uredinia paraphysate,
         pores equatorial.......18. <u>Puccinia arthraxonis-ciliaris</u>
1. Uredinia aparaphysate, pores
                        scattered......199. <u>Puccinia arthraxonis</u>
<u>Arundinaria</u> (Bambusoideae)
1. Teliospores mostly 38-65μ long....148. <u>Puccinia arundinariae</u>
1. Teliospores mostly 20-28μ long......125. <u>Puccinia bambusarum</u>

<u>Arundinellae</u> (Festucoideae:Arundinelleae)
1. Teliospores with apical digitations....54. <u>Puccinia coronata</u>
1. Teliospores without digitations (2)
2. Uredinia paraphysate.......16. <u>Puccinia arundinellae-setosae</u>
2. Uredinia aparaphysate (3)
3. Urediniospores pores equatorial...145. <u>Puccinia arundinellae</u>
3. Urediniospore pores
         scattered..........201. <u>Puccinia arundinellae-anomalae</u>

<u>Arundo</u> (Festucoideae:Festuceae)
1. Uredinia paraphysate (2)
1. Uredinia aparaphysate (3)
2. Urediniospore pores
            equatorial.........24. <u>Puccinia arundinis-donacis</u>
2. Urediniospore pores scattered........75. <u>Puccinia magnusiana</u>
3. Teliospores wall 2.5-3μ at sides,
         6-8μ apically....................155. <u>Puccinia torosa</u>
3. Teliospores wall mostly 5-7μ at sides,
         10-12μ apically.................156. <u>Puccinia trabutii</u>

<u>Asperella</u> see <u>Hystrix</u>

<u>Astrebla</u> (Festucoideae:Festuceae)
1. Urediniospores echinulate, pores
                   scattered........13. <u>Uromyces tripogonicola</u>

<u>Avellinia</u> (Festucoideae:Aveneae)
1. Urediniospores echinulate, pores
                   scattered.............186. <u>Puccinia hordei</u>

<u>Avena</u> (Festucoideae:Aveneae)
1. Telia exposed; germ pores equatorial...98. <u>Puccinia graminis</u>
1. Telia covered; germ pores scattered (2)
2. Teliospores with apical digitations....54. <u>Puccinia coronata</u>
2. Teliospores without digitations (3)
3. Uredinia in chlorotic streaks.......58. <u>Puccinia striiformis</u>
3. Uredinia without such streaks (4)
4. Urediniospore wall pale yellowish......186. <u>Puccinia hordei</u>
4. Urediniospore wall brownish..........187. <u>Puccinia recondita</u>

<u>Avenochloa</u> (Festucoideae:Aveneae) also see <u>Helictotrichon</u>
1. Teliospores with apical digitations....54. <u>Puccinia coronata</u>
1. Teliospores without digitations (2)
2. Urediniospore pores equatorial.........98. <u>Puccinia graminis</u>
2. Teliospore pores scattered (3)
3. Teliospore wall echinulate-verrucose, telia exposed (4)
3. Teliospore wall smooth, telia covered (5)
4. Teliospores 42-60μ long; pores
         mostly 10-12.....................170. <u>Puccinia pratensis</u>

5

4. Teliospores 37-48µ long; pores
                    mostly 6 or 7......171. Puccinia bromoides
5. Teliospores mostly 40-60 x
                    17-22µ.........173. Puccinia helictotrichi
5. Teliospores mostly 37-62 x
                    12-19µ............277. Puccinia lavroviana

Axonopus (Panicoideae:Paniceae)
1. Teliospores sessile in chains.......12. Physopella compressa
1. Teliospore pedicellate..................109. Puccinia levis

Bambusa (Bambusoideae)
1. Teliospores pedicellate (2)
1. Teliospores sessile, in erumpent crusts (4)
2. Teliospores with 3 pores per
                    cell..........1. Stereostratum corticioides
2. Teliospores with 1 pore per cell (3)
3. Urediniospores mostly 27-37µ
                    long........43. Puccinia xanthosperma
3. Urediniospores 18-20µ diam.........77. Puccinia kwanhsiensis
4. Telia mostly 150-200µ thick..............2. Dasturella divina
4. Telia mostly less than 100µ thick....1. Dasturella bambusina

Bambusoideae (undetermined)
1. Teliospore wall unilaterally thickened,
                    verrucose.....288. Puccinia tenella

Beckeropsis (Panicoideae:Paniceae)
1. Teliospores 2-celled; pores
                    equatorial.........121. Puccinia substriata
1.. Teliospores 1-celled; pores
                    scattered...........28. Uromyces pegleriae

Beckmannia (Festucoideae:Beckmanniae)
1. Teliospores 1-celled; pores
                    scattered.........30. Uromyces beckmanniae
1. Teliospores 2-celled (2)
2. Urediniospore pores scattered (3)
2. Urediniospore pores equatorial.........98. Puccinia graminis
3. Teliospores with apical digitations....54. Puccinia coronata
3. Teliospores without digitations.....58. Puccinia striiformis

Bewsia (Eragrostoideae:Eragrosteae)
1. Teliospores mostly 34-39 x 24-29µ......283. Puccinia bewsiae

Boissiera (Festucoideae:Festuceae)
1. Teliospores 1-celled; pores
                    scattered.............5. Uromyces turcomanicum
1. Teliospores 2-celled; pores scattered (2)
2. Uredinia in chlorotic streaks;
                    telia seriate...........58. Puccinia striiformis
2. Uredinia not in streaks; telia not seriate (3)
3. Urediniospore wall pale yellowish......186. Puccinia hordei
3. Urediniospore wall brownish..........187. Puccinia recondita

Bothriochloa (Andropogonoideae:Andropogoneae)
1. Teliospores 1-celled; urediniospores
                    echinulate...........38. Uromyces clignyi

6

1. Teliospores 2-celled (2)
2. Urediniospores verrucose (3)
2. Urediniospores echinulate (5)
3. Teliospores pedicels
        thick-walled.........260. Puccinia pseudocesatii
3. Teliospore pedicels thin-walled (4)
4. One-celled teliospores common,
        amphispores none.............247. Puccinia infuscans
4. One-celled teliospores rare, amphi-
        spores usually common.......251. Puccinia cesatii
5. Uredinia paraphysate (7)
5. Uredinia aparaphysate.................98. Puccinia graminis
6. Urediniospore pores equatorial (7)
6. Urediniospore pores scattered (8)
7. Teliospore pedicels brown,
        thick-walled.....37. Puccinia nakanishikii
7. Teliospore pedicels yellowish,
        thin-walled...........50. Puccinia duthiae
8. Lumen of urediniospore stellate.....197. Puccinia versicolor
8. Lumen not stellate (9)
9. Teliospores with apical digitations....54. Puccinia coronata
9. Teliospores without digitations.....70. Puccinia kenmorensis

Bouteloua (Eragrostoideae:Chlorideae)
1. Urediniospores verrucose, pores
        scattered..........274. Puccinia opuntiae
1. Urediniospores echinulate (2)
2. Urediniospore pores equatorial.......142. Puccinia cacabata
2. Urediniospore pores scattered (3)
3. Apically thick-walled amphispores
        abundant...........166. Puccinia vexans
3. Amphispores not produced (4)
4. Teliospore pedicels thick-walled, not
        collapsing......219. Puccinia exasperans
4. Teliospore pedicels thin-walled, usually collapsing (5)
5. Teliospores mostly diorchidioid.....206. Puccinia boutelouae
5. Teliospores typically puccinioid (6)
6. Urediniospores mostly 18-23μ long....214. Puccinia chloridis
6. Urediniospores mostly 22-26μ long...221. Puccinia diplachnis

Brachiaria (Panicoideae:Paniceae)
1. Teliospores in sessile chains.........9. Physopella africana
1. Teliospores pedicellate (2)
2. Teliospores 1-celled..........11. Uromyces setariae-italicae
2. Teliospores 2-celled (3)
3. Teliospores typically puccinioid..205. Puccinia nyasalandica
3. Teliospores diorchidioid (4)
4. Urediniospore pores basal............25. Puccinia orientalis
4. Urediniospore pores equatorial...........109. Puccinia levis

Brachyelytrum (Festucoideae:Festuceae)
1. Teliospores with apical digitations....3. Uromyces halstedii

Brachypodium (Festucoideae:Festuceae)
1. Urediniospore pores equatorial.........98. Puccinia graminis
1. Urediniospore pores scattered (2)

2. Teliospores commonly 3- or 4-
                      celled...163. Puccinia agropyricola
2. Teliospores predominantly or only 2-celled (3)
3. Teliospores with apical digitations....54. Puccinia coronata
3. Teliospores without digitations (4)
4. Uredinia in chlorotic streaks.......58. Puccinia striiformis
4. Uredinia not in chlorotic streaks (5)
5. Uredinia aparaphysate................187. Puccinia recondita
5. Uredinia with paraphyses (6)
6. Paraphyses uniformly thin-walled.....79. Puccinia corteziana
6. Paraphysis wall thick (7)
7. Paraphysis wall uniformly
                      thick-walled........63. Puccinia brachypodii
7. Paraphysis wall abruptly thicker
                  at apex.....62. Puccinia brachypodii-phoenicoidis

Brachystachyum (Bambusoideae)
1. Teliospores narrowly
                  ellipsoid.......291. Puccinia brachystachyicola

Briza (Festucoideae:Festuceae)
1. Urediniospore pores equatorial; telia
                      exposed..............98. Puccinia graminis
1. Urediniospore pores scattered; telia covered (2)
2. Teliospores with apical digitations....54. Puccinia coronata
2. Teliospores without digitations (3)
3. Uredinia in chlorotic streaks.......58. Puccinia striiformis
3. Uredinia not in such streaks (4)
4. Teliospores 2-celled.................187. Puccinia recondita
4. Teliospores 1-celled.................32. Uromyces dactylidis

Bromus (Festucoideae:Festuceae)
1. Teliospores typically 1-celled (2)
1. Teliospores more than 1-celled (3)
2. Teliospores mostly 33-34 x 27-31µ.....176. Puccinia cryptica
2. Teliospores 19-28 x 14-23µ...........23. Uromyces brominus
3. Teliospores typically multicellular...162. Puccinia tomipara
3. Teliospores typically 2-celled (4)
4. Teliospores with apical digitations....54. Puccinia coronata
4. Teliospores without digitations (5)
5. Uredinia in chlorotic streaks.......58. Puccinia striiformis
5. Uredinia not in such streaks (6)
6. Urediniospore pores equatorial.........98. Puccinia graminis
6. Urediniospore pores scattered (7)
7. Uredinia with paraphyses (8)
7. Uredinia aparaphysate (10)
8. Telia exposed......................80. Puccinia decolorata
8. Telia covered (9)
9. Paraphysis wall thick throughout....63. Puccinia brachypodii
9. Paraphysis wall abruptly thicker
                  above....62. Puccinia brachypodii-phoenicoidis
10. Urediniospore wall pale yellow (11)
10. Urediniospore wall brownish (12)
11. Telia with abundant brown paraphyses....186. Puccinia hordei
11. Telia with few or no
                  paraphyses.........184. Puccinia tsinlingensis

12. Teliospores mostly 16-23μ wide.......187. Puccinia recondita
12. Teliospores mostly 20-30μ wide........176. Puccinia cryptica

Buchloë (Eragrostoideae:Chlorideae)
1. Telia exposed; germ pores scattered..190. Puccinia kansensis

Calamagrostis (Festucoideae:Aveneae)
1. Teliospores 1-celled............33. Uromyces calamagrostidis
1. Teliospores 2-celled (2)
2. Teliospores with apical digitations....54. Puccinia coronata
2. Teliospores without digitations (3)
3. Urediniospore pores equatorial........98. Puccinia graminis
3. Urediniospore pores scattered (4)
4. Uredinia in chlorotic streaks.......58. Puccinia striiformis
4. Uredinia not in such streaks (5)
5. Uredinia with paraphyses (6)
5. Uredinia aparaphysate (7)
6. Paraphyses clavate-capitate.........63. Puccinia brachypodii
6. Paraphyses capitate....................60. Puccinia pygmaea
7. Urediniospore wall nearly colorless.....185. Puccinia poarum
7. Urediniospore wall brownish.........187. Puccinia recondita

Calamovilfa (Festucoideae:Festuceae)
1. Urediniospore pores scattered........237. Puccinia amphigena
1. Urediniospore pores around the
                    hilum...........93. Puccinia sporoboli

Capillipedium (Andropogonoideae:Andropogoneae)
1. Uredinia paraphysate (2)
1. Uredinia aparaphysate (4)
2. Urediniospore pores scattered.......71. Puccinia eritreensis
2. Urediniospore pores equatorial (3)
3. Teliospores mostly 33-44μ long,
                pedicels brown......37. Puccinia nakanishikii
3. Teliospores mostly 29-36μ long,
                pedicels pale yellowish.........31. Puccinia pusilla
4. Urediniospores echinulate (5)
4. Urediniospores verrucose (6)
5. Urediniospore wall very unevenly
                        thick.......197. Puccinia versicolor
5. Urediniospore wall quite or nearly
                        uniform...........200. Puccinia agrophila
6. Teliospore pedicels thick-walled;
                no amphispores........249. Puccinia miyoshiana
6. Teliospore pedicel thin-walled,
        amphispores usually abundant.........251. Puccinia cesatii

Catabrosa (Festucoideae:Festuceae)
1. Urediniospore pores equatorial; telia
                        exposed.......98. Puccinia graminis
1. Urediniospore pores, scattered; telia covered (2)
2. Uredinia in chlorotic streaks.......58. Puccinia striiformis
2. Uredinia not in such streaks (3)
3. Teliospores with apical digitations....54. Puccinia coronata
3. Teliospores without digitations.....63. Puccinia brachypodii

9

Cathestecum (Eragrostoideae:Chlorideae)
1. Urediniospore pores scattered......206. Puccinia boutelouae
1. Urediniospore pores equatorial.......142. Puccinia cacabata

Cenchrus (Panicoideae:Paniceae)
1. Urediniospore pores equatorial.........89. Puccinia cenchri

Centotheca (Festucoideae:Festuceae)
1. Urediniospore pores equatorial......113. Puccinia lophatheri

Chaetii (Panicoideae:Paniceae)
1. Urediniospores echinulate, pores
                        scattered...........87. Puccinia chaetii

Chimonobambusa (Bambusoideae)
1. Teliospore with 3 pores per
                        cell........1. Stereostratum corticioides

Chloris (Eragrostoideae:Chlorideae)
1. Teliospores 1-celled (2)
1. Teliospores 2-celled (3)
2. Urediniospores verrucose............55. Uromyces archerianus
2. Urediniospores echinulate.............50. Uromyces kenyensis
3. Uredinia in chlorotic streaks.......58. Puccinia striiformis
3. Uredinia not in streaks (4)
4. Urediniospore pores equatorial........142. Puccinia cacabata
4. Urediniospore pores scattered (5)
5. Uredinia paraphysate..............65. Puccinia enteropogonis
5. Uredinia aparaphysate (6)
6. Urediniospore wall thickened
                apically...........202. Puccinia dietelii
6. Urediniospore wall uniformly thin....214. Puccinia chloridis

Chrysopogon (Andropogonoideae:Andropogoneae)
1. Urediniospore pores equatorial (2)
1. Urediniospore pores scattered (3)
2. Urediniospores mostly oblong-
                        ellipsoid.......98. Puccinia graminis
2. Urediniospores mostly obovoid......139. Puccinia kawandensis
3. Urediniospores verrucose........260. Puccinia pseudocesatii
3. Urediniospores echinulate (4)
4. Teliospores with apical digitations....54. Puccinia coronata
4. Teliospores without digitations (5)
5. Uredinia paraphysate..................72. Puccinia purpurea
5. Uredinia aparaphysate..............198. Puccinia chrysopogi

Cinna (Festucoideae:Festuceae)
1. Teliospores with apical digitations....54. Puccinia coronata
1. Teliospores without digitations (2)
2. Urediniospore pores equatorial........98. Puccinia graminis
2. Urediniospore pores scattered.......187. Puccinia recondita

Cleistogenes (Festucoideae:Arundineae)
1. Urediniospore wall 1-1.5µ thick; teliospore
            wall 3-5µ apically.....212. Puccinia diplachnicola
1. Urediniospore wall 2-3µ thick; teliospore wall
                        more than 5µ (2)
2. Teliospores mostly 36-43 x 24-27µ;
                        apex 5-8µ....213. Puccinia permixta

10

2. Teliospores mostly 30-40 x 21-24μ,
                    apex 7-10μ..........211. Puccinia australis

Coix (Andropogonoideae:Maydeae)
1. Urediniospores echinulate, sori
                paraphysate..............48. Puccinia operta

Coleanthus (Festucoideae:Festuceae)
1. Urediniospores echinulate, pores
                equatorial.........98. Puccinia graminis

Colpodium (Festucoideae:Festuceae)
1. Urediniospores echinulate, pores
                scattered........187. Puccinia recondita

Corynephorus (Eragrostoideae:Aveneae)
1. Urediniospores echinulate, pores
                equatorial.......98. Puccinia graminis

Cutandia (Festucoideae:Festuceae)
1. Telia covered, loculate with brown
                paraphyses..........186. Puccinia hordei

Cymbopogon (Andropogonoideae:Andropopgoneae)
1. Teliospores 1-celled (2)
1. Teliospores 2-celled (3)
2. Urediniospores pores equatorial.....15. Uromyces schoenanthi
2. Urediniospore pores scattered..........38. Uromyces clignyi
3. Urediniospores echinulate (4)
3. Urediniospores verrucose (7)
4. Uredinia paraphysate (5)
4. Uredinia aparaphysate..............197. Puccinia versicolor
5. Urediniospore pores equatorial (6)
5. Urediniospore pores scattered......71. Puccinia eritraeensis
6. Teliospores mostly 40-56μ long,
                apex 9-12μ......46. Puccinia andropogonicola
6. Teliospores mostly 33-44μ long,
                apex 4-8μ..........37. Puccinia nakanishikii
7. Teliospore pedicels thin-walled,
                to 80μ long..........250. Puccinia cymbopogonis
7. Teliospore pedicels thick-walled,
                to 130μ long...........261. Puccinia schoenanthi

Cynodon (Eragrostoideae:Chlorideae)
1. Urediniospores echinulate..............98. Puccinia graminis
1. Urediniospores verrucose...........256. Puccinia cynodontis

Cynosurus (Festucoideae:Festuceae)
1. Teliospores 1-celled................32. Uromyces dactylidis
1. Teliospores 2-celled, with apical
                digitations............54. Puccinia coronata
1. Teliospores 2-celled, without
                digitations........98. Puccinia graminis

Cypholepis (Eragrostoideae:Eragrosteae)
1. Teliospores 1-celled; urediniospores
                echinulate......54. Uromyces eragrostidis

Cyrtococcum (Panicoideae:Paniceae)
1. Teliospores in sessile chains......11. Physopella clemensiae
                                                        11

1. Teliospores pedicellate (2)
2. Teliospores 1-celled.........11. _Uromyces setariae-italicae_
2. Teliospores 2-celled (3)
3. Urediniospore pores at the hilum.....25. _Puccinia orientalis_
3. Urediniospore pores equatorial......115. _Puccinia taiwaniana_

Dactylis (Festucoideae:Festuceae)
1. Teliospores 1-celled; germ pores
                    scattered.......32. _Uromyces dactylidis_
1. Teliospores 2-celled (2)
2. Urediniospore pores equatorial........98. _Puccinia graminis_
2. Urediniospore pores scattered (3)
3. Teliospores with apical digitations....54. _Puccinia coronata_
3. Teliospores without digitations (4)
4. Uredinia in chlorotic streaks.......58. _Puccinia striiformis_
4. Uredinia not in such streaks........187. _Puccinia recondita_

Dactyloctenium (Eragrostoideae:Eragrosteae)
1. Teliospores 1-celled; germ pores
                    equatorial.......9. _Uromyces dactyloctenii_
1. Teliospores 2-celled; germ pores
                    scattered..........202. _Puccinia dietelii_

Danthonia (Festucoideae:Aveneae)
1. Teliospores 1-celled (2)
1. Teliospores 2-celled (3)
2. Urediniospores mostly 30µ long or
                    less......41. _Uromyces danthoniae_
2. Urediniospores mostly 35µ or
                    longer...........39. _Uromyces mcnabbii_
3. Urediniospores echinulate.............98. _Puccinia graminis_
3. Urediniospores verrucose...........262. _Puccinia danthoniae_

Danthoniopsis (Festucoideae:Arundinelleae)
1. Urediniospores echinulate, pores
                    equatorial..........8. _Puccinia angusii_

Dendrocalamus (Bambusoideae)
1. Telia erumpent, spores in sessile
                    chains..........2. _Dasturella divina_

Deschampsia (Festucoideae:Aveneae)
1. Teliospores 1-celled (2)
1. Teliospores 2-celled (3)
2. Telia loosely covered but spores
                    loose........24. _Uromyces fragilipes_
2. Telia tightly covered, spores
                    attached..27. _Uromyces airae-flexuosae_
3. Urediniospore pores equatorial........98. _Puccinia graminis_
3. Urediniospore pores scattered (4)
4. Teliospores with apical digitations (5)
4. Teliospores without digitations (6)
5. Uredinia with cylindrical paraphyses...54. _Puccinia coronata_
5. Uredinia aparaphysate.............161. _Puccinia praegracilis_
6. Uredinia paraphysate (7)
6. Uredinia aparaphysate (8)
7. Uredinia with clavate-capitate
                    paraphyses........63. _Puccinia brachypodii_

12

7. Uredinia with capitate paraphyses.......60. Puccinia pygmaea
8. Urediniospore wall pale yellowish.......186. Puccinia hordei
8. Urediniospore wall brownish.........187. Puccinia recondita

Desmazeria (Festucoideae:Festuceae)
1. Teliospores with apical digitations....54. Puccinia coronata

Desmostachya (Festucoideae:Festuceae)
1. Teliospores 2-celled; uredinia in
                chlorotic streaks....58. Puccinia striiformis
1. Teliospores 1-celled; uredinia
                not in streaks......54. Uromyces eragrostidis

Deyeuxia (Festucoideae:Aveneae)
1. Teliospores with apical digitations....54. Puccinia coronata
1. Teliospores without digitations (2)
2. Urediniospore pores equatorial........98. Puccinia graminis
2. Urediniospore pores scattered (3)
3. Telia exposed; urediniospore wall
                        2.5-5µ...232. Puccinia changtuensis
3. Telia covered; urediniospore wall less than 2.5µ (4)
4. Urediniospore wall pale yellowish.......186. Puccinia hordei
4. Urediniospore wall brownish.........187. Puccinia recondita

Diarrhena (Festucoideae:Festuceae)
1. Teliospores with apical digitations..84. Puccinia diarrhenae
1. Teliospores without digitations........98. Puccinia graminis

Dichanthium (Andropogonoideae:Andropogoneae)
1. Teliospores 1-celled; urediniospores
                        echinulate.......38. Uromyces clignyi
1. Teliospores 2-celled (2)
2. Urediniospores echinulate...............50. Puccinia duthiae
2. Urediniospores verrucose...............251. Puccinia cesatii

Dichelachne (Festucoideae:Stipeae)
1. Uredinia aparaphysate, pores
                equatorial............98. Puccinia graminis
1. Uredinia paraphysate, pores
                scattered.............61. Puccinia crinitae

Digitaria (Panicoideae:Paniceae)
1. Teliospores in sessile, subepidermal
                        crusts........6. Physopella digitariae
1. Teliospores pedicellate (2)
2. Teliospores 1-celled (3)
2. Teliospores 2-celled (4)
3. Urediniospores echinulate.............28. Uromyces pegleriae
3. Urediniospores verrucose...........252. Puccinia esclavensis
                                var. unicellula
4. Urediniospores verrucose (5)
4. Urediniospores echinulate (6)
5. Urediniospore pores equatorial.....252. Puccinia esclavensis
5. Urediniospore pores scattered.......271. Puccinia pseudoatra
6. Uredinia paraphysate (7)
6. Uredinia aparaphysate (8)
7. Urediniospore pores equatorial........3. Puccinia oahuensis

7. Urediniospore pores
                scattered..........66. Puccinia digitaria-velutinae
8. Teliospores diorchidioid, pedicel
                long...............109. Puccinia levis
8. Teliospores puccinioid, pedicel
                short.........121. Puccinia substriata

Dimeria (Andropogonoideae:Andropogoneae)
1. Teliospores sessile, in subepidermal
                crusts.....2. Phakopsora incompleta

Distichlis (Festucoideae:Festuceae)
1. Teliospores 2-celled.................273. Puccinia subnitens
1. Teliospores 1-celled.................62. Uromyces peckianus

Eccoilopus (Andropogonoideae:Andropogoneae)
1. Urediniospores verrucose, pores
                equatorial............249. Puccinia miyoshiana

Echinaria (Festucoideae:Festuceae)
1. Telia covered; urediniospores
                echinulate...............186. Puccinia hordei

Echinochloa (Panicoideae:Paniceae)
1. Urediniospore pores equatorial.........98. Puccinia graminis
1. Urediniospore pores scattered.........193. Puccinia abnormis

Echinopogon (Festucoideae:Festuceae)
1. Urediniospore pores equatorial.........98. Puccinia graminis

Ehrharta (Festucoideae:Phalarideae)
1. Telia deep cushions, felt-like;
                uredinia unknown.....64. Uromyces ehrhartae-giganteae
1. Telia not deep and felt-like;
                uredinia brownish...............44. Uromyces ehrhartae

Elymus (Festucoideae:Triticeae)
1. Teliospores with apical digitations....54. Puccinia coronata
1. Telia without digitations (2)
2. Uredinia in chlorotic streaks.......58. Puccinia striiformis
2. Uredinia not in such streaks (3)
3. Uredinia paraphysate (4)
3. Uredinia aparaphysate (5)
4. Uredinia with capitate paraphyses...59. Puccinia montanensis
4. Uredinia with clavate-capitate
                paraphyses........63. Puccinia brachypodii
5. Urediniospore pores equatorial.........98. Puccinia graminis
5. Urediniospore with scattered pores (6)
6. Teliospores finely striately
                ridged........168. Puccinia pattersoniana
6. Teliospores smooth (7)
7. Teliospores mostly 3- or 4-celled........164. Puccinia elymi
7. Teliospores typically 2-celled (8)
8. Urediniospores mostly 32-44μ long......175. Puccinia procera
8. Urediniospores mostly 24-32μ long....187. Puccinia recondita

Elytrigia see Agropyron

Enneapogon (Festucoideae:Pappophoreae)

14

1. Urediniospores echinulate, pores
                    equatorial...........123. Puccinia enneapogonis

Enteropogon (Eragrostoideae:Chlorideae)
1. Teliospores 2-celled; urediniospores
                echinulate...............65. Puccinia enteropogonis
1. Teliospores 1-celled; urediniospores
                verrucose...............55. Uromyces archerianus

Entolasia (Panicoideae:Paniceae)
1. Urediniospores dark brown, echinulate....109. Puccinia levis

Eragrostis (Eragrostoideae:Eragrosteae)
1. Teliospores in sessile chains.......8. Physopella hiratsukae
1. Teliospores pedicellate (2)
2. Teliospores 1-celled...............54. Uromyces eragrostidis
2. Teliospores 2-celled (3)
3. Urediniospores echinulate (5)
3. Urediniospore verrucose (4)
4. Urediniospores pores
                equatorial.......253. Puccinia eragrostidis-arundinaceae
4. Urediniospore pores scattered........272. Puccinia morigera
5. Uredinia paraphysate (6)
5. Uredinia aparaphysate (7)
6. Urediniospore pores scattered...73. Puccinia eragrostidicola
6. Urediniospore pores
                equatorial.........49. Puccinia eragrostidis-superbae
7. Urediniospore wall colorless, spores
                    mostly 20-25µ long.222. Puccinia eragrostidis
7. Urediniospore wall golden, spores
                    mostly 25-28µ long........261. Puccinia pogonarthriae

Eremopogon (Andropogonoideae:Andropogoneae)
1. Urediniospores echinulate, pores
                scattered.....................38. Uromyces clignyi

Erianthus (Andropogonoideae:Andropogoneae)
1. Urediniospores verrucose, pores
                equatorial.................259. Puccinia daniloi
1. Urediniospores echinulate, pores equatorial (2)
2. Uredinia paraphysate (3)
2. Uredinia aparaphysate (4)
3. Urediniospores mostly 28-33µ
                    long............17. Puccinia melanocephala
3. Urediniospores mostly 23-27µ long....14. Puccinia microspora
4. Telia covered; urediniospore wall
                golden...............92. Puccinia polysora
4. Telia exposed; urediniospores brown (5)
5. Urediniospore pores 2............124. Puccinia erianthicola
5. Urediniospore pores 3 or 4.........101. Puccinia erythropus

Eriochloa (Panicoideae:Paniceae)
1. Telia exposed, spores 2-celled...........109. Puccinia levis
1. Telia covered, spores
                1-celled.......11. Uromyces setariae-italicae

Euchlaena (Andropogonoideae:Maydeae)
1. Teliospores is sessile chains (2)

1. Teliospores pedicellate (3)
2. Urediniospores mostly 18-24µ long...2. Physopella pallescens
2. Urediniospores mostly 24-30µ long.........3. Physopella zeae
3. Telia exposed; teliospore apex
                   4-9µ thick...........140. Puccinia sorghi
3. Telia covered; teliospore wall
                   uniform.............92. Puccinia polysora

Eulalia (Andropogonoideae:Andropogoneae)
1. Teliospore wall 3-3.5µ thick at sides,
           5-7µ apically....267. Puccinia polliniae-quadrinervis
1. Teliospore wall 1.5-2µ at sides,
           2.5-5µ apically.............284. Puccinia phaeopoda

Exotheca (Andropogonoideae:Andropogoneae)
1. Teliospores pedicellate, 1-celled......38. Uromyces clignyi
1. Teliospores sessile, in subepidermal
                   crusts.........2. Phakopsora imcompleta

Festuca (Festucoideae:Festuceae), also see Vulpia
1. Teliospores 1-celled (2)
1. Teliospores 2-celled (6)
2. Telia exposed, compact (3)
2. Telia covered (4)
3. Teliospore wall 4-8µ thick
                   apically............37. Uromyces cuspidatus
3. Teliospore wall 15-21µ apically........65. Uromyces procerus
4. Telia loosely covered, spores
                   loose........5. Uromyces turcomanicum
4. Telia firmly covered, spores attached (5)
5. Telia paraphyses abundant............32. Uromyces dactylidis
5. Telial paraphyses few.................34. Uromyces hordeinus
6. Urediniospore pores equatorial (7)
6. Urediniospore pores scattered (8)
7. Teliospores mostly 40-60µ long.........98. Puccinia graminis
7. Teliospores mostly 28-43µ
                   long...........286. Puccinia festucae-ovinae
8. Teliospores with apical digitations (9)
8. Teliospores without digitations (10)
9. Telia early exposed...................159. Puccinia festucae
9. Telia usually covered.................54. Puccinia coronata
10. Uredinia in chlorotic streaks.......58. Puccinia striiformis
10. Uredinia not in such streaks (11)
11. Uredinia with paraphyses (12)
11. Uredinia aparaphysate (13)
12. Paraphyses clavate-capitate.........63. Puccinia brachypodii
12. Paraphyses capitate....................60. Puccinia pygmaea
13. Telia early exposed (14)
13. Telia covered (15)
14. Teliospores 28-43µ long........286. Puccinia festucae-ovinae
14. Teliospores mostly 40-60µ
                   long..............238. Puccinia crandallii
15. Teliospores with longitudinal
                   ridges...............167. Puccinia piperi
15. Teliospores without ridges (16)
16. Urediniospore wall pale yellowish (17)

16

16. Urediniospore wall brownish (18)
17. Teliospores mostly 40-58μ................185. Puccinia poarum
17. Teliospores mostly 60-80μ........182. Puccinia cockerelliana
18. Telia with numerous paraphyses.......187. Puccinia recondita
18. Telia with few or no paraphyses.......183. Puccinia sessilis

Garnotia (Eragrostoideae:Festuceae)
1. Telia exposed; urediniospores
                        verrucose......7. Puccinia garnotiae

Gastridium (Festucoideae:Aveneae)
1. Telia exposed; urediniospores
                        echinulate...........98. Puccinia graminis

Gaudinia (Festucoideae:Aveneae)
1. Uredinia in chlorotic streaks.......58. Puccinia striiformis
1. Uredinia not in such streaks (2)
2. Urediniospore wall yellowish............186. Puccinia hordei
2. Urediniospore wall brownish.........187. Puccinia recondita

Glyceria (Festucoideae:Festuceae)
1. Teliospores 1-celled; urediniospores
                        echinulate......42. Uromyces amphidymus
1. Teliospores 2-celled (2)
2. Urediniospore pores equatorial.........98. Puccinia graminis
2. Urediniospore pores scattered (3)
3. Teliospores with apical digitations....54. Puccinia coronata
3. Teliospores without digitations (4)
4. Uredinia in chlorotic streaks.......58. Puccinia striiformis
4. Uredinia not in such streaks (5)
5. Uredinia with clavate-capitate
                        paraphyses........63. Puccinia brachypodii
5. Uredinia aparaphysate (6)
6. Urediniospore wall 1μ, colorless.....181. Puccinia glyceriae
6. Urediniospore wall 1.5-2μ,
                        brownish...........187. Puccinia recondita

Gouinia (Festucoideae:Arundineae)
1. Telia exposed; urediniospores
                        echinulate........127. Puccinia guaranitica

Gymnopogon (Festucoideae:Festuceae)
1. Teliospores mostly diorchidioid.....206. Puccinia boutelouae
1. Teliospores typically
        puccinioid............204. Puccinia gymnopogonicola

Hackelochloa (Andropogonoideae:Andropogoneae)
1. Urediniospore pores basal............25. Puccinia orientalis
1. Urediniospore pores equatorial (2)
2. Germ pores 2, uredinia aparaphysate......109. Puccinia levis
2. Germ pores 3, uredinia paraphysate........11. Puccinia cacao
2. Germ pores 4, uredinia paraphysate.....38. Puccinia pappiana

Haynaldia (Festucoideae:Festuceae)
1. Urediniospore pores equatorial; telia
                        exposed............98. Puccinia graminis
1. Urediniospore pores scattered; telia covered (2)
2. Uredinia in chlorotic streaks.......58. Puccinia striiformis

2. Uredinia not in such streaks........187. Puccinia recondita

Helictotrichon (Festucoideae:Festuceae) also see Avenochloa
1. Teliospores with apical digitations....54. Puccinia coronata
1. Teliospores without digitations.....63. Puccinia brachypodii

Hemarthria (Andropogonoideae:Andropogoneae)
1. Teliospores 1-celled; germ pores
                     scattered..........38. Uromyces clignyi
1. Teliospores 2-celled; germ pores equatorial (2)
2. Uredinia with paraphyses (3)
2. Uredinia aparaphysate...................109. Puccinia levis
3. Paraphyses nearly cylindrical.............11. Puccinia cacao
3. Paraphyses capitate.................14. Puccinia microspora

Hesperochloa (Festucoideae:Festuceae)
1. Telia exposed; urediniospore wall
                     brown.......238. Puccinia crandallii
1. Telia covered; urediniospore wall
                     yellowish.......58. Puccinia striiformis

Heteranthelium (Festucoideae:Triticeae)
1. Telia exposed; germ pores
                     equatorial..........98. Puccinia graminis
1. Telia covered; germ pores
                     scattered..........58. Puccinia striiformis

Heteropogon (Andropogonoideae:Andropogoneae)
1. Teliospores 2-celled; uredinia
                     orange.......197. Puccinia versicolor
1. Teliospores 1-celled; uredinia
                     brown.............38. Uromyces clignyi

Hierochloë (Festucoideae:Phalarideae)
1. Telia exposed; germ pores
                     equatorial...........98. Puccinia graminis
1. Telia covered; germ pores scattered
2. Teliospores with apical digitations (3)
2. Teliospores without digitations (4)
3. Uredinia paraphysate...................54. Puccinia coronata
3. Uredinia aparaphysate.............161. Puccinia praegracilis
4. Uredinia paraphysate...............63. Puccinia brachypodii
4. Uredinia aparaphysate...............187. Puccinia recondita

Hilaria (Eragrostoideae:Lappagineae)
1. Urediniospores echinulate, wall
                     colorless..........210. Puccinia hilariae
1. Urediniospores verrucose, wall
                     brown.............263. Puccinia aristidae

Holcus (Festucoideae:Aveneae)
1. Teliospores 1-celled; germ pores
                     scattered............51. Uromyces holci
1. Teliospores 2-celled; germ pores
                     equatorial...........98. Puccinia graminis
1. Teliospores 2-celled; germ pores scattered (2)
2. Uredinia in chlorotic streaks.......58. Puccinia striiformis
2. Uredinia not in such streaks (3)

18

3. Teliospores with apical
                    digitations.........54. Puccinia coronata
3. Teliospores without digitations.........186. Puccinia hordei

Hordelymus see Elymus

Hordeum (Festucoideae:Festuceae)
1. Teliospores 1-celled (2)
1. Teliospores 2-celled (4)
2. Telia firmly covered, spores
                    attached...........34. Uromyces hordeinus
2. Telia loosely covered, spores loose (3)
3. Teliospores mostly 18-24 x
                    14-20µ..........5. Uromyces turcomanicum
3. Teliospores mostly 24-30 x
                    20-25µ...........24. Uromyces fragilipes
4. Uredinia in chlorotic streaks.......58. Puccinia striiformis
4. Uredinia not in such streaks (5)
5. Teliospores with apical digitations....54. Puccinia coronata
5. Teliospores without digitations (6)
6. Uredinia with paraphyses (7)
6. Uredinia aparaphysate (8)
7. Uredinial paraphyses clavate-
                    capitate.........63. Puccinia brachypodii
7. Uredinial paraphyses capitate.......59. Puccinia montanensis
8. Urediniospore pores equatorial (9)
8. Urediniospore pores scattered (10)
9. Telia exposed.........................98. Puccinia graminis
9. Telia covered........................85. Puccinia hordeina
10. Telia exposed.......................194. Puccinia tornata
10. Telia covered (11)
11. Telia not loculate...................176. Puccinia cryptica
11. Telia obviously loculate (12)
12. Urediniospore wall yellowish............186. Puccinia hordei
12. Urediniospore wall brownish..........187. Puccinia recondita

Hyparrhenia (Andropogonoideae:Andropogoneae)
1. Teliospores 1-celled; uredinia
                    aparaphysate............38. Uromyces clignyi
1. Teliospores 2-celled (2)
2. Uredinia aparaphysate (3)
2. Uredinia paraphysate (5)
3. Urediniospore wall brown, uniformly
                    thin............109. Puccinia levis
3. Urediniospore wall colorless (4)
4. Urediniospore wall thick apically
                    only.....103. Puccinia hyparrheniae
4. Urediniospore wall thick laterally
                    also........197. Puccinia versicolor
5. Urediniospore pores equatorial..46. Puccinia andropogonicola
5. Urediniospore pores scattered (6)
6. Teliospore pedicel thick-
                    walled.......68. Puccinia andropogonis-hirti
6. Teliospore pedicel thin-walled (7)
7. Teliospores mostly 33-40 x
                    20-27µ..........71. Puccinia eritraeensis

7. Teliospores mostly 33-40 x
                    16-19μ.......69. Puccinia hyparrheniicola

Hystrix (Festucoideae:Triticeae)
1. Teliospores with apical digitations (2)
1. Teliospores without digitations (3)
2. Teliospore pedicels about 100μ
                    long....157. Puccinia asperellae-japonicae
2. Teliospore pedicels about 20μ long.....54. Puccinia coronata
3. Uredinia paraphysate (5)
3. Uredinia aparaphysate (4)
4. Urediniospore pores equatorial........98. Puccinia graminis
4. Urediniospore pores scattered.......187. Puccinia recondita
5. Uredinospore pores equatorial........33. Puccinia kiusiana
5. Urediniospore pores scattered (6)
6. Paraphyses capitate, obvious........59. Puccinia montanensis
6. Paraphyses sack-like, collapsing....58. Puccinia striiformis

Ichnanthus (Panicoideae:Paniceae)
1. Teliospores diorchidioid................109. Puccinia levis
1. Teliospores puccinioid (2)
2. Teliospores delicate, colorless; uredinio-
        spore wall brown...................117. Puccinia ichnanthi
2. Teliospores robust, brown; urediniospore
        wall colorless....................135. Puccinia inclita

Imperata (Andropogonoideae:Andropogoneae)
1. Uredinia aparaphysate, spore wall thick
                        apically..108. Puccinia imperatae
1. Uredinia paraphysate (2)
2. Teliospore pedicels about 70-90μ long...30. Puccinia rufipes
2. Teliospore pedicels short (3)
3. Urediniospore wall thickened
                    apically.......23. Puccinia fragosoana
3. Urediniospore wall uniform (4)
4. Urediniospores commonly 30μ or more long (5)
4. Urediniospores less than 30μ long....14. Puccinia microspora
5. Teliospores mostly 40-60μ long........20. Puccinia miscanthi
5. Teliospores mostly 36-50μ long.......21. Puccinia posadensis

Isachne (Panicoideae:Isachneae)
1. Teliospores mostly 35-43μ long.........36. Puccinia isachnes
1. Teliospores mostly 24-28μ long.........12. Puccinia sublesta

Ischaemum (Andropogonoideae:Andropogoneae)
1. Uredinia paraphysate; teliospores
                    sessile......2. Phakopsora incompleta
1. Uredinia aparaphysate; teliospores
                    pedicellate...197. Puccinia versicolor

Ischurochloa (Bambusoideae)
1. Uredinia paraphysate; teliospores in
                    chains.........2. Dasturella divina

Ixophorus (Panicoideae:Paniceae)
1. Uredinia paraphysate, spores
                    echinulate.......1. Puccinia chaetochloaè

Koeleria (Festucoideae:Aveneae)
1. Teliospores 1-celled; urediniospores
                          echinulate....31. Uromyces koeleriae
1. Teliospores 2-celled (2)
2. Urediniospores pores equatorial........98. Puccinia graminis
2. Urediniospore pores scattered (3)
3. Teliospores with apical digitations....54. Puccinia coronata
3. Teliospores without digitations (4)
4. Uredinia on chlorotic streaks.......58. Puccinia striiformis
4. Uredinia without such streaks (5)
5. Uredinia with clavate-capitate
                          paraphyses........63. Puccinia brachypodii
5. Uredinia aparaphysate (6)
6. Telia exposed (7)
6. Telia covered (8)
7. Teliospores mostly 70-100µ long.....191. Puccinia longissima
7. Teliospores mostly 40-51µ long........217. Puccinia monoica
8. Telia with.scant paraphyses, rarely
                          loculate   185. Puccinia poarum
8. Telia with numerous paraphyses, typically loculate (9)
9. Urediniospore wall pale yellowish.......186. Puccinia hordei
9. Urediniospore wall brownish..........187. Puccinia recondita
9. Urediniospore color unknown.......188. Puccinia koeleriicola

Lagurus (Festucoideae:Festuceae)
1. Teliospores with apical digitations....54. Puccinia coronata
1. Teliospores without digitations (2)
2. Telia covered; germ pores scattered.....186. Puccinia hordei
2. Telia exposed; germ pores
                          equatorial..........98. Puccinia graminis

Lamarckia (Festucoideae:Festuceae)
1. Teliospores with apical digitations....54. Puccinia coronata
1. Teliospores without digitations (2)
2. Uredinia in chlorotic streaks.......58. Puccinia striiformis
2. Uredinia not in such streaks (3)
3. Uredinia with clavate-capitate
                          paraphyses....63. Puccinia brachypodii
3. Uredinia aparaphysate, pores
                          equatorial....98. Puccinia graminis

Lasiagrostis see Stipa

Lasiacis (Panicoideae:Paniceae)
1. Teliospores sessile, in chains....5. Physopella lenticularis
1. Teliospores pedicellate,
                          1-celled.......12. Uromyces costaricensis
1. Teliospores pedicellate, 2-celled...126. Puccinia lasiacidis

Leersia (Oryzoideae:Oryzeae)
1. Teliospores 1-celled; uredinia
                          paraphysate.......3. Uromyces halstedii
1. Teliospores 2-celled (2)
2. Telia covered (3)
2. Telia exposed (4)
3. Uredinia in chlorotic streaks.......58. Puccinia striiformis
3. Uredinia not in such streaks.........187. Puccinia recondita

21

4. Teliospores mostly 40-60µ long.........98. _Puccinia graminis_
4. Teliospores mostly less than 36µ long (5)
5. Teliospores mostly 22-30µ long.....285. _Puccinia fushunensis_
5. Teliospores mostly 29-36µ long..........28. _Puccinia ekmanii_

Leleba see _Bambusa_

Leptochloa (Eragrostoideae:Eragrosteae)
1. Teliospores 1-celled; urediniospores
                            echinulate..49. _Uromyces leptochloae_
1. Teliospores 2-celled (2)
2. Urediniospores verrucose...........269. _Puccinia leptochloae_
2. Urediniospores echinulate (3)
3. Urediniospores about cinnamon-
                    brown........226. _Puccinia leptochloae-uniflorae_
3. Urediniospore wall colorless (4)
4. Urediniospores 16-18µ long; teliospores
                        23-31µ long.....207. _Puccinia subtilipes_
4. Urediniospores 22-26µ long; teliospores
                        32-40µ long.....221. _Puccinia diplachnis_

Leptoloma see _Digitaria_

Lepturus (Eragrostoideae:Chlorideae)
1. Telia exposed; urediniospores
                        echinulate.........228. _Puccinia lepturi_

Limnodea (Festucoideae:Aveneae)
1. Telia exposed; germ pores equatorial...98. _Puccinia graminis_
1. Telia covered; germ pores scattered..179. _Puccinia limnodeae_

Lolium (Festucoideae:Festuceae)
1. Teliospores with apical digitations....54. _Puccinia coronata_
1. Teliospores without digitations (2)
2. Urediniospore pores equatorial.........98. _Puccinia graminis_
2. Urediniospore pores scattered (3)
3. Uredinia in chlorotic streaks.......58. _Puccinia striiformis_
3. Uredinia not in such streaks (4)
4. Uredinia with clavate-capitate
                    paraphyses......63. _Puccinia brachypodii_
4. Uredinia aparaphysate (5)
5. Urediniospore wall pale yellowish.......186. _Puccinia hordei_
5. Urediniospore wall brownish.........187. _Puccinia recondita_

Lophatherum (Festucoideae:Festuceae)
1. Telia exposed; urediniospores
                    echinulate........113. _Puccinia lophatheri_

Lophochloa see _Koeleria_

Loudetia (Festucoideae:Arundinelleae) Also see _Tristachya_
1. Teliospores sessile, in subepidermal
                        crusts.......4. _Phakopsora loudetiae_
1. Teliospores pedicellate..............10. _Puccinia loudetiae_

Lycurus (Eragrostoideae:Eragrosteae)
1. Telia exposed; urediniospores
                    echinulate..225. _Puccinia schedonnardi_

<u>Lygeum</u> (Festucoideae:Lygeeae)
1. Telia covered; urediniospores
                    echinulate..........32. <u>Uromyces</u> <u>dactylidis</u>

<u>Melica</u> (Festucoideae:Festuceae)
1. Teliospores 1-celled; germ pores scattered (2)
1. Teliospores 2-celled (3)
2. Urediniospores closely echinulate, pore caps
                    small.......47. <u>Uromyces</u> <u>epicampis</u>
2. Urediniospores sparsely echinulate, pore
                    caps large.........46. <u>Uromyces</u> <u>graminis</u>
3. Urediniospore pores equatorial (4)
3. Urediniospore pores scattered (5)
4. Urediniospores mostly 22-25µ wide....243. <u>Puccinia</u> <u>trebouxii</u>
4. Urediniospores mostly 16-22µ wide......98. <u>Puccinia</u> <u>graminis</u>
5. Teliospores with apical digitations....54. <u>Puccinia</u> <u>coronata</u>
5. Teliospores without digitations (6)
6. Teliospores verrucose...............56. <u>Puccinia</u> <u>paradoxica</u>
6. Teliospores smooth (7)
7. Uredinial paraphyses clavate-
                    capitate......63. <u>Puccinia</u> <u>brachypodii</u>
7. Uredinia aparaphysate (8)
8. Telia exposed....................225. <u>Puccinia</u> <u>schedonnardi</u>
8. Telia covered (9)
9. Urediniospores echinulate...............185. <u>Puccinia</u> <u>poarum</u>
9. Urediniospores verrucose..........265. <u>Puccinia</u> <u>abramoviana</u>

<u>Melinis</u> (Panicoideae:Paniceae)
1. Teliospores in sessile chains.......13. <u>Physopella</u> <u>melinidis</u>
1. Teliospores pedicellate,
                    1-celled.........11. <u>Uromyces</u> <u>setariae-italicae</u>

<u>Microchloa</u> (Eragrostoideae:Chlorideae)
1. Urediniospore pores equatorial.....9. <u>Uromyces</u> <u>dactyloctenii</u>
1. Urediniospore pores scattered......26. <u>Uromyces</u> <u>microchloae</u>

<u>Microlaena</u> (Eragrostoideae:Phalarideae)
1. Urediniospores echinulate, pores
                    scattered............44. <u>Uromyces</u> <u>ehrhartae</u>

<u>Microstegium</u> (Andropogonoideae:Andropogoneae)
1. Teliospores sessile, in subepidermal
                    crusts......2. <u>Phakopsora</u> <u>incompleta</u>
1. Teliospores pedicellate (2)
2. Urediniospores pores equatorial...133. <u>Puccinia</u> <u>polliniicola</u>
2. Urediniospore pores scattered (3)
3. Teliospores golden, germinating without
                    dormancy..........6. <u>Puccinia</u> <u>aestivalis</u>
3. Teliospores chestnut-brown, requiring dormancy (4)
4. Teliospore pedicels 25µ or less
                    long........13. <u>Puccinia</u> <u>benguetensis</u>
4. Teliospore pedicels exceeding 25µ (5).35. <u>Puccinia</u> <u>polliniae</u>

<u>Milium</u> (Festucoideae:Stipeae)
1. Teliospores 1-celled; germ pores
                    scattered...........32. <u>Uromyces</u> <u>dactylidis</u>
1. Teliospores 2-celled (2)

2. Urediniospore pores equatorial.........98. Puccinia graminis
2. Urediniospore pores scattered (3)
3. Teliospores with apical digitations....54. Puccinia coronata
3. Teliospores without digitations (4)
4. Uredinia with capitate-clavate
                            paraphyses......63. Puccinia brachypodii
4. Uredinia without such paraphyses (5)
5. Uredinia in chlorotic streaks.......58. Puccinia striiformis
5. Uredinia not in such streaks........187. Puccinia recondita

Miscanthus (Andropogonoideae:Andropogoneae)
1. Teliospore 2-4-celled............276. Puccinia miscanthicola
1. Teliospores only 2-celled (2)
2. Uredinia paraphysate (4)
2. Uredinia aparaphysate (3)
3. Urediniospore wall brown............101. Puccinia erythropus
3. Urediniospore wall colorless......136. Puccinia miscanthidii
4. Teliospore wall 4-6μ thick apically...20. Puccinia miscanthi
4. Teliospore wall 7-13μ thick
                            apically.......22. Puccinia daisenensis

Molinia (Festucoideae:Arundineae)
1. Teliospores with apical digitations....54. Puccinia coronata
1. Teliospores without digitations (2)
2. Urediniospores mostly 16-22μ wide......98. Puccinia graminis
2. Urediniospores mostly 22-26μ wide.....151. Puccinia moliniae

Moliniopsis (Festucoideae:Arundineae)
1. Telia covered; urediniospores
                            echinulate....180. Puccinia ishikariensis

Monanthochloë (Festucoideae:Festuceae)
1. Urediniospores verrucose, pores
                            scattered......273. Puccinia aristidae

Monocymbium (Andropogonoideae:Andropogoneae)
1. Teliospores 2-celled; uredinia
                            orange.......197. Puccinia versicolor
1. Teliospores 1-celled; uredinia
                            brown............38. Uromyces clignyi

Muhlenbergia (Eragrostoideae:Eragrosteae)
1. Teliospores 1-celled (2)
1. Teliospores 2-celled (5)
2. Urediniospore pores equatorial (3)
2. Urediniospores scattered (4)
3. Teliospores mostly 23-28 x 22-26μ.........20. Uromyces major
3. Teliospores mostly 22-27 x
                            16-18μ.......17. Uromyces muhlenbergiae
4. Teliospores mostly 28-32 x 22-25μ.....47. Uromyces epicampis
4. Teliospores mostly 19-24 x 14-17μ.......45. Uromyces minimus
5. Uredinia in chlorotic streaks.......58. Puccinia striiformis
5. Uredinia not in such streaks (6)
6. Urediniospores verrucose, pores
                            scattered......270. Puccinia chihuahuana
6. Urediniospores echinulate (7)
7. Urediniospore pores equatorial.........98. Puccinia graminis

7. Urediniospore pores scattered (8)
8. Urediniospore wall brownish (9)
8. Urediniospore wall colorless (10)
9. Teliospores mostly 26-30 x 22-25µ......220. Puccinia dochmia
9. Teliospores mostly 28-36 x
                        18-26µ.....225. Puccinia schedonnardi
10. Urediniospores mostly 22-27µ long...218. Puccinia sierrensis
10. Urediniospores mostly 14-19µ long.......208. Puccinia sinica

Nardurus (Festucoideae:Festuceae)
1. Uredinia with clavate-capitate
                        paraphyses....63. Puccinia brachypodii

Nassella (Festucoideae:Stipeae)
1. Teliospores 1-celled; urediniospores echinulate (2)
1. Teliospores 2-celled (3)
2. Urediniospores mostly 30-35µ long, wall golden
            to cinnamon brown...............36. Uromyces nassellae
2. Urediniospores mostly less than 30µ long,
                    wall colorless.........35. Uromyces pencanus
3. Urediniospores verrucose..............266. Puccinia pazensis
3. Urediniospores echinulate or not formed (4)
4. Uredinia lacking; teliospores mostly 53-60µ long,
            pedicels to 200µ long (5)
4. Uredinia formed; teliospores pedicels less than
            30µ long (6)
5. Telia associated with aecia,
                    autoecious.....281. Puccinia graminella
5. Telia separated from aecia,
                    heteroecious....282. Puccinia interveniens
6. Paraphysis wall uniformly 1-1.5µ thick....82. Puccinia digna
6. Paraphysis wall 2.5-4µ thick (7)
7. Urediniospores mostly 23-26µ wide, telio-
            spores mostly 21-25µ wide........74. Puccinia nassellae
7. Urediniospores mostly 16-20µ wide; telio-
            spores mostly 16-22µ wide........78. Puccinia saltensis

Neostapfia (Festucoideae:Festuceae)
1. Urediniospores echinulate, pores
                    equatorial.......98. Puccinia graminis

Neyraudiae (Eragrostoideae:Eragrosteae)
1. Urediniospores echinulate, pores
                    scattered.....224. Puccinia neyraudiae

Nipponobambusa (Bambusoideae)
1. Uredinia paraphysate; teliospore apex
                    prolonged....51. Puccinia longicornis
1. Uredinia aparaphysate; teliospore apex
                    rounded.........149. Puccinia kusanoi

Olyra (Olyroideae:Olyreae)
1. Teliospores in sessile chains...7. Physopella phakopsoroides
1. Teliospores pedicellate (2)
2. Uredinia paraphysate, pores
                    equatorial...34. Puccinia obliquo-septata
2. Uredinia aparaphysate, pores equatorial (3)

3. Urediniospores mostly 34-46µ long...100. Puccinia belizensis
3. Urediniospores less than 34µ long (4)
4. Urediniospores mostly 27-32µ long....112. Puccinia deformata
4. Urediniospores mostly 23-26µ long.......134. Puccinia faceta

Oplismenus (Panicoideae:Paniceae)
1. Teliospores sessile in crusts........1. Phakopsora oplismeni
1. Teliospores pedicellate (2)
2. Urediniospore pores at the hilum.........95. Puccinia advena
2. Urediniospore pores equatorial (3)
3. Urediniospore wall brown.............110. Puccinia flaccida
3. Urediniospore wall colorless (4)
4. Urediniospores mostly 27-34µ long; telio-
            spores chestnut-brown.....135. Puccinia inclita
4. Urediniospores mostly 31-40µ long; telio-
        spores opaque chestnut-brown......131. Puccinia opipara

Orcuttia (Festucoideae;Festuceae)
1. Telia exposed; urediniospores
            echinulate.............98. Puccinia graminis

Oryza (Oryzoideae:Oryzeae)
1. Telia erumpent; urediniospores
            echinulate.............98. Puccinia graminis

Oryzopsis (Festucoideae:Stipeae)
1. Uredinia with clavate-capitate
        paraphyses......62. Puccinia brachypodii-phoenicoidis
1. Uredinia aparaphysate (2)
2. Urediniospore pores equatorial (3)
2. Urediniospore pores scattered (4)
3. Amphispores predominant, wall mostly 3.5-4.5µ
                thick..........165. Puccinia substerilis
3. Only urediniospores formed, wall mostly
        3-3.5µ thick...............146. Puccinia burnettii
4. Teliospores germinating without
                dormancy........217. Puccinia monoica
4. Teliospores requiring dormancy (5)
5. Telia covered......................187. Puccinia recondita
5. Telia exposed (6)
6. Teliospore wall mostly 1-1.5µ at
            sides........215. Puccinia micrantha
6. Teliospore wall mostly 2.5-3.5µ
            at sides..........287. Puccinia oryzopsidis

Ottochloa (Panicoideae:Paniceae)
1. Teliospores 1-celled; germ pores
            equatorial...11. Uromyces setariae-italicae
1. Teliospores 2-celled; germ pores
        at the hilum..............25. Puccinia orientalis

Oxytenanthera (Bambusoideae)
1. Teliospores in sessile chains...........2. Dasturella divina

Panicum (Panicoideae:Paniceae)
1. Teliospores in sessile chains (2)
1. Teliospores pedicellate (3)
2. Uredinia paraphysate.................10. Physopella cameliae
26

2. Uredinia aparaphysate....................1. _Physopella_ _aurea_
3. Teliospores 1-celled (4)
3. Teliospores 2-celled (8)
4. Uredinia paraphysate................1. _Uromyces_ _niteroyensis_
4. Uredinia aparaphysate (5)
5. Urediniospores verrucose................56. _Uromyces_ _vossiae_
5. Urediniospores echinulate (6)
6. Telia covered.................11. _Uromyces_ _setariae-italicae_
6. Telia exposed (7)
7. Urediniospore wall 1.5-2μ thick.....18. _Uromyces_ _graminicola_
7. Urediniospore wall 2.5-3μ thick........22. _Uromyces_ _linearis_
8. Urediniospores verrucose..........252. _Puccinia_ _esclavensis_
8. Urediniospores echinulate (9)
9. Uredinia paraphysate (10)
9. Uredinia aparaphysate (11)
10. Urediniospore pores basal............25. _Puccinia_ _orientalis_
10. Urediniospore pores equatorial............5. _Puccinia_ _dolosa_
11. Teliospore pedicels less than 30μ long (12)
11. Teliospore pedicels 35-80μ long (15)
12. Pore depressed in lower teliospore
                    cell..........97. _Puccinia_ _subcentripora_
12. Pore at septum in lower cell (13)
13. Apical wall of teliospores much paler
                externally..............118. _Puccinia_ _puttemansii_
13. Apical wall nearly uniformly brown (14)
14. Urediniospores 32-40μ long.........121. _Puccinia_ _substriata_
14. Urediniospores 24-27μ long..............119. _Puccinia_ _huberi_
15. Teliospore pedicels 35-80μ long (16)
15. Teliospore pedicels exceeding 100μ.......109. _Puccinia_ _levis_
16. Teliospores typically puccinioid (17)
16. Teliospores diorchidioid or tending so (19)
17. Teliospores pale golden with paler
                umbo................129. _Puccinia_ _millegranae_
17. Teliospores nearly uniformly chestnut-brown (18)
18. Urediniospores nearly globoid........138. _Puccinia_ _emaculata_
18. Urediniospores oblong-ellipsoid........98. _Puccinia_ _graminis_
19. Teliospore pedicels less than 70μ long (20)
19. Teliospore pedicels exceeding 100μ.......109. _Puccinia_ _levis_
20. Urediniospores dark chestnut-brown..111. _Puccinia_ _nyasaensis_
20. Urediniospores golden or cinnamon-brown (21)
21. Teliospores mostly 24-26μ long.......114. _Puccinia_ _negrensis_
21. Teliospores mostly 25-44μ long........110. _Puccinia_ _flaccida_

Pappophorum (Festucoideae:Pappophoreae)
1. Telia exposed; urediniospores
                echinulate.........132. _Puccinia_ _pappophori_

Paspalidium (Panicoideae:Paniceae)
1. Telia covered; urediniospores
                echinulate...11. _Uromyces_ _setariae-italicae_

Paspalum (Panicoideae:Paniceae)
1. Teliospores in sessile chains.......12. _Physopella_ _compressa_
1. Teliospores pedicellate (2)
2. Teliospores 1-celled...............25. _Uromyces_ _paspalicola_
2. Teliospores 2-celled (3)

3. Teliospores with apical digitations....54. Puccinia coronata
3. Teliospores without digitations (4)
4. Uredinia paraphysate (5)
4. Uredinia aparaphysate (7)
5. Paraphyses capitate....................15. Puccinia thiensis
5. Paraphyses cylindrical (6)
6. Urediniospores mostly 32-40μ
long.........1. Puccinia chaetochloae
6. Urediniospores mostly less than 30μ.......5. Puccinia dolosa
7. Urediniospores verrucose (14)
7. Urediniospores echinulate (8)
8. Urediniospore pores scattered...........229. Puccinia macra
8. Urediniospore pores equatorial (9)
9. Teliospore pedicels less than 25μ long (10)
9. Teliospores exceeding 25μ (13)
10. Telia covered.......................90. Puccinia dolosoides
10. Telia exposed (11)
11. Teliospores yellowish.................88. Puccinia paspalina
11. Teliospores brown (12)
12. Urediniospore wall pale brown, pores
obscure...........120. Puccinia araguata
12. Urediniospore cinnamon-brown, pores
obvious.........121. Puccinia substriata
13. Teliospore pedicels to 80μ long......138. Puccinia emaculata
13. Teliospore pedicels 100μ or more.........109. Puccinia levis
14. Teliospores typically
diorchidioid..........252. Puccinia esclavensis
14. Teliospores typically puccinioid....271. Puccinia pseudoatra

Pennisetum (Panicoideae:Paniceae)
1. Teliospores sessile, in crusts...........5. Phakopsora apoda
1. Teliospores pedicellate (2)
2. Teliospores 1-celled (3)
2. Teliospores 2-celled (4)
3. Telia covered.................11. Uromyces setariae-italicae
3. Telia exposed.......................19. Uromyces penniseti
4. Urediniospores verrucose...........252. Puccinia esclavensis
4. Urediniospores echinulate (5)
5. Uredinia paraphysate; telia covered (6)
5. Uredinia aparaphysate (7)
6. Teliospores mostly 30-42μ long......1. Puccinia chaetochloae
6. Teliospores mostly 44-60μ long.......2. Puccinia stenotaphri
7. Urediniospore pores scattered.178. Puccinia penniseti-lanati
7. Urediniospore pores equatorial (8)
8. Teliospore pedicels exceeding 100μ
long...........109. Puccinia levis
8. Teliospore pedicels less than 100μ (9)
9. Teliospore pedicels 50-90μ long (11)
9. Teliospore pedicels 25μ or less (10)
10. Teliospores mostly 25-34μ long....97. Puccinia subcentripora
10. Teliospores mostly 34-50μ long......121. Puccinia substriata
11. Urediniospore pores 3 or 4.......130. Puccinia gymnothrichis
11. Urediniospore pores 4-6................141. Puccinia arthuri

Pereilema (Eragrostoideae:Eragrosteae)
1. Urediniospores echinulate, pores
                    scattered............220. Puccinia dochmia

Perotis (Eragrostoideae:Lappagineae)
1. Urediniospores echinulate, pores
                    scattered..........227. Puccinia perotidis

Peyritschia (Festucoideae:Aveneae)
1. Urediniospores echinulate, pores
                    scattered............185. Puccinia poarum

Phacelurus (Andropogonoideae:Andropogoneae)
1. Urediniospores verrucose...............56. Uromyces vossiae

Phaenosperma (Eragrostoideae:Phaenospermeae)
1. Urediniospores echinulate, pores
                    scattered.......234. Puccinia phaenospermae

Phalaris (Festucoideae:Festuceae)
1. Teliospores 1-celled...............6. Uromyces phalaridicola
1. Teliospores otherwise (2)
2. Teliospores mostly 3-celled..............55. Puccinia addita
2. Teliospores typically 2-celled (3)
3. Teliospores with apical digitations....54. Puccinia coronata
3. Teliospores without digitations (4)
4. Uredinia in chlorotic streaks.......58. Puccinia striiformis
4. Uredinia not in such streaks (5)
5. Urediniospore pores equatorial.........98. Puccinia graminis
5. Urediniospore pores scattered (6)
6. Uredinia paraphysate................63. Puccinia brachypodii
6. Uredinia aparaphysate................183. Puccinia sessilis

Phippsia (Festucoideae:Festuceae)
1. Uredinia with clavate-capitate
                    paraphyses.......63. Puccinia brachypodii

Phleum (Festucoideae:Festuceae)
1. Teliospores 1-celled................32. Uromyces dactylidis
1. Teliospores 2-celled (2)
2. Teliospores with apical digitations....54. Puccinia coronata
2. Teliospores without digitations (3)
3. Uredinia in chlorotic streaks.......58. Puccinia striiformis
3. Uredinia not in such streaks (4)
4. Urediniospore pores equatorial.........98. Puccinia graminis
4. Urediniospore pores scattered (5)
5. Uredinia paraphysate................63. Puccinia brachypodii
5. Uredinia aparaphysate.................185. Puccinia poarum

Phragmites (Festucoideae:Arundineae)
1. Teliospores 1-celled...................21. Uromyces blandus
1. Teliospores 2-celled (2)
2. Urediniospores verrucose.........246. Puccinia cagayanensis
2. Urediniospores echinulate (3)
3. Uredinia aparaphysate (7)
3. Uredinia paraphysate (4)
4. Urediniospore pores scattered........75. Puccinia magnusiana
4. Urediniospore pores equatorial (5)
5. Teliospores less than 50µ long........29. Puccinia invenusta

29

5. Teliospores more than 50µ long (6)
6. Teliospores mostly 14-21µ wide.....42. Puccinia moriokaensis
6. Teliospores mostly 20-23µ wide.........41. Puccinia tepperi
7. Urediniospore pores mostly
                 4 or 5.........153. Puccinia phragmitis
7. Urediniospore pores mostly 3 (8)
8. Teliospores mostly 37-48µ long.........154. Puccinia isiacae
8. Teliospores mostly 48-60µ long........156. Puccinia trabutii

Phyllostachys (Bambusoideae)
1. Teliospores mostly 25-29µ
               long......1. Stereostratum corticioides
1. Teliospores exceeding 40µ long (2)
2. Teliospore apex long-acuminate (3)
2. Teliospore apex rounded (4)
3. Teliospores mostly 65-100µ long.....51. Puccinia longicornis
3. Teliospores mostly 70-85µ long...290. Puccinia nigroconoidea
4. Uredinia paraphysate............40. Puccinia phyllostachydis
4. Uredinia aparaphysate................149. Puccinia kusanoi

Piptochaetium (Festucoideae:Stipeae)
1. Teliospore pedicels to 200µ long....281. Puccinia graminella
1. Teliospore pedicels much shorter (2)
2. Urediniospore pores scattered (3)
2. Urediniospore pores equatorial (4)
3. Teliospore apex commonly with a few projections
              apically.....158. Puccinia neocoronata
3. Teliospore apex without
            projections........241. Puccinia durangensis
4. Teliospores with a pale conical
           apex...........128. Puccinia piptochaetii
4. Teliospores apex rounded or
           obtuse......137. Puccinia chisosensis

Pleioblastus (Bambusoideae)
1. Teliospores mostly 25-29µ
             long........1. Stereostratum corticioides
1. Teliospores mostly more than 50µ long (2)
2. Teliospores 65-100µ long, apex
           elongate.........51. Puccinia longicornis
2. Teliospores 50-78µ long, apex
           narrowly rounded...........149. Puccinia kusanoi

Poa (Festucoideae:Festuceae)
1. Teliospores 1-celled (2)
1. Teliospores 2-celled (4)
2. Telia covered......................32. Uromyces dactylidis
2. Telia exposed (3)
3. Urediniospores mostly 25-30µ long........43. Uromyces otakou
3. Urediniospores mostly 30-40µ long....37. Uromyces cuspidatus
4. Teliospores with apical digitations....54. Puccinia coronata
4. Teliospores without digitations (5)
5. Urediniospores pores equatorial........98. Puccinia graminis
5. Urediniospores pores scattered (6)
6. Uredinia in chlorotic streaks.......58. Puccinia striiformis
6. Uredinia not in such streaks (7)
7. Uredinia paraphysate...............63. Puccinia brachypodii

7. Uredinia aparaphysate (8)
8. Telia exposed (10)
8. Telia covered (9)
9. Urediniospore wall colorless or
nearly so..............185. Puccinia poarum
9. Urediniospore wall brownish..........187. Puccinia recondita
10. Teliospores pale yellowish, 9-12µ
wide....189. Puccinia agropyri-ciliaris
10. Teliospores brown, more than 16µ wide (11)
11. Teliospores germinating without
dormancy.........217. Puccinia monoica
11. Teliospores requiring dormancy......238. Puccinia crandallii

Pogonarthria (Eragrostoideae:Eragrosteae)
1. Telia exposed; germ pores
scattered...216. Puccinia pogonarthriae

Pogonatherum (Andropogonoideae:Andropogoneae)
1. Telia exposed; uredinia
paraphysate........39. Puccinia pogonatheri

Polypogon (Festucoideae:Aveneae)
1. Teliospores with apical digitations....54. Puccinia coronata
1. Teliospores without digitations (2)
2. Urediniospore pores scattered......236. Puccinia polypogonis
2. Urediniospore pores equatorial........98. Puccinia graminis

Polytrias (Andropogonoideae:Andropogoneae)
1. Teliospores 1-celled; urediniospores
echinulate.................15. Uromyces schoenanthi

Psammochloa (Festucoideae:Stipeae)
1. Telia exposed; urediniospores
echinulate......245. Puccinia psammochloae

Pseudoraphis (Panicoideae:Paniceae)
1. Teliospores verrucose..............96. Puccinia brachycarpa

Pseudosasa (Bambusoideae)
1. Teliospores mostly 25-29µ
long.........1. Stereostratum corticioides
1. Teliospores mostly 50-78µ long.........149. Puccinia kusanoi
1. Teliospores mostly 65-100µ long.....51. Puccinia longicornis

Psilurus (Festucoideae:Festuceae)
1. Telia exposed; germ pores equatorial...98. Puccinia graminis
1. Telia covered; germ pores scattered.....186. Puccinia hordei

Puccinellia (Festucoideae:Festuceae)
1. Teliospores 1-celled.................32. Uromyces dactylidis
1. Teliospores 2-celled (2)
2. Teliospores with apical digitations....54. Puccinia coronata
2. Teliospores without digitations (3)
3. Uredinia in chlorotic streaks, telia
covered...........58. Puccinia striiformis
3. Uredinia not in such streaks; telia
exposed..........98. Puccinia graminis

Redfieldia (Eragrostoideae:Eragrosteae)

1. Telia exposed; urediniospores
                 verrucose......254. Puccinia redfieldiae

Relchella (Festucoideae:Festuceae)
1. Telia covered; uredinia
                 paraphysate.........63. Puccinia brachypodii

Reimarochloa (Panicoideae:Paniceae)
1. Teliospores diorchidioid; germ pores
                 equatorial...............109. Puccinia levis

Rhynchelytrum (Panicoideae:Paniceae) also see Tricholaena
1. Teliospores diorchioid; germ pores
                 equatorial...............109. Puccinia levis

Roegneria see Agropyron

Rottboellia (Andropogonoideae:Andropogoneae)
1. Uredinia paraphysate; teliospores
                 puccinioid...........14. Puccinia microspora
1. Uredinia aparaphysate; teliospores
                 diorchidioid................109. Puccinia levis

Saccharum (Andropogonoideae:Andropogoneae)
1. Urediniospore wall usually thickened
                 apically............9. Puccinia kuehnii
1. Urediniospore wall uniformly thin (2)
2. Teliospore pedicels more than 100μ
                 long..........44. Puccinia pugiensis
2. Teliospore pedicels less than 30μ long (3)
3. Teliospores mostly 30-43μ long....17. Puccinia melanocephala
3. Teliospores mostly 40-60μ long........20. Puccinia miscanthi

Sacciolepis (Panicoideae:Paniceae)
1. Telia exposed; urediniospores
                 echinulate............138. Puccinia emaculata

Sasa (Bambusoideae)
1. Teliospores mostly 25-29μ
                 long.........1. Stereostratum corticioides
1. Teliospores exceeding 50μ long (2)
2. Teliospore apex rounded; uredinia
                 aparaphysate............149. Puccinia kusanoi
2. Teliospore apex long-acuminate (3)
3. Teliospore side wall uniformly 2μ thick (4)
3. Teliospore side wall unilaterally thickened (5)
4. Teliospores mostly 60-100 x
                 14-19μ...........51. Puccinia longicornis
4. Teliospores mostly 90-125 x 16-22μ.....52. Puccinia sasicola
5. Teliospores smooth or minutely
                 rugose.......53. Puccinia mitriformis
5. Teliospores obviously rugose....289. Puccinia flammuliformis

Sasaella (Bambusoideae)
1. Uredinia paraphysate.................51. Puccinia longicornis
1. Uredinia aparaphysate.................149. Puccinia kusanoi

Sasamorpha (Bambusoideae)
1. Uredinia paraphysate, pores
                 equatorial.......45. Puccinia hikawaensis

Schedonnardus (Eragrostoideae:Chlorideae)
1. Uredinia aparaphysate, pores
                    scattered.....225. Puccinia schedonnardi

Schizachyrium (also see Andropogon)
1. Uredinia paraphysate, pores
                    equatorial........23. Puccinia fragosoana

Schismus (Festucoideae:Aveneae)
1. Teliospores 2-celled; telia covered.....186. Puccinia hordei
1. Teliospores 1-celled; telia exposed.......51. Uromyces holci

Schizachne (Festucoideae:Festuceae)
1. Teliospores with apical digitations....54. Puccinia coronata

Sclerochloa (Festucoideae:Festuceae)
1. Telia covered; germ pores
                    scattered.........32. Uromyces dactylidis

Scleropoa (Festucoideae:Festuceae)
1. Teliospores 2-celled; germ pores
                    equatorial........98. Puccinia graminis
1. Teliospores 2-celled; germ pores
                    scattered.......32. Uromyces dactylidis

Scleropogon (Eragrostoideae:Eragrosteae)
1. Telia exposed; germ pores
                    scattered.....209. Puccinia scleropogonis

Sclerostachya (Andropogonoideae:Andropogoneae)
1. Uredinia paraphysate, pores
                    equatorial..........9. Puccinia kuehnii

Scolochloa (Festucoideae:Festuceae)
1. Teliospores with apical digitations....54. Puccinia coronata
1. Teliospores without digitations......187. Puccinia recondita

Scribnera (Festucoideae:Monermeae)
1. Telia covered; germ pores scattered...34. Uromyces hordeinus

Secale (Festucoideae:Triticeae)
1. Teliospores 1-celled (2)
1. Teliospores 2-celled (3)
2. Teliospores mostly 18-24 x
                    14-20µ............5. Uromyces turcomanicum
2. Teliospores mostly 24-30 x
                    20-25µ.............24. Uromyces fragilipes
3. Uredinia in chlorotic streaks.......58. Puccinia striiformis
3. Uredinia not in such streaks (4)
4. Telia covered; germ pores
                    scattered.........187. Puccinia recondita
4. Telia exposed; germ pores
                    equatorial..........98. Puccinia graminis

Semiarundinaria (Bambusoideae)
1. Teliospores yellow, mostly
                    25-29µ long...1. Stereostratum corticioides
1. Teliospores brown, exceeding 50µ
                    long...........149. Puccinia kusanoi

<u>Sesleria</u> (Festucoideae:Festuceae)
1. Teliospores with apical digitations....54. <u>Puccinia</u> <u>coronata</u>
1. Teliospores without digitations........98. <u>Puccinia</u> <u>graminis</u>
                            ...99. <u>Puccinia</u> <u>sesleriae</u>
<u>Setaria</u> (Panicoideae:Paniceae)
1. Teliospores sessile (2)
1. Teliospores pedicellate (3)
2. Teliospores irregularly arranged......3. <u>Phakopsora</u> <u>setariae</u>
2. Teliospores in chains................10. <u>Physopella</u> <u>cameliae</u>
3. Teliospores 1-celled (4)
3. Teliospores 2-celled (5)
4. Uredinia paraphysate................1. <u>Uromyces</u> <u>niteroyensis</u>
4. Uredinia aparaphysate........11. <u>Uromyces</u> <u>setariae-italicae</u>
5. Urediniospores verrucose.............268. <u>Puccinia</u> <u>setariae</u>
5. Urediniospores echinulate (6)
6. Uredinia paraphysate (7)
6. Uredinia aparaphysate (8)
7. Urediniospores mostly 30-42μ long...1. <u>Puccinia</u> <u>chaetochloae</u>
7. Urediniospores mostly 23-29μ long.........5. <u>Puccinia</u> <u>dolosa</u>
8. Urediniospores wall colorless, thick
                   above...........106. <u>Puccinia</u> <u>wiehei</u>
8. Urediniospore wall brown (9)
9. Teliospore pedicels exceeding
                      100μ..152. <u>Puccinia</u> <u>setariae-longisetae</u>
9. Teliospore pedicels less than 100μ long (10)
10. Teliospore pedicels usually about
                       50μ........98. <u>Puccinia</u> <u>graminis</u>
10. Teliospore pedicels less than 25μ long (11)
11. Telia covered..............91. <u>Puccinia</u> <u>setariae-forbesianae</u>
11. Telia exposed (12)
12. Teliospores brown, wall thicker
                 apically.....121. <u>Puccinia</u> <u>substriata</u>
12. Teliospores colorless, wall
              uniform........116. <u>Puccinia</u> <u>panici-montani</u>

<u>Sieglingia</u> (Festucoideae:Aveneae)
1. Telia covered; uredinia
              paraphysate.........63. <u>Puccinia</u> <u>brachypodii</u>

<u>Sinobambusa</u> (Bambusoideae)
1. Uredinia aparaphysate, pores
              equatorial...............149. <u>Puccinia</u> <u>kusanoi</u>

<u>Sitanion</u> (Festucoideae:Triticeae)
1. Uredinia in chlorotic streaks.......58. <u>Puccinia</u> <u>striiformis</u>
1. Uredinia not in such streaks (2)
2. Uredinia with capitate paraphyses...59. <u>Puccinia</u> <u>montanensis</u>
2. Uredinia aparaphysate (3)
3. Urediniospore pores equatorial.........98. <u>Puccinia</u> <u>graminis</u>
3. Urediniospore pores scattered (4)
4. Telia covered, spores smooth.........187. <u>Puccinia</u> <u>recondita</u>
4. Telia exposed, spores striate....168. <u>Puccinia</u> <u>pattersoniana</u>

<u>Snowdenia</u> (Panicoideae:Arthropogoneae)
1. Uredinia aparaphysate, pores
              scattered......52. <u>Uromyces</u> <u>snowdeniae</u>

34

<u>Sorghastrum</u> (Andropogonoideae:Andropogoneae)
1. Teliospores 2-celled; uredinia
                    paraphysate..........19. <u>Puccinia</u> <u>virgata</u>
1. Teliospores 1-celled; uredinia
                    aparaphysate..........38. <u>Uromyces</u> <u>clignyi</u>

<u>Sorghum</u> (Andropogonoideae:Andropogoneae)
1. Uredinia aparaphysate...................109. <u>Puccinia</u> <u>levis</u>
1. Uredinia paraphysate (2)
2. Urediniospore pores equatorial.....37. <u>Puccinia</u> <u>nakanishikii</u>
2. Urediniospore pores scattered.........72. <u>Puccinia</u> <u>purpurea</u>

<u>Spartina</u> (Festucoideae:Phalarideae)
1. Teliospores 1-celled (2)
1. Teliospores 2-celled (3)
2. Urediniospore pores scattered........40. <u>Uromyces</u> <u>acuminatus</u>
2. Urediniospore pores equatorial..........14. <u>Uromyces</u> <u>argutus</u>
3. Urediniospore pores scattered.....240. <u>Puccinia</u> <u>distichlidis</u>
3. Urediniospore pores equatorial (4)
4. Urediniospore wall thick at
                    apex........105. <u>Puccinia</u> <u>sparganioides</u>
4. Urediniospore wall irregularly thickened at
                    sides and apex...........102. <u>Puccinia</u> <u>seymouriana</u>

<u>Sphenophlis</u> (Festucoideae:Aveneae)
1. Germ pores equatorial.................98. <u>Puccinia</u> <u>graminis</u>
1. Germ pores scattered.................172. <u>Puccinia</u> <u>eatoniae</u>

<u>Spodiopogon</u> (Andropogonoideae:Andropogoneae)
1. Urediniospores echinulate..............81. <u>Puccinia</u> <u>pachypes</u>
1. Urediniospores verrucose (2)
2. Teliospore wall mostly 6-10μ
                    apically......249. <u>Puccinia</u> <u>miyoshiana</u>
2. Teliospore wall mostly 10-16μ
                    apically.....258. <u>Puccinia</u> <u>crassapicalis</u>

<u>Sporobolus</u> (Eragrostoideae:Eragrosteae)
1. Teliospores 1-celled (2)
1. Teliospores 2-celled (4)
2. Urediniospore pores mostly 2......10. <u>Uromyces</u> <u>sporobolicola</u>
2. Urediniospore pores mostly 4 or 5 (3)
3. Telia exposed; urediniospores mostly
                    36-40μ long............16. <u>Uromyces</u> <u>sporoboli</u>
3. Telia covered; urediniospores mostly
                    24-30μ long............7. <u>Uromyces</u> <u>tenuicutis</u>
4. Urediniospore pores equatorial (6)
4. Urediniospore pores scattered (5)
5. Urediniospores mostly 21-26μ
                    long.........225. <u>Puccinia</u> <u>schedonnardi</u>
5. Urediniospores mostly 26-33μ
            long.........244. <u>Puccinia</u> <u>cryptandri</u> <u>var.</u> <u>luxurians</u>
6. Urediniospore pores basal.............93. <u>Puccinia</u> <u>sporoboli</u>
6. Urediniospore pores equatorial (7)
7. Urediniospore wall colorless (8)
7. Urediniospore wall brown (9)
8. Urediniospore wall thick apically.......107. <u>Puccinia</u> <u>vilfae</u>
8. Urediniospore wall uniform.......144. <u>Puccinia</u> <u>kakamariensis</u>

9. Teliospore pedicel mostly about 50µ
                    long.................98. Puccinia graminis
9. Teliospore pedicel exceeding 100µ...150. Puccinia cryptandri

Stapfiola see Desmostachya

Stenotaphrum (Panicoideae:Paniceae)
1. Uredinia paraphysate; teliospores
                    2-celled........2. Puccinia stenotaphri
1. Uredinia aparaphysate; teliospores
                    1-celled.....11. Uromyces setariae-italicae

Stereochlaena (Panicoideae:Paniceae)
1. Uredinia paraphysate; telia
                    covered........2. Puccinia stenotaphri

Stipa (Festucoideae:Stipeae)
1. Teliospores 1-celled (2)
1. Teliospores 2-celled (5)
2. Teliospores with a pale, differentiated
                    apical umbo......63. Uromyces stipinus
2. Teliospores without such an apex (3)
3. Urediniospores verrucose..........60. Uromyces mussooriensis
3. Urediniospores echinulate (4)
4. Teliospore apex mostly 4-6µ thick...48. Uromyces ferganensis
4. Teliospore apex mostly 6-10µ thick.....35. Uromyces pencanus
5. Teliospores finely rugose, pale
                    golden........169. Puccinia wolgensis
5. Teliospores smooth (6)
6. Uredinia in chlorotic streaks.......58. Puccinia striiformis
6. Uredinia not in such streaks (7)
7. Teliospore pedicels less than 40µ long (8)
7. Teliospore pedicels exceeding 40µ (11)
8. Teliospore apex 2.5-5µ
                    thick........278. Puccinia achnatheri-sibiricae
8. Teliospore apex exceeding 5µ (9)
9. Teliospore apex mostly 6-9µ thick (10)
9. Teliospore apex 20-60µ,
                    rostroid.......279. Puccinia longirostroides
10. Teliospore usually with a few apical
                    projections..........158. Puccinia neocoronata
10. Teliospores without projections.....192. Puccinia mexicensis
11. Telia several mm long; opis-forms (12)
11. Telia small (14)
12. Aecia associated with telia;
                    autoecious..........281. Puccinia graminella
12. Aecia not associated; heteroecious (13)
13. Teliospore pedicel thick-walled,
                    persistent........282. Puccinia interveniens
13. Teliospore pedicel thin-walled,
                    collapsing...........280. Puccinia avocensis
14. Uredinia paraphysate (15)
14. Uredinia aparaphysate (17)
15. Paraphyses incurved, thick-walled.....74. Puccinia nassellae
15. Paraphyses straight, capitate (16)
16. Paraphysis wall 2.5-4µ thick..........78. Puccinia saltensis

16. Paraphysis wall uniformly 1μ..............82. Puccinia digna
17. Urediniospore pores scattered (18)
17. Urediniospore pores equatorial (24)
18. Amphispores produced, usually
                       predominant...165. Puccinia substerilis
18. Amphispores not produced (19)
19. Teliospores germinating without
                       dormancy........217. Puccinia monoica
19. Teliospores requiring dormancy (20)
20. Apical wall of teliospores less than 8μ thick (21)
20. Apical wall of teliospores more than 8μ (22)
21. Teliospore pedicels to 120μ,
                       persistent....223. Puccinia malalhuensis
21. Teliospore pedicels to 85μ, usually broken
                       much shorter...235. Puccinia flavescens
22. Teliospores mostly 40-50μ long.........231. Puccinia stipae
22. Teliospores mostly 50-70μ long (23)
23. Apical wall of teliospores mostly 5-12μ
                       thick......242. Puccinia lasiagrostis
23. Apical wall mostly 12-20μ thick.......233. Puccinia harryana
24. Amphispores predominant............165. Puccinia substerilis
24. Amphispores not produced (25)
25. Urediniospores oblong-ellipsoid........98. Puccinia graminis
25. Urediniospores obovoid or broadly ellipsoid (26)
26. Teliospores mostly 36-41 x
                       25-28μ...........146. Puccinia burnettii
26. Teliospores mostly 38-52 x
                       16-22μ.........147. Puccinia entrerriana

Taeniatherum (Festucoideae:Triticeae)
1. Uredinia in chlorotic streaks.......58. Puccinia striiformis
1. Uredinia not in such streaks (2)
2. Urediniospore pores equatorial........98. Puccinia graminis
2. Urediniospore pores scattered..........186. Puccinia hordei

Tetrarrhena (Festucoideae:Phalarideae)
1. Uredinia aparaphysate, pores
                       scattered.........44. Uromyces ehrhartiae

Themeda (Andropogonoideae:Andropogoneae)
1. Teliospores sessile, in crusts......2. Phakopsora incompleta
1. Teliospores pedicellate (2)
2. Teliospores 1-celled...................38. Uromyces clignyi
2. Teliospores 2-celled (3)
3. Lumen of urediniospores strongly
                       stellate........197. Puccinia versicolor
3. Lumen not or only slightly
                       stellate........198. Puccinia chrysopogi

Thraysia (Panicoideae:Paniceae)
1. Urediniospore pores equatorial..........109. Puccinia levis

Trachypogon (Andropogonoideae:Andropogoneae)
1. Uredinia paraphysate, pores
                       scattered......71. Puccinia eritraeensis
1. Uredinia aparaphysate, pores
                       scattered.......197. Puccinia versicolor

37

<u>Tragus</u> (Eragrostoideae:Lappagineae)
1. Uredinia aparaphysate, pores
                    scattered.............29. <u>Uromyces</u> <u>tragi</u>

<u>Trichachne</u> see <u>Digitaria</u>

<u>Tricholaena</u> (Panicoideae:Paniceae) also see <u>Rhynchelytrum</u>
1. Teliospores in sessile chains.......13. <u>Physopella</u> <u>melinidis</u>

<u>Trichloris</u> (Eragrostoideae:Chlorideae)
1. Uredinia aparaphysate, pores
                    scattered.........214. <u>Puccinia</u> <u>chloridis</u>

<u>Trichoneura</u> (Eragrostoideae:Eragrosteae)
1. Uredinia aparaphysate, pores
                    equatorial.........8. <u>Uromyces</u>˙<u>trichoneurae</u>

<u>Tridens</u> (Eragrostoideae:Eragrosteae)
1. Urediniospores echinulate..............98. <u>Puccinia</u> <u>graminis</u>
1. Urediniospores verrucose (2)
2. Urediniospore wall 1.5-2µ thick.... 257. <u>Puccinia</u> <u>windsoriae</u>
2. Urediniospore wall 3.5-4µ thick......263. <u>Puccinia</u> <u>aristidae</u>

<u>Triodia</u> see <u>Tridens</u>

<u>Triplacis</u> (Eragrostoideae:Eragrosteae)
1. Uredinia aparaphysate, pores
                    scattered....225. <u>Puccinia</u> <u>schedonnardi</u>

<u>Tripogon</u> (Eragrostoideae:Eragrosteae)
1. Uredinia aparaphysate, pores
                    scattered........13. <u>Uromyces</u> <u>tripogonicola</u>

<u>Tripsacum</u> (Andropogonoideae:Maydeae)
1. Teliospores in sessile chains (2)
1. Teliospores pedicellate (3)
2. Urediniospores mostly 28-38µ long.....4. <u>Physopella</u> <u>mexicana</u>
2. Urediniospores mostly 18-24µ long...2. <u>Physopella</u> <u>pallescens</u>
3. Urediniospores smooth, pore 1,
                    basal.........94. <u>Puccinia</u> <u>tripsacicola</u>
3. Urediniospores echinulate, pores equatorial (4)
4. Telia covered; urediniospores pale
                    brownish.........92. <u>Puccinia</u> <u>polysora</u>
4. Telia exposed; urediniospores cinnamon-brown (5)
5. Teliospores mostly 30-40 x 22-27µ.....122. <u>Puccinia</u> <u>tripsaci</u>
5. Teliospores mostly 40-54 x
                    18-22µ.......143. <u>Puccinia</u> <u>pattersoniae</u>

<u>Trisetum</u> (Festucoideae:Aveneae)
1. Teliospores 1-celled................32. <u>Uromyces</u> <u>dactylidis</u>
1. Teliospores 2-celled (2)
2. Teliospores with apical digitations (3)
2. Teliospores without such digitations (4)
3. Teliospores usually less than 90µ
                    long.........54. <u>Puccinia</u> <u>coronata</u>
3. Teliospores mostly 85-140µ long.....160. <u>Puccinia</u> <u>leptospora</u>
4. Uredinia in chlorotic streaks.......58. <u>Puccinia</u> <u>striiformis</u>
4. Uredinia not in such streaks (5)
5. Uredinia paraphysate (6)
5. Uredinia aparaphysate (7)
38

6. Uredinia with thick-walled
                  paraphyses..........63. Puccinia brachypodii
6. Uredinia with thin-walled
                  paraphyses..............67. Puccinia azteca
7. Urediniospore pores equatorial (8)
7. Urediospore pores scattered (9)
8. Telia covered; germ pores
                  subequatorial......86. Puccinia triseticola
8. Telia exposed; germ pores
                  equatorial..........98. Puccinia graminis
9. Telia exposed.........................217. Puccinia monoica
9. Telia covered (10)
10. Urediniospore wall pale yellowish (11)
10. Urediniospore wall brownish (12)
11. Telia with abundant brown
                  paraphyses............186. Puccinia hordei
11. Telia with few or no paraphyses.........185. Puccinia poarum
12. Teliospores 36μ long.........177. Puccinia austroussuriensis
12. Teliospores mostly exceeding 40μ
                  long............187. Puccinia recondita

Tristachya (Festucoideae:Arundinelleae)
1. Teliospores dark chestnut-
                  brown.....26. Puccinia loudetiae-superbae
1. Teliospore clear chestnut or
                  golden..............27. Puccinia tristachyae

Triticum (Festucoideae:Triticeae)
1. Uredinia in chlorotic streaks.......58. Puccinia striiformis
1. Uredinia without such streaks (2)
2. Urediniospore pores scattered........187. Puccinia recondita
2. Urediniospore pores equatorial.........98. Puccinia graminis

Urochloa (Panicoideae:Paniceae)
1. Urediniospores echinulate, pores
                  equatorial...11. Uromyces setariae-italicae

Ventenata (Festucoideae:Festuceae)
1. Uredinia aparaphysate, pores
                  equatorial.......98. Puccinia graminis

Vulpia (Festucoideae:Festuceae) also see Festuca
1. Teliospores 1-celled (2)
1. Teliospores 2-celled (3)
2. Teliospores dusty beneath the
                  epidermis..........24. Uromyces fragilipes
2. Teliospores firmly attached in
                  locules............32. Uromyces dactylidis
3. Teliospores with apical digitations....54. Puccinia coronata
3. Teliospores without digitations (4)
4. Uredinia in chlorotic streaks......58. Puccinia striiformis
4. Uredinia not in such streaks (5)
5. Uredinia with thick-walled paraphyses (6)
5. Uredinia aparaphysate (7)
6. Paraphyses clavate-capitate.........63. Puccinia brachypodii
6. Paraphyses capitate.....................64. Puccinia mellea
7. Urediniospore pores equatorial.........98. Puccinia graminis

39

7. Urediniospore pores scattered (8)
8. Urediniospore wall brownish..........187. <u>Puccinia</u> <u>recondita</u>
8. Urediniospore wall pale yellowish (9)
9. Telia loculate with abundant brown
                paraphyses..........186. <u>Puccinia</u> <u>hordei</u>
9. Telia not loculate....................167. <u>Puccinia</u> <u>piperi</u>

<u>Zea</u> (Andropogonoideae:Maydeae)
1. Teliospores in sessile chains.............3. <u>Physopella</u> <u>zeae</u>
1. Teliospores pedicellate (2)
2. Telia covered; urediniospores mostly
              29-36μ long..............92. <u>Puccinia</u> <u>polysora</u>
2. Telia exposed; urediniospores mostly
              26-31μ long..............140. <u>Puccinia</u> <u>sorghi</u>

<u>Zerna</u> see <u>Bromus</u>

<u>Zizania</u> (Oryzoideae:Oryzeae)
1. Uredinia paraphysate; pores
                equatorial........2. <u>Uromyces</u> <u>coronatus</u>

<u>Zoysia</u> (Eragrostoideae:Lappagineae)
1. Uredinia aparaphysate, pores
                scattered..........203. <u>Puccinia</u> <u>zoysiae</u>

40

1. DASTURELLA Mundkur & Kheswala

Mycologia 35:202-203. 1943

Type species: <u>Dasturella</u> <u>divina</u> (Syd.) Mundk. & Khes.

Key to Species

1. Telia mostly less than 100µ thick...............1. <u>bambusina</u>
1. Telia mostly 150-200µ thick.......................2. <u>divina</u>

1. DASTURELLA BAMBUSINA Mundk. & Khes. Mycologia 35:203. 1943.

Aecia unknown. Uredinia on abaxial leaf surface, small, yellowish brown, with abundant, incurved, colorless to golden paraphyses, then ventral wall 1-1.5µ thick, dorsal wall 2-6µ thick, the terminal portion of the paraphysis commonly solid for 20-30µ; spores (24-)28-36(-40) x (17-)19-24(-28)µ, mostly obovoid, wall (1-)1.5-2µ thick, echinulate, yellow or slowly becoming golden brown, germ pores (4)5(6), equatorial. Telia on abaxial surface, exposed, erumpent, blackish, compact, the telium mostly less than 100µ thick, 3 or 4(5) spores deep; spores (12-)14-30 x 10-15(-17)µ, wall 1-1.5µ thick except apical wall of terminal spores 5-7µ, golden brown to chestnut-brown.

Hosts and distribution: Bambusa sp.: India and Singapore.

Type: Ajrekar, on Bambusa sp., Mahableshwar, India, Mar. 1917 (HC10).

A photograph of the telia was published with the diagnosis.

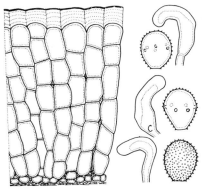

Figure 1

2. DASTURELLA DIVINA (Syd.) Mundk. & Khes. Mycologia 35:203. 1943. Fig. 1.

Uredo inflexa Ito J. Agr. Coll. Tohoku Imp. Univ. 3:247. 1909.

Puccinia inflexa Hori ex Fujik. in Bot. Mag. Tokyo 32:360. 1918 (nomen nudum).

Kuehneola bambusae Fujik. ex Sawada in Descr. Cat. Formosan Fungi 4:71. 1928 (nomen nudum).

Angiopsora divina Syd. Ann. Mycol. 34:71. 1936.

Dasturella oxytenantherae Sathe Sydowia 19:149. 1965.

Aecia occur on species of Randia, locally systemic and forming witches' brooms, cupulate; spores 18-24 x 15-19μ, polygonal or globoid, wall 1.5μ thick, verrucose. Uredinia yellowish brown, with hyaline or yellowish, incurved, thick-walled (especially apically and dorsally) paraphyses, 40-75 x 8-11μ; spores (20-)25-30(-34) x (16-)18-23(-25)μ, ellipsoid, obovoid or nearly globoid, wall 1.5-2μ thick, golden to brownish, echinulate, pores indistinct, 4-6, equatorial. Telia blackish brown, erumpent, pulvinate, crustose, mostly 150-200μ thick; spores 13-28 x 10-16μ, mostly cuboid or oblong, in chains of mostly 3-6 spores, wall 1-1.5μ thick at sides, 3-12μ at apex, chestnut-brown or darker.

Hosts and distribution: Bambusa multiplex Raeusch, B. oldhami (Munro) Nakai, B. shimadai Hayata, B. vulgaris Schrad., Dendrocalamus latiflorus Munro, D. strictus Nees, Ischurochloa stenostachya (Hack.) Nakai, Oxytenanthera sp., Sasa (?) sp.; India, Taiwan, and Japan.

Type: Tandon No. 188, on Bambusa sp. (=Dendrocalamus sp.); Majhgawan, India (Isotypes HC10, PUR).

Thirumalachar, Narasimhan, and Gopalkrishnan (Bot. Gaz. 108:371-379. 1947) proved the life cycle by inoculation. They used Randia dumetorum Lam. and Dendrocalamus strictus as hosts. Mundkur and Kheswalla (loc. cit.) published photographs of telia of the type.

The species differs from D. bambusina mainly in the number of spores per chain and the depth of the telia. The uredinio-spores are not distinguishable.

Uredo ignava Arth. is similar and perhaps synonymous.

2. PHAKOPSORA Dietel

Ber. Deut. Bot. Ges. 13:333. 1895

Type species: Phakopsora punctiformis (Diet. & Barcl.) Diet.

Key to species

1. Wall of teliospore uniformly thin (2)
1. Wall of teliospore thickened apically (3)
2. Telia becoming erumpent; urediniospores
        22-26 x 17-21µ.............................1. oplismeni
2. Telia covered; urediniospores 21-29 x
                        15-21µ.................2. incompleta
3. Side wall of teliospore 1-1.5µ thick, apical
        wall 1.5-2.5µ; urediniospores 22-27µ
                            long...................3. setariae
3. Side wall of teliospore exceeding 1.5µ (4)
4. Apical wall of teliospore 2-3µ thick; uredinio-
        spore wall pale cinnamon-brown.................4. loudetiae
4. Apical wall of teliospores 2.5-5µ thick,
        urediniospore wall pale yellowish.................5. apoda

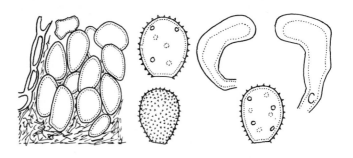

Figure 2

1. PHAKOPSORA OPLISMENI Cumm. Bull. Torrey Bot. Club
83:223. 1956 Fig. 2.

Phakopsora oplismeni Cumm. Mycologia 33:143. 1941
(nomen nudum).

Uredo oplismeni Arth. & Cumm. Phil. J. Sci. 59:442.
1936.

Aecia unknown. Uredinia on abaxial leaf surface, with
hyaline to golden, incurved, apically and dorsally thick-
walled paraphyses, 30-45 x 8-15μ; spores mostly 22-26 x
17-21μ, obovoid or ellipsoid, wall 1.5μ thick, hyaline to
yellowish, echinulate, pores obscure, scattered, probably
6-8. Telia becoming erumpent, crustose, 3-8 spores deep,
waxy-golden in appearance, spores 15-23 x 10-15μ; cuboid,
oblong, or ellipsoid, wall uniformly 0.5-1μ thick, hyaline
to yellowish, germinating at once.

Hosts and distribution: Oplismenus compositus (L.) P.
Beauv., O. hirtellus (L.) Beauv., O. undulatifolius (Ard.)
P. Beauv.: New Guinea, the Phillippine Islands and Mauritius.

Type; Clemens No. 10568, on O. compositus, Kajabit Mission,
Morobe, New Guinea (PUR).

A photograph of teliospores of the type was published by
Cummins (loc. cit., 1941).

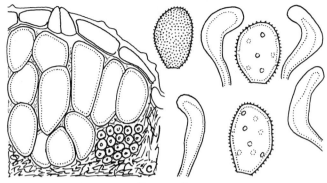

Figure 3

2. PHAKOPSORA INCOMPLETA (Syd.) Cumm. Mycologia 42:786.
1950 Fig. 3.

Puccinia incompleta Syd. Ann. Mycol. 10:261. 1912.

Uredo paraphysata Karst. Rev. Mycol. 12:127. 1890.

Uredo polliniae-imberbis Ito J. Coll. Agr. Tohoku Imp.
Univ. 3:246. 1909.

Aecia unknown. Uredinia mostly on abaxial leaf surface,
with hyaline to golden, incurved paraphyses, the wall apically
and dorsally thickened, 35-45 x 8-13µ; spores mostly 21-29 x
15-21µ, mostly ellipsoid or obovoid, wall 1-1.5µ thick, hyaline
to pale brownish, echinulate, germ pores 7-10, obscure,
scattered. Telia blackish, covered by the epidermis, 2-4
spores deep; spores mostly oblong or ellipsoid, 19-26 x 8-15µ,
wall uniformly (1-)1.5-2µ thick, golden.

Hosts and distribution: Andropogon appendiculatus Nees,
A. dummeri Stapf, A. eucomus Nees, Dimeria filiformis (Roxb.)
Hochst., Exotheca abyssinica (Hochst.) Anders., Ischaemum ari-
statum L., I. arundinaceum F. Muell, I. ciliare Retz, I.
crassipes (Steud.) Thell., Microstegium biaristatum (Steud.)
Keng, M. ciliatum (Trin.) A. Camus, M. vimineum (Trin.)
A. Camus, Themeda triandra Forsk.: Africa to India, Indo
China, New Guinea, the Phillipine Islands, Taiwan, and China.

Type: Mc Rae (Butler No. 1600), on Ischaemum ciliare var.
wallichii, Panora, Wynaad, India (HC10).

Telia, when forming and perhaps occasionally when mature,
consist of a single layer of spores.

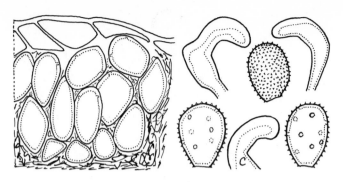

Figure 4

3. PHAKOPSORA SETARIAE Cumm. Bull. Torrey Bot. Club 83:223.
1956. Fig. 4.

Aecia unknown. Uredinia amphigenous, with yellowish to
golden, incurved, apically and dorsally thick-walled paraphyses, 25-40 x 8-14μ; spores 22-27 x 14-19μ, ellipsoid or
obovoid, wall 1-1.5μ thick, hyaline to very pale yellowish,
echinulate, pores obscure, about 8-10 scattered. Telia
blackish brown, covered by the epidermis; crustose, 2-4
spores deep; spores oblong, ellipsoid or nearly globoid,
18-26 x 10-16μ, wall 1-1.5μ at sides, 1.5-2.5μ at apex
golden.

Hosts and distribution: Setaria aequalis Stapf, S.
lancea Stapf, S. sphacelata (Schum.) Stapf & C.E. Hubb.:
Sudan, Uganda, and Nyasaland.

Type: Tarr No. 1908, on S. lancea, Juba, Sudan (PUR;
isotype IMI).

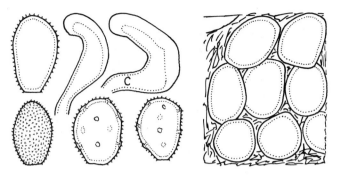

Figure 5

4. PHAKOPSORA LOUDETIAE Cumm. Bull. Torrey Bot. Club
83:223. 1956. Fig. 5.

Aecia unknown. Uredinia amphigenous, yellowish brown,
with peripheral, incurved paraphyses 35-55 x 8-11μ, wall
yellow to golden, 1-2μ thick basally and ventrally, 3-5μ
thick dorsally, to 12μ apically; spores (24-)26-32(-34) x
(15-)18-21(-23)μ, obovoid or ellipsoid, wall pale cinnamon-
brown, echinulate, germ pores obscure, scattered (5)6-9.
Telia mostly abaxial, covered by epidermis, dark brown;
spores irregularly arranged in crusts 2-4 spores deep, 16-28
x 14-18μ, ellipsoid or more or less oblong, wall 2μ thick
at sides, 2-3μ at apex, smooth.

Hosts and distribution: Loudetia arundinacea (Hochst.)
Steud., L. kagerensis (K. Schum.) C.E. Hubb.: Kenya and
Uganda.

Type: Liebmbing No. 23, on Loutedia arundinacea,
Uganda (PUR F15755; isotype BPI).

The brown uredinia and urediniospores distinguish this
fungus from most species of Phakopsora and from Uredo
arundinellae-nepalensis.

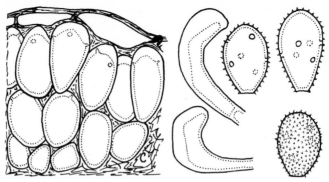

Figure 6

5. PHAKOPSORA APODA (Har. & Pat.) Mains Mycologia 30:45.
1938. Fig. 6.

Puccinia apoda Har. & Pat. Bull. Mus. Hist. Nat. Paris
15:199. 1909.

Aecia unknown. Uredinia amphigenous, with hyaline to
golden, incurved paraphyses, the wall apically and dorsally
thickened, 40-60 x 8-10μ; spores 24-30(-34) x 18-23μ,
ellipsoid or obovoid, wall 1-1.5μ thick, yellowish, echinulate,
pores obscure, 5-8, scattered or tending to be equatorial.
Telia blackish, covered by the epidermis; crustose, 2-4 spores
deep; spores mostly 16-32 x 14-20μ, mostly ellipsoid or
obovoid, side wall 1.5-2μ, apical wall 2.5-5μ thick, golden
or chestnut, with 1-3 fairly obvious germ pores near the
apex.

Hosts and distribution: Pennisetum pedicellatum Trin.,
P. polystachyon (L.) Schult., P. setosum (Sw.) Rich.: Sudan
and Abyssinia to Uganda, Nyasaland, and French Congo.

Type: Chevalier, on P. setosum (probably = P. polystachyon),
Fort Lamy, Chari, French Congo (PC; isotypes: Vestergr.
Micromy. rar. sel. No. 1565).

A photograph of telia of the type was published by
Mains (loc. cit.).

3. PHYSOPELLA Arthur

Result. Sci. Congr. Internat. Bot. Wien p. 338. 1906

Type species: Physopella vitis Arth.

Key to species

1. Uredinia aparaphysate (2)
1. Uredinia with paraphyses (5)
2. Teliospore wall uniformly 1μ thick.................1. aurea
2. Teliospore wall thickened apically (3)
3. Urediniospores mostly 18-24μ long.............2. pallescens
3. Urediniospores larger (4)
4. Urediniospores mostly 24-30 x 15-20μ................3. zeae
4. Urediniospores mostly 28-38 x 18-23μ............4. mexicana
5. Paraphyses short, thin-walled,
               inconspicuous...............5. lenticularis
5. Paraphyses conspicuous, thickened apically
       and usually dorsally (6)
6. Teliospore wall uniformly thin (7)
6. Teliospore wall thickened apically in terminal spore (9)
7. Telia only 1 spore thick......................6. digitariae
7. Telia more than 1 spore thick (8)
8. Urediniospores mostly 28-34 x 20-24μ.......7. phakopsoroides
8. Urediniospores mostly 18-28 x 15-22μ...........8. hiratsukae
9. Urediniospore pores equatorial, 5 or 6...........9. africana
9. Urediniospore pores scattered, very obscure (10)
10. Apical wall of terminal teliospores 4-8μ
                    thick............10. cameliae
10. Apical wall of teliospores mostly 4μ or less (11)
11. Apical wall of teliospores 2-3μ, side wall
                  1-1.5μ.........11. clemensiae
11. Apical wall of teliospores mostly 3-4μ thick (12)
12. Urediniospore wall colorless or pale yellowish,
               1-1.5μ thick........12. compressa
12. Urediniospore wall tending to be brownish,
               1.5-2μ thick.........13. melinidis

1. PHYSOPELLA AUREA (Cumm.) Cumm. & Ramachar Mycologia 50:742. 1958.

Angiopsora aurea Cumm. Bull. Torrey Bot. Club. 83:221. 1956.

Aecia unknown. Uredinia amphigenous or mostly on adaxial leaf surface, spores 22-29 x 14-19μ, wall 1μ thick, hyaline or very pale yellowish, echinulate, pores obscure, probably several and scattered. Telia golden to brown, covered by the epidermis; spores 14-24(-28) x 8-13μ, oblong or cuboid, in chains of 3 or 4(-6), wall uniformly 1μ thick, hyaline or pale yellowish.

Hosts and distribution: Panicum olivaceum Hitchc. & Chase, P. sphaerocarpon Ell.: Honduras.

Type: Müller No. 419, on P. olivaceum, Uyaca, Honduras (PUR).

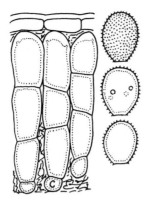

Figure 7

2. PHYSOPELLA PALLESCENS (Arth.) Cumm. & Ramachar Mycologia
50:743. 1958. Fig. 7.

Angiopsora pallescens (Arth.) Mains Mycologia 26:128. 1934.

Uredo pallida Diet. & Holw. in Holway Bot. Gaz. 24:37. 1897.

Puccinia pallescens Arth. Bull. Torrey Bot. Club 46:111.
1919.

Aecia unknown. Uredinia amphigenous, yellowish, without
paraphyses or these hyphoid if present; spores (16-)18-24(-26)
x (12-)14-18µ, ellipsoid or obovoid, wall 1µ thick, colorless
or very pale yellowish, echinulate, pores obscure, probably
about 5 in the equatorial zone. Telia blackish brown, covered
by the epidermis; spores 12-28(-33) x (7-)10-14(-18)µ cuboid
or oblong, in chains of 2-4 spores, wall 1-1.5µ thick at sides,
2-3.5µ at apex of apical spore, golden to light chestnut-brown.

Hosts and distribution: Euchlaena mexicana Schrad.,
Tripsacum fasciculatum Trin., T. lanceolatum Rupr., T.
latifolium Hitchc., T. laxum Nash, T. pilosum Scribn. &
Merrill: Mexico to Columbia, and in Florida.

Lectotype: Hitchcock No. 8720, on Tripsacum latifolium,
Jinotepe, Nicaragua (PUR).

Mains (loc. cit.) published a photograph of teliospore but
did not indicate the source.

53

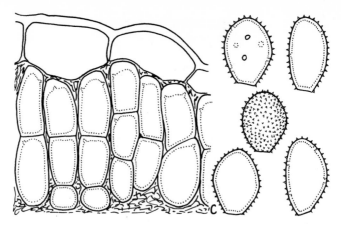

Figure 8

3. PHYSOPELLA ZEAE (Mains) Cumm. & Ramachar Mycologia 50:743.
1958. Fig. 8.

Angiopsora zeae Mains Mycologia 30:42. 1938.

Aecia unknown. Uredinia amphigenous, yellow, without
paraphyses; spores (22-)24-30(-33) x 15-20(-22)μ, ellipsoid
or obovoid, wall 1.5(-2)μ thick, hyaline or very pale yellowish,
echinulate, pores very obscure, probably 7 or 8 and scattered.
Telia blackish, covered by the epidermis; spores 16-36 x
12-18μ, in chains of 2 or 3, usually 2, spores, mostly oblong,
wall 1.5-2μ thick at sides 2.5-4(-6)μ at apex of apical spores,
golden to chestnut-brown.

Hosts and distribution: Euchlaena mexicana Schrad., E.
perennis Hitchc., Zea mays L.; Trinidad to Puerto Rico,
Florida, Mexico, Guatemala, and Venezuela.

Type: Johnston, Alameda, Guatemala (PUR; isotypes BPI,
K, LE, MICH).

Mains published a photograph of the teliospores with the
original diagnosis as did Cummins (Phytopathology 31: 856-857.
1941).

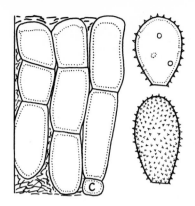

Figure 9

4. PHYSOPELLA MEXICANA Cumm. Southw. Nat. 12:71. 1967.
Fig. 9.

Aecia unknown. Uredinia mostly on abaxial leaf surface,
yellow, without paraphyses; spores (24-)28-38(-44) x
(16-)18-23(-25)µ wall hyaline, echinulate, 1.5-2µ thick,
pores 5-7, scattered, obscure. Telia covered by the epidermis,
blackish brown; spores 12-32 x 11-18(-20)µ, in chains of 2
or 3, oblong, wall (1-)1.5-2µ thick, golden or yellowish,
apex of terminal spores 2.5-4.5µ thick, chestnut-brown,
smooth.

Hosts and distribution: Tripsacum lanceolatum Rupr:
Mexico.

Type: Cummins 63-550, on Tripsacum lanceolatum, Durango,
Mexico (PUR).

P. mexicana has longer urediniospores than other gramini-
colous species.

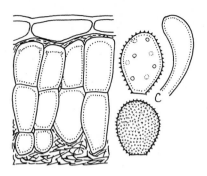

Figure 10

5. PHYSOPELLA LENTICULARIS (Mains) Cumm. & Ramachar Mycologia
50:743. 1958. Fig. 10.

Angiopsora lenticularis Mains Mycologia 26:127. 1934.

Aecia unknown. Uredinia amphigenous, with inconspicuous,
hyaline, uniformly thin-walled paraphyses; spores 22-27 x
15-20μ, ellipsoid or obovoid, wall 1-1.5μ thick, hyaline to
yellowish, echinulate, pores obscure, 7 or 8, scattered.
Telia blackish brown, covered by the epidermis; spores 16-30
x 11-16μ, in chains of 2 to 4, mostly oblong, wall 1-1.5μ
thick at sides, 2-4μ at apex of apical spore, golden to
nearly chestnut-brown.

Hosts and distribution: Lasiacis divaricata (L.)
Hitchc., L. ligulata Hitchc. & Chase, L. procerrima (Hack.)
Hitchc., L. ruscifolia (H.B.K.) Hitchc. & Chase, L.
sorghoidea (Desv.) Hitchc., Panicum arundinariae Trin.;
Trinidad to Mexico, Guatemala, Venezuela, and Ecuador.

Type: Holway No. 801, on Lasiacis ruscifolia, Guayaquil,
Ecuador (PUR; isotypes Reliq. Holw. No. 95).

Mains published photographs of spores of the type with the
diagnosis.

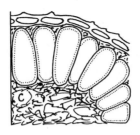

Figure 11

6. PHYSOPELLA DIGITARIAE (Cumm.) Cumm. & Ramachar Mycologia
50:742. 1958. Fig. 11.

Angiopsora digitariae Cumm. Bull. Torrey Bot. Club
83:222. 1956.

Melampsora syntherismae Saw. Taiwan Agr. Res. Inst.
Rept. 87:41. 1944 (nom. nud.)

Aecia unknown. Uredinia mostly on abaxial leaf surface,
with hyaline to golden paraphyses, incurved, the wall apically
and dorsally thickened, 25-40 x 8-11μ; spores (18-)21-26(-28)
x 15-20μ, wall 1-1.5μ thick, hyaline to yellowish, echinulate,
pores obscure, probably several and scattered. Telia
blackish, covered by the epidermis, as seen only one spore
deep; spores (16-)20-25(-30) x (7-)9-11(-13)μ mostly oblong
to ellipsoid, wall uniformly 1-2μ thick, yellowish to
golden.

Hosts and distribution: Digitaria chinensis Hornem., D.
ischaemum (Schreb.) Schreb.; Taiwan.

Type: Sawada, on Syntherisma formosana (=D. chinensis),
Taipeh, Taiwan (PUR).

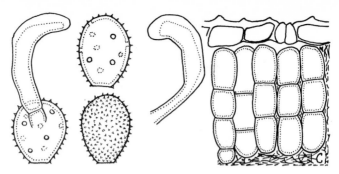

Figure 12

7. PHYSOPELLA PHAKOPSOROIDES (Arth. & Mains) Cumm. &
Ramachar Mycologia 50:743. 1958. Fig. 12.

Angiopsora phakopsoroides (Arth. & Mains) Mains Mycologia
26:128. 1934.

Puccinia phakopsoroides Arth. & Mains Bull. Torrey Bot.
Club 46:412. 1919.

Aecia unknown. Uredinia on abaxial leaf surface, with
abundant, yellowish to brownish, incurved paraphyses, the
wall apically and dorsally thickened, 35-50 x 10-12μ; spores
(25-)28-34(-38) x (18-)20-24(-26)μ ellipsoid or obovoid,
wall 1-1.5μ thick, hyaline to yellow, echinulate, pores
obscure, 7-11, scattered. Telia brownish to blackish,
covered by the epidermis; spores 12-21 x 8-14μ, in chains
of 2 or 3, cuboid to oblong, wall uniformly 1-1.5μ thick,
yellow to golden.

Hosts and distribution: Olyra cordifolia H.B.K., O.
latifolia L.: Cuba and Puerto Rico to Ecuador and Brazil.

Type: Johnston No. 1028, on O. latifolia, Guantanamo, Cuba
(PUR).

Mains (loc. cit.) published a photograph of the teliospores but did not indicate the source.

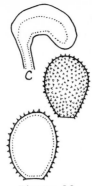

Figure 13

8. PHYSOPELLA HIRATSUKAE (Syd.) Cumm. & Ramachar Mycologia
50:742. 1958. Fig. 13.

Angiopsora hiratsukae Syd. Ann. Mycol. 34:70. 1936.

Aecia unknown. Uredinia amphigenous, with abundant
hyaline to brownish, incurved paraphyses, the wall dorsally
and apically thickened, 35-50 x 8-12µ; spores 18-28 x 15-22µ,
ellipsoid or obovoid, wall 1-1.5µ thick, hyaline to pale
brownish, echinulate, pores obscure, probably scattered.
Telia blackish brown, covered by the epidermis, spores 15-20
x 13-16µ, mostly cuboid or oblong, in chains of 2 or 3,
wall 1µ thick, yellowish to pale brownish.

Hosts and distribution: Eragrostis sp.: Taiwan.

Type: Hashioka No. 686, Kuraru, Prov. Takao (Herb.
Hiratsuka; isotype PUR).

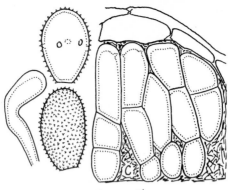

Figure 14

9. PHYSOPELLA AFRICANA (Cumm.) Cumm. & Ramachar Mycologia
50:742. 1958. Fig. 14.

Angiopsora africana Cumm. Bull. Torrey Bot. Club 83:221.
1956.

Aecia unknown. Uredinia amphigenous, with hyaline or
yellowish, incurved, peripheral paraphyses, the wall slightly
apically and dorsally thickened; spores (23-)26-33(-36)
x (14-)16-20(-23)µ, ellipsoid or obovoid, wall 1.5µ thick,
echinulate, pale golden brown, germ pores obscure, (4?)
5 or 6, approximately equatorial. Telia amphigenous,
covered by the epidermis, blackish brown; spores (16-)20-28(-33)
x 10-16µ, in chains of 2 or 3 spores, mostly oblong, wall
2µ thick at sides, 3-4µ thick apically, especially in apical
spore, golden or pale chestnut-brown, smooth.

Hosts and distribution: Brachiaria brizantha (Hochst.)
Stapf, B. decumbens Stapf: Kenya and Uganda.

Type: Hansford No. 2178, on Brachiaria decumbens,
Kabale, Kigesi, Uganda (PUR; isotype IMI).

60

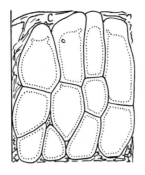

Figure 15

10. PHYSOPELLA CAMELIAE (Arth.) Cumm. & Ramachar Mycologia
50:742. 1958. Fig. 15.

Uredo cameliae Mayor Mem. Soc. Neuchatel. Sci. Nat.
5:578. 1913 (telia present but not described).

Puccinia cameliae Arth. Mycologia 7:227. 1915.

Angiopsora cameliae (Arth.) Mains Papers Michigan Acad.
Sci. Arts, Letters 22:154. 1936 (1937).

Aecia unknown. Uredinia amphigenous, with colorless or
golden, inconspicuous, mostly apically and dorsally somewhat thickened, or uniformly thin-walled, peripheral paraphyses, 25-35 x 8-14µ; spores (18-)20-25(-28) x (13-)15-18(-21)µ,
ellipsoid or obovoid, wall 1-1.5µ thick, colorless or golden,
echinulate, germ pores 7-9, scattered, obscure. Telia
amphigenous, blackish brown, covered by the epidermis;
spores (16-)20-28(32) x 10-15(-18)µ, mostly oblong, in
chains of 2-4 spores, wall 1.5-2µ thick at sides, 4-8µ
apically in the apical spores, golden or chestnut-brown.

Hosts and distribution: Species of Panicum and Setaria:
U.S.A. (Texas) to Puerto Rico, Trinidad, Brazil, and
Columbia.

Type: Mayor, on Setaria scandens (Jacq.) Schrad.,
Cafetal La Camelia, near Angelopolis, Columbia (PUR).

Arthur described the telia from a portion of the type of
Uredo cameliae but Mayor gave no indication that he recognized
their presence.

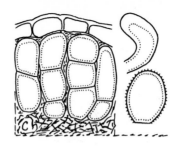

Figure 16

11. PHYSOPELLA CLEMENSIAE (Arth. & Cumm.) Cumm. & Ramachar
Mycologia 50:742. 1958. Fig. 16.

Angiopsora clemensiae Arth. & Cumm. Philippine J. Sci.
59:438. 1936.

Angiopsora cyrtococci T. S. Ramak. & Sund. Indian
Phytopathology 7:143-144. 1954.

Aecia unknown. Uredinia amphigenous, with colorless or
brownish, incurved paraphyses, the wall apically and dorsally
thickened, 25-40 x 7-12µ; spores 20-26 x 16-19µ, obovoid or
ellipsoid, wall 1-1.5µ thick, colorless or pale brownish,
echinulate, germ pores obscure, scattered or possibly
equatorial. Telia blackish, covered by the epidermis; spores
16-29 x 10-15µ, cuboid or oblong, in chains of 2 or 3, wall
1-1.5µ thick at sides, 2-3µ apically, golden or pale chestnut-
brown.

Hosts and distribution: Cyrtococcum patens (L.) A. Camus,
C. warburgii (Mez) Stapf, Ottochloa nodosa (Kunth) Dandy,
Panicum montanum Roxb.: India and the Philippines.

Type: Clemens No. 6946, on Panicum warburgii (=Cyrtococcum
warburgii), Anda, Anda Island, the Philippines (PUR).

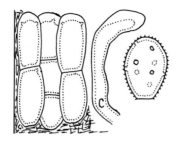

Figure 17

12. PHYSOPELLA COMPRESSA (Mains) Cumm. & Ramachar
Mycologia 50:742. 1958. Fig. 17.

Angiopsora compressa Mains Mycologia 26:129. 1934.

Uredo paspalicola P. Henn. Hedwigia 44:57. 1905.

Uredo stevensiana Arth. Mycologia 7:326. 1915.

Puccinia compressa Arth. & Holw. in Arthur Proc. Amer.
Phil. Soc. 64:157. 1925; not P. compressa Diet. 1907.

Aecia unknown. Uredinia amphigenous, with abundant
hyaline, incurved paraphyses, the wall apically and dorsally
thickened, 26-50 x 8-14µ; spores 20-27(-30) x 15-19µ,
ellipsoid or obovoid, wall 1-1.5µ thick, hyaline or
yellowish, closely echinulate, pores 6-9, obscure, scattered.
Telia blackish brown, covered by the epidermis; spores 20-32
x 12-14µ, in chains of 2 or 3, mostly oblong, wall 1.5µ
thick at sides, 3-4(-6)µ at apex of apical spore, golden
to chestnut-brown.

Hosts and distribution: Axonopus compressus (Swartz) P.
Beauv., species of Paspalum: Southern U.S.A. to Brazil and
Bolivia.

Type: Holway No. 331½, on Paspalum elongatum Griseb.,
Cochabamba, Bolivia (PUR).

Arthur published a photograph of teliospores of the type
with the diagnosis.

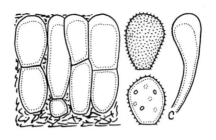

Figure 18

13. PHYSOPELLA MELINIDIS Cumm. & Ramachar Mycologia
50:743. 1958. Fig. 18.

Angiopsora hansfordii Cumm. Bull. Torrey Bot. Club
72:206. 1945, not Thirumalachar & Kern 1949.

Aecia unknown. Uredinia on abaxial side of leaf, with
hyaline to pale yellowish, incurved paraphyses, the wall
apically and dorsally thickened, 25-40 x 9-15μ; spores
20-27 x 14-19μ, mostly ellipsoid or obovoid, wall 1.5-2μ
thick, hyaline to pale brownish, echinulate, pores obscure,
about 7-9, scattered. Telia blackish brown, covered by
the epidermis; spores 18-30 x 9-17μ, in chains of 2 or 3,
cuboid or oblong, wall 1.5μ thick at sides, 2-5μ at apex
of apical spore, golden to chestnut-brown.

Hosts and distribution: Melinis tenuissima Stapf,
Tricholaena sp.; Uganda and Angola.

Type: Hansford No. 1714, Kyasoweri, Elgon, Uganda
(PUR; isotype IMI).

4. STEREOSTRATUM Magnus

Ber. Deut. Bot. Ges. 17:181. 1899

Type (and only) species: Stereostratum corticioides (Berk. & Br.) Magnus.

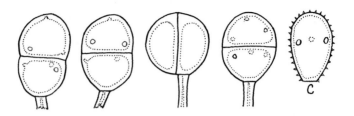

Figure 19

1. STEREOSTRATUM CORTICIOIDES (Berk. & Br.) Magn. Ber. Deut. Bot. Ges. 17:181. 1899. Fig. 19.

Puccinia corticioides Berk. & Br. J. Linn. Soc. 16:52. 1877.

Puccinia schottmuelleri P. Henn. Hedwigia 32:61. 1893.

Sori in long series on stems. Uredinia brownish, without paraphyses; spores 16-28 x 14-20μ, narrowly to broadly ellipsoid, wall 1.5-2μ thick, yellowish to pale brownish, echinulate, pores equatorial, 2 or 3. Telia cinnamon-brown, erumpent, pulvinate, large and Stereum-like; spores (23-)25-29(-33) x 19-25(-28)μ, ellipsoid to broadly ellipsoid, wall nearly uniformly 2.5-3.5μ thick, yellowish to golden, smooth; germ pores 3 in each cell near septum or one often is apical in upper cell, small, very obscure; pedicels hyaline, slender and tapering, attaining 350μ in length but usually broken near spore. 1-celled spores occasionally occur.

Hosts and distribution: On species of Bambusa, Chimono-bambusa, Phyllostachys, Pleioblastus, Pseudosasa, Sasa, and Semiarundinaria: China and Japan.

Type: Challenger Expedition, on Arundinaria (?), Kobe, Japan (K).

5. PUCCINIA Persoon

Synopsis Methodica Fungorum p. 225. 1801

Type species: <u>Puccinia graminis</u> Pers.

GROUP I: Uredinia paraphysate, urediniospores echinulate, germ pores equatorial or basal.

1. Telia covered or only tardily exposed, not erumpent (2)
1. Telia exposed early, erumpent (6)
2. Urediniospores mostly 30-40µ long (3)
2. Urediniospores seldom more than 30µ long (4)
3. Teliospores mostly 32-40µ long...............1. <u>chaetochloae</u>
3. Teliospores mostly 44-60µ long................2. <u>stenotaphri</u>
4. Teliospores regularly clavate or oblong-clavate, mostly 35-45µ long, not brittle (5)
4. Teliospores irregular but mostly oblong angular, rarely 40µ long, brittle........................5. <u>dolosa</u>
5. Telia with some peripheral paraphyses but not loculate......................3. <u>oahuensis</u>
5. Telia aparaphysate...............................4. <u>chaseana</u>
6. Teliospore pedicels typically less than 30µ long, hence sori not conspicuously pulvinate (7)
6. Teliospore pedicels typically more than 40µ long (23)
7. Teliospore pale golden, germinating without a dormant period (8)
7. Teliospores darker brown, requiring dormancy (9)
8. Teliospores mostly 29-40µ long; paraphysis wall thick..................................6. <u>aestivalis</u>
8. Teliospores mostly 43-68µ long; paraphysis wall thin...................................7. <u>garnotiae</u>
9. Amphispores produced, the apical wall thick.......8. <u>angusii</u>
9. Only ordinary urediniospores produced (10)
10. Urediniospore wall thickened apically.............9. <u>kuehnii</u>
10. Urediniospore wall uniformly thin (11)
11. Teliospore wall uniformly thin (12)
11. Teliospore wall thickened apically (13)
12. Uredinial paraphyses capitate..................10. <u>loudetiae</u>
12. Uredinial paraphyses more or less cylindrical......11. <u>cacao</u>
13. Apical wall of teliospore rarely more than 2-3µ thick (14)
13. Apical wall of teliospore more than 3µ (17)
14. Wall of uredinial paraphysis uniformly thick....12. <u>sublesta</u>
14. Wall of uredinial paraphysis thickened apically (15)
15. Wall of uredinial paraphysis 4-9µ apically..13. <u>benguetensis</u>
15. Wall of uredinial paraphysis 6-15-18µ apically (16)
16. Urediniospores mostly 23-27µ long, teliospores mostly 29-34µ long................14. <u>microspora</u>
16. Urediniospores mostly 26-32µ long, teliospores mostly 30-40µ long..................15. <u>thiensis</u>
17. Apical wall of teliospore 3-5µ thick (18)
17. Apical wall of teliospore more than 5µ thick (20)
18. Urediniospore wall 2-2.5µ thick.....16. <u>arundinellae-setosae</u>
18. Urediniospore 1.5µ thick (19)

GROUP II: Uredinia paraphysate, urediniospores echinulate,
            germ pores scattered

3. Teliospores verrucose, wall uniformly
        1.5-2.5μ........................56. paradoxica
3. Teliospores smooth (4)
4. Telia covered (5)
4. Telia exposed, mostly obviously erumpent (12)
5. Teliospores mostly 3- or 4-celled.....57. Puccinia naumovii
5. Teliospores typically 2-celled (6)
6. Uredinial paraphyses large, saccate, thin-
    walled and collapsing, uredinia in conspicuous
    chlorotic streaks........................58. striiformis
6. Uredinial paraphyses not saccate, uredinia not in
    chlorotic streaks even if seriate (7)
7. Paraphyses capitate, the wall uniformly 1-1.5μ thick (8)
7. Paraphyses not capitate or if capitate with thick
                            wall (10)
8. Teliospores mostly 22-32μ wide..............59. montanensis
8. Teliospores mostly less than 23μ wide (9)
9. Urediniospores mostly cinnamon-brown...........60. pygmaea
9. Urediniospores mostly pale yellow..............61. crinitae
10. Paraphysis wall thin in the stipe, abruptly
        thickened 3-7μ apically....62. brachypodii-phoenicoidis
10. Paraphysis wall 2-5μ thick in the stipe (11)
11. Paraphysis wall thicker in the head than in the
        stipe; teliospores brown...................63. brachypodii
11. Paraphysis wall uniformly 2-4μ thick; telio-
            spores pale golden......................64. mellea
12. Uredinial paraphyses 1-septate............65. enteropogonis
12. Uredinial paraphyses aseptate (13)
13. Teliospore pedicels mostly less than 30μ long (14)
13. Teliospore pedicels mostly more than 40μ long (18)
14. Paraphyses thin-walled; urediniospore golden
        or pale cinnamon-brown (15)
14. Paraphyses thick-walled; urediniospores
        cinnamon-brown or darker (16)
15. Teliospores mostly 39-50μ long.....66. digitariae-velutinae
15. Teliospores 50-95μ or more long................67. azteca
16. Urediniospores mostly 29-35μ long....68. andropogonis-hirti
16. Urediniospores mostly less than 30μ long (17)
17. Teliospores mostly 33-40 x 16-19μ.......69. hyparrheniicola
17. Teliospores mostly 23-30 x 18-22μ...........70. kenmorensis
18. Uredinial paraphyses typically clavate (19)
18. Uredinial paraphyses typically capitate (23)
19. Urediniospore wall cinnamon-brown or darker (20)
19. Urediniospore wall colorless or pale golden (21)
20. Urediniospores mostly 24-32μ long; teliospores
            mostly 33-40μ long...................71. eritraeensis
20. Urediniospores mostly 30-40μ long; teliospores
            mostly 40-50μ long....................72. purpurea
21. Apical wall of teliospores 4-6μ thick apically,
        spores mostly 29-33μ long.........73. eragrostidicola
21. Apical wall of teliospores 5-10μ or more, spores
            exceeding 35μ long (22)
22. Urediniospores mostly 26-30 x 23-26μ; teliospores
            mostly 36-44μ long.....................74. nassellae

22. Urediniospores mostly 26-35 x 15-19μ; teliospores
        mostly 42-56μ long........................75. magnusiana
23. Teliospores commonly diorchidioid, wall from 2-7μ
        thick basally to 8-12μ apically..............76. eylesii
23. Teliospores rarely or not diorchidioid, wall not
        progressively thickened (24)
24. Teliospore pedicels commonly more than 100μ long (29)
24. Teliospore pedicels rarely or not 100μ long (25)
25. Urediniospores 18-20μ diam. globoid, wall 2.8μ
                    thick......................77. kwanhsiensis
25. Urediniospores larger (26)
26. Teliospore pedicel thin-walled, mostly collapsing (27)
26. Teliospore pedicel thick-walled, mostly terete (28)
27. Wall of uredinial paraphyses nearly uniformly
                    2.5-4μ thick...................78. saltensis
27. Wall of uredinial paraphyses uniformly 0.5-1μ
                    thick.........................79. corteziana
28. Teliospores mostly ellipsoid, mostly 29-34
            x 18-22μ........................80. decolorata
28. Teliospores mostly broadly ellipsoid, mostly
            31-37 x 23-26μ..........................81. pachypes
29. Wall of uredinial paraphyses uniformly 1-1.5μ
        thick, urediniospore wall yellow or golden........82. digna
29. Wall of paraphyses thickened apically, uredinio-
            spore wall cinnamon-brown or darker.........83. unica

GROUP III: Uredinia paraphysate, urediniospore verrucose,
            germ pores equatorial: no species.

GROUP IV: Uredinia paraphysate, urediniospore verrucose,
            germ pores scattered: no species, but see Uredo.

GROUP V: Uredinia aparaphysate; urediniospores echinulate,
            germ pores equatorial or basal.

1. Teliospores with apical digitations...........84. diarrhenae
1. Teliospores without such digitations (2)
2. Telia covered by the epidermis (3)
2. Telia exposed (10)
3. Telia with brown paraphyses (4)
3. Telia without such paraphyses (5)
4. Teliospores 37-89μ long; urediniospores 16-27
                x 16-22μ.....................85. hordeina
4. Teliospores 35-49μ long; urediniospores 19-22
                x 16-18μ..........................86. triseticola
5. Amphispores produced, the pores subequatorial....87. chaetii
5. No amphispores, only ordinary urediniospores (6)
6. Teliospore wall yellowish or pale golden; germ
                pores 3...................88. paspalina
6. Teliospore wall more or less chestnut-brown (7)
7. Urediniospore wall 2-3μ thick, cinnamon-brown....89. cenchri
7. Urediniospore wall 1-1.5μ, yellowish or pale
                cinnamon-brown (8)
8. Teliospores typically clavate or oblong.......90. dolosoides

8. Teliospores very irregular and angular, brittle and
easily broken (9)
9. Teliospores mostly 25-31 x 16-24µ; urediniospores
mostly 27-31 x 20-24µ...........91. setariae-forbesianae
9. Teliospores mostly 29-41 x 20-27µ; urediniospores
mostly 29-36 x 23-29µ......................92. polysora
10. Urediniospores with germ pores next to the hilum (11)
10. Urediniospores with germ pores in the equator (13)
11. Urediniospores mostly broadly obovoid, mostly
less than 30µ long.......................93. sporoboli
11. Urediniospores ellipsoid or oblong-ellipsoid,
smooth, mostly more than 40µ long (12)
12. Urediniospore pores 2; teliospore wall mostly 5-7µ
thick apically...............94. tripsacicola
12. Urediniospore pore 1; teliospore wall nearly
uniformly 2-3µ thick...................95. advena
13. Germ pore in lower cell of teliospore depressed
toward the hilum (14)
13. Germ pore in lower cell at the septum (15)
14. Teliospore punctate-verrucose................96. brachycarpa
14. Teliospore smooth........................97. subcentripora
15. Wall of urediniospore typically thickened apically (16)
15. Wall of urediniospore uniform (26)
16. Urediniospore wall yellowish brown or darker (17)
16. Urediniospore wall colorless or essentially so (19)
17. Urediniospores mostly oblong-ellipsoid or ellipsoid,
wall yellowish brown....................98. graminis
...................99. sessleriae
17. Urediniospores mostly broadly ellipsoid or obovoid,
wall cinnamon-brown or darker (18)
18. Urediniospores mostly 34-46µ long, coarsely
echinulate......100. belizensis
18. Urediniospores mostly 25-33µ long, moderately to
finely echinulate..........101. erythropus
19. Urediniospore wall unevenly thickened, the lumen
somewhat stellate.........102. seymouriana
19. Urediniospore wall thickened only apically (20)
20. Apical wall of urediniospore 10-19µ thick; teliospores mostly 38-40µ long.103. hyparrheniae
20. Apical wall mostly 10µ or less thick (21)
21. Teliospores mostly broadly ellipsoid, typically less than
twice as long as wide (22)
21. Teliospores mostly ellipsoid or oblong-ellipsoid,
typically twice as long as wide (23)
22. Teliospores mostly 35-44 x 24-30µ................104. eucomi
22. Teliospores mostly 31-40 x 22-26µ..107. vilfae var. mexicana
23. Urediniospores mostly 30-43µ long.........105. sparganioides
23. Urediniospores mostly 33µ or less long (24)
24. Apical wall of teliospore 8-12µ thick, urediniospores mostly 25-28µ long.......106. wiehei
24. Apical wall of teliospore mostly 8µ or less;
urediniospores longer (25)
25. Urediniospores mostly 26-33 x 22-26µ.............107. vilfae
25. Urediniospores mostly 23-30 x 18-22µ..........108. imperatae

26. Teliospores tending to be diorchidioid, the septum
    at least typically oblique (26)
26. Teliospores typically puccinioid, septum not or
    only occasionally oblique (34)
27. Urediniospore pores 2, wall dark cinnamon-brown..109. <u>levis</u>
27. Urediniospore pores mostly or only 3, wall color
                    various (28)
28. Teliospore pedicels thick-walled, to 150μ or
                   more long.............109. <u>levis</u>
28. Teliospore pedicels thin-walled (29)
29. Urediniospore wall dark cinnamon- or chestnut-brown (30)
29. Urediniospores yellowish to cinnamon-brown (31)
30. Teliospores with a pale umbo over the pores, the
    septum usually only oblique.................110. <u>flaccida</u>
30. Teliospores without such an umbo, the septum
    mostly vertical..........................111. <u>nyasaensis</u>
31. Urediniospores mostly 27-32 x 21-27μ; telio-
        spore septum mostly only oblique.......112. <u>deformata</u>
31. Urediniospores mostly 24-28 x 17-22μ or less;
        teliospore septum commonly vertical (32)
32. Apical wall of teliospore mostly 6-8μ thick;
               urediniospores mostly 24-28 x
                     20-23μ............113. <u>lophatheri</u>
32. Apical wall of teliospore mostly less than 5μ;
    urediniospores narrower (33)
33. Urediniospores mostly 24-27 x 17-21μ; telio-
    spore side wall 1-1.5μ thick...............114. <u>negrensis</u>
33. Urediniospores mostly 20-24 x 17-19μ; telio-
    spore side wall 1.5-2.5μ thick............115. <u>taiwaniana</u>
34. Teliospores thin-walled, colorless, delicate;
        urediniospores mostly 32-38μ long, brown (35)
34. Teliospores not colorless and delicate; urediniospores
        similar or smaller (36)
35. Teliospores 26-31 x 15-19μ; urediniospores mostly
    27-31μ wide........................116. <u>panici-montani</u>
35. Teliospores 28-34 x 12-14; urediniospores mostly
    23-27μ wide............................117. <u>ichnanthi</u>
36. Teliospore pedicels typically 30μ or less long (37)
36. Teliospore pedicels typically more than 40μ long (40)
37. Teliospores less than 40μ long; urediniospores less than
    28μ long (38)
37. Teliospores more than 40μ long; urediniospores mostly
    more than 28μ long (39)
38. Apical wall of teliospore 4-7μ thick, paler
               externally............118. <u>puttemansii</u>
38. Apical wall of teliospore 3-5μ thick, uniformly
               brown....................119. <u>huberi</u>
39. Apical wall of teliospore paler externally;
    urediniospore wall golden................120. <u>araguata</u>
39. Apical wall of teliospore uniformly brown...121. <u>substriata</u>
40. Teliospore pedicels typically less than 100μ long (41)
40. Teliospore pedicels typically more than 100μ long (63)
41. Amphispores produced, usually commoner than ordinary
    urediniospores (42)

41. Amphispores not produced (43)
42. Amphispores ellipsoid or obovoid, mostly
    28-36 x 22-28μ..........165. substerilis var. oryzopsidis
42. Amphispores globoid, mostly 26-30μ diam........122. tripsaci
43. Urediniospore pores 2 (44)
43. Urediniospore pores 3 or more (45)
44. Teliospore wall nearly uniformly thick;
    urediniospores 25-33μ long...............123. enneapogonis
44. Teliospore wall much thicker apically;
    urediniospores 23-28μ long...............124. erianthicola
45. Teliospores mostly 30μ or less long (46)
45. Teliospores typically more than 30μ long (48)
46. Teliospores mostly 12-15μ wide; uredinio-
    spores mostly 24-32μ long................125. bambusarum
46. Teliospores wider; urediniospores shorter (47)
47. Teliospores mostly 18-20μ wide, side wall
    1.5-2μ thick............................126. lasiacidis
47. Teliospores mostly 20-23μ wide, side wall
    2-3μ thick.............................127. guaranitica
48. Teliospores with a conspicuously paler, rather
    narrowly conical umbo (49)
48. Teliospores without such an umbo (51)
49. Urediniospores mostly 19-25 x 17-21μ.......128. piptochaetii
49. Urediniospores larger (50)
50. Urediniospores colorless or pale yellowish;
    teliospores yellowish to golden...........129. millegranae
50. Urediniospores golden to cinnamon-brown;
    teliospores golden to chestnut-brown.....130. gymnothrichis
51. Urediniospores commonly 31-38μ long (52)
51. Urediniospores typically shorter (53)
52. Urediniospore wall colorless or yellowish.......131. opipara
52. Urediniospore wall cinnamon-brown............132. pappophori
53. Teliospores 36μ or less long (54)
53. Teliospores mostly more than 36μ long (55)
54. Teliospore side wall 3-3.5μ thick; uredinio-
    spore wall yellowish, pores 3.............133. polliniicola
54. Teliospore side wall 2-2.5μ thick; uredinio-
    spore wall cinnamon, pores 4..................134. faceta
55. Urediniospore wall colorless, pores obscure (56)
55. Urediniospore wall brown, pores obvious (57)
56. Apical wall of teliospore mostly 3-5μ thick,
    spores broadly ellipsoid.....................135. inclita
56. Apical wall of teliospore mostly 5-9μ, spores
    ellipsoid or oblong-ellipsoid...........136. miscanthidii
57. Teliospore pedicels mostly thick-walled, usually
    terete (61)
57. Teliospore pedicels mostly thin-walled, collapsing (58)
58. Teliospores commonly more than 45μ long.....137. chisosensis
58. Teliospores typically less than 45μ long (59)
59. Urediniospores mostly 21-27μ long, germ pores
    typically 3.............................138. emaculata
59. Urediniospores mostly 26-31μ or longer (60)
60. Teliospores mostly 25-29μ wide; germ pores
    4 or 5....................139. kawandensis

74

60. Teliospores mostly 18-23µ wide; germ pores
        3 or 4...........................140. sorghi
61. Urediniospore pores 4-6, usually 5..............141. arthuri
61. Urediniospore pores 3 or occasionally 4 (62)
62. Teliospores broadly ellipsoid or broadly obovoid,
    mostly 34-40 x 20-24µ....................142. cacabata
62. Teliospores oblong-ellipsoid, mostly 40-54 x
             18-22µ...................143. pattersoniae
63. Urediniospore wall colorless or essentially so,
    2µ or less thick (64)
63. Urediniospore wall golden to cinnamon-brown (65)
64. Urediniospores mostly 20-24 x 17-20µ; telio-
    spores mostly 32-42 x 22-26µ............144. kakamariensis
64. Urediniospores mostly 23-30 x 20-24µ; telio-
    spores mostly 40-56 x 22-30µ.............145. arundinellae
65. Teliospores mostly 40µ or less long (66)
65. Teliospores mostly more than 40µ long (67)
66. Urediniospore wall cinnamon-brown, 1.5-2µ
             thick.................142. cacabata
66. Urediniospore wall yellow or golden, 3-3.5µ
             thick.................146. burnettii
67. Urediniospores mostly 20-24µ long, wall
          1.5-2µ thick................147. entrerriana
67. Urediniospores longer, or wall thicker, or both (68)
68. Walls of at least some teliospores finely punctate-
                verrucose (69)
68. Wall of all teliospores smooth (70)
69. Teliospores mostly 38-68 x 20-26µ, wall
           mostly 5-7µ apically.......148. arundinariae
69. Teliospores mostly 50-78 x 17-21µ, wall
           mostly 6-12µ apically..............149. kusanoi
70. Side wall of teliospore 1.5-2.5µ thick;
    germ pores 4-6........................150. cryptandri
70. Side wall of teliospore 2.5µ or more; germ pores
    mostly 3 or 4 (71)
71. Teliospore pedicels thin-walled, mostly
            collapsing................151. moliniae
71. Teliospore pedicel thick-walled, terete (72)
72. Urediniospore wall cinnamon-brown, 2-2.5µ
             thick....152. setariae-longisetae
72. Urediniospore wall yellow, yellowish brown, or
    golden 2.5-3µ or thicker (73)
73. Teliospores mostly less than 24µ wide with pale
   and umbonate apical thickening..............153. phragmitis
73. Teliospore typically 24µ or wider (74)
74. Teliospores mostly 37-48µ long; urediniospores
   mostly 24-29µ long............................154. isiacae
74. Teliospores mostly more than 50µlong; uredinio-
    spores mostly longer (75)
75. Side wall of teliospore 2.5-3µ thick; germ
         pores 4.............................155. torosa
75. Side wall of teliospore 3-4µ or thicker; pores
         typically 3.......................156. trabutii

GROUP VI: Uredinia without paraphyses, urediniospores
echinulate, germ pores scattered.

1. Teliospores with apical digitations (2)
1. Teliospores without such digitations (8)
2. Teliospore pedicels long, to 150μ..157. asperellae-japonicae
2. Teliospore pedicels short, less than 40μ (3)
3. Urediniospores cinnamon-brown...............158. neocoronata
3. Urediniospores yellowish or colorless (4)
4. Teliospores commonly 3- or 4-celled...............55. addita
4. Teliospores typically 2-celled (5)
5. Telia early exposed (6)
5. Telia covered or tardily exposed (7)
6. Teliospores mostly 42-58μ long................159. festucae
6. Teliospores mostly 85-140μ long.............160. leptospora
7. Teliospores commonly exceeding 50μ long.........54. coronata
7. Teliospores mostly less than 50μ long......161. praegracilis
8. Teliospores commonly with 3 or more cells (9)
8. Teliospores typically 2-celled (11)
9. Teliospores mostly muriformly septate, 3- to
7-celled........................162. tomipara
9. Teliospores mostly 3-celled (10)
10. Urediniospores mostly 20-25μ long; teliospores mostly 30-46μ long..............163. agropyricola
10. Urediniospores mostly 28-35μ long; teliospores mostly 55-85μ long....................164. elymi
11. Amphispores predominant (12)
11. Amphispores lacking (13)
12. Amphispore wall nearly uniformly thick......165. substerilis
12. Amphispore wall much thicker apically............166. vexans
13. Teliospores with a few conspicuous surface
ridges...................167. piperi
13. Teliospore surface otherwise (14)
14. Teliospores finely striate longitudinally.168. pattersoniana
14. Teliospore surface otherwise (15)
15. Teliospore echinulate-verrucose or rugose (16)
15. Teliospores smooth or at least without conspicuous
ridges (18)
16. Teliospore wall mostly 8-10μ thick............169. wolgensis
16. Teliospore wall mostly 3-5μ thick (17)
17. Teliospores mostly 42-60μ long................170. pratensis
17. Teliospores mostly 37-48μ long................171. bromoides
18. Telia typically covered, not erumpent (19)
18. Telia typically erumpent (35)
19. Uredinia seriate in conspicuous chlorotic
streaks...............58. striiformis
19. Uredinia not in such streaks (20)
20. Teliospores germinating without dormancy.......172. eatoniae
20. Teliospores requiring a dormant period (21)
21. Urediniospores typically more than 30μ long (22)
21. Urediniospores typically 30μ or less long (24)
22. Urediniospore wall 1-1.5μ thick, nearly or
colorless..............173. helictotrichi
22. Urediniospore wall thicker (23)

23. Urediniospores mostly 28-34µ long; teliospores
    mostly 38-60µ long........................174. ammophilae
23. Urediniospores mostly 32-44µ long; teliospores
    mostly 50-70µ long........................175. procera
24. Teliospores typically less than 42µ long (25)
24. Teliospores typically more than 42µ long (28)
25. Teliospores mostly 20-30µ wide................176. cryptica
25. Teliospores mostly less than 20µ wide (26)
26. Urediniospores 20-30µ long (or
            globoid)..........177. austroussuriensis
26. Urediniospores mostly 19-24µ long (27)
27. Telia without paraphyses, spore wall 3-5µ
          apically......178. penniseti-lanatae
27. Telia with brown paraphyses, spore wall
    2-3µ apically........................179. limnodeae
28. Telia with no or few paraphyses, the sori scarcely
        or not loculate (29)
28. Telia typically with brown paraphyses, the sori
        typically loculate (34)
29. Telia tending to be or typically early exposed (30)
29. Telia typically long covered (32)
30. Telia without paraphyses; urediniospore wall colorless
        or pale yellowish (31)
30. Telia with few paraphyses; urediniospore near
        cinnamon-brown..................180. ishikariensis
31. Urediniospores 23-27 x 18-22µ; teliospores
    40-65 x 14-19µ........................181. glyceriae
31. Urediniospores 27-32 x 22-25µ; teliospores
    60-80 x 14-18µ........................182. cockerelliana
32. Urediniopsore wall near cinnamon-brown.........183. sessilis
32. Urediniospore wall colorless or yellowish (33)
33. Urediniospore wall 2-2.5µ thick..........184. tsinglingensis
33. Urediniospore wall 1.5µ thick....................185. poarum
34. Teliospores typically more than 20µ wide,
    1-celled spores often abundant..................186. hordei
34. Teliospores typically 20µ or less wide,
    1-celled spores uncommon...................187. recondita
                ....188. koeleriicola
35. Teliospore pedicels 30µ or less long (36)
35. Teliospore pedicels typically 40µ or more (39)
36. Teliospores germinating without dormancy; telia
    waxy in appearance..................189. agropyri-ciliaris
36. Teliospores requiring a dormant period; telia not waxy (37)
37. Teliospores 23-30µ long; urediniospores mostly
        17-22µ long............................190. kansensis
37. Teliospores and urediniospores much larger (38)
38. Teliospores mostly 60-80 x 14-18µ, apical
    wall mostly 4-6µ thick..............182. cockerelliana
38. Teliospores mostly 70-100 x 17-22µ, apical
    wall mostly 7-12µ thick.................191. longissima
39. Teliospore pedicels rarely more than 50µ long (40)
39. Teliospore pedicels mostly exceeding 50µ (44)
40. Urediniospores mostly 19-23µ long (41)
40. Urediniospores mostly exceeding 24µ (42)

41. Teliospore septum horizontal, spores
    chestnut-brown.....................192. mexicensis
41. Teliospore septum oblique or vertical,
    spores golden........................193. abnormis
42. Apical wall of teliospore 4μ or less thick......194. tornata
42. Apical wall exceeding 5μ (43)
43. Teliospores mostly 36-56 x 17-27μ.........195. agrostidicola
43. Teliospores mostly 27-31 x 23-28μ...........196. aegopogonis
44. Lumen of urediniospore stellate or tending
  so due to irregularly thickened (colorless) wall (45)
44. Lumen of urediniospore not stellate (48)
45. Urediniospore wall mostly 3-6μ thick.........197. versicolor
45. Urediniospore wall mostly 2-3μ thick (46)
46. Apical wall of teliospore 6-10μ thick........198. chrysopogi
46. Apical wall mostly less than 6μ thick (47)
47. Teliospores mostly 27-33μ wide, side wall mostly
    3-4μ thick......................199. arthraxonis
47. Teliospores mostly 21-26μ wide, side wall mostly
    3μ thick..........................200. agrophila
48. Urediniospore wall (colorless) thickened apically
   in at least some spores (49)
48. Urediniospore wall uniform in thickness (51)
49. Teliospores mostly 38-54 x 19-24μ.201. arundinellae-anomalae
49. Teliospores larger (50)
50. Urediniospores 17-26μ long; teliospore pedicel
    thin walled..........................202. dietelii
50. Urediniospores 17-22μ long; teliospore pedicel
    thick-walled.........................203. zoysiae
51. Urediniospores mostly 22μ or less long (52)
51. Urediniospores mostly exceeding 22μ (66)
52. Teliospore wall golden brown or paler (53)
52. Teliospores chestnut-brown (55)
53. Teliospores mostly diorchidioid...............193. abnormis
53. Teliospores puccinioid (54)
54. Germ pores of lower teliospore cell at the
    septum...............204. gymnopogonicola
54. Germ pore of lower cell near pedicel.......205. nyasalandica
55. Teliospores typically diorchidioid...........206. boutelouae
55. Teliospores rarely or not diorchidioid (56)
56. Teliospores mostly 30μ or less long..........207. subtilipes
56. Teliospores typically longer (57)
57. Teliospores mostly 35μ or less long (58)
57. Teliospores typically longer (60)
58. Teliospores 12-17μ wide, apical wall
    3-3.5μ thick....................208. sinica
58. Teliospores wider, apical wall thicker (59)
59. Teliospores mostly 28-35 x 17-20μ.........209. scleropogonis
59. Teliospores mostly 28-35 x 22-25μ..............210. hilariae
60. Teliospore pedicels thick-walled, mostly remaining
    terete (61)
60. Teliospore pedicels thin-walled, commonly collapsing
    laterally (64)
61. Teliospores mostly 16-22μ wide (62)
61. Teliospores more than 20μ wide (63)

62. Urediniospore wall 1.5-2µ thick, often
                 thicker apically.....................203. zoysiae
62. Urediniospore wall uniformly 1-1.5µ
                       thick...........212. diplachnicola
63. Teliospores mostly 21-24µ wide, apical wall
     mostly 7-10µ thick.........................211. australis
63. Teliospores mostly 24-27µ wide, apical wall
     mostly 5-8µ thick............................213. permixta
64. Urediniospore wall colorless.................214. chloridis
64. Urediniospore wall golden or cinnamon-brown (65)
65. Teliospore pedicels 90µ or less long;
             urediniospores cinnamon-brown.............215. micrantha
65. Teliospore pedicels commonly exceeding 100µ;
     urediniospores golden.....231. stipae var. stipae-sibiricae
66. Germ pore of lower teliospore cell midway or
     more toward pedicel....................216. pogonarthriae
66. Germ pore of lower cell at the septum (67)
67. Teliospores germinating without dormancy, the
     telia cinereous with basidia..................217. monoica
67. Teliospores requiring or presumably requiring a
                  dormant period (68)
68. Teliospores typically or commonly diorchidioid (69)
68. Teliospores typically puccinioid (71)
69. Urediniospore wall colorless.................218. sierrensis
69. Urediniospore wall about cinnamon-brown (70)
70. Apical wall of teliospore 4-10µ thick........219. exasperans
70. Apical wall of teliospores 4-7µ thick...........220. dochmia
71. Teliospores mostly in the range of 30-40µ long (72)
71. Teliospores usually exceeding 40µ long (81)
72. Urediniospore wall colorless or pale yellowish (73)
72. Urediniospore wall about cinnamon-brown (77)
73. Urediniospores mostly about 20µ long............203. zoysiae
73. Urediniospores mostly more than 20µ long (74)
74. Teliospore pedicels typically thin-walled (75)
74. Teliospore pedicels typically thick-walled (76)
75. Urediniospores mostly 22-26 x 20-24µ.........221. diplachnis
75. Urediniospores mostly 20-25 x 18-20µ.......222. eragrostidis
76. Urediniospores mostly 22-26µ wide; teliospores
             mostly 18-22µ wide.....................223. malalhuensis
76. Urediniospores mostly 19-22µ wide; teliospores
             mostly 23-28µ wide......................224. neyraudiae
77. Side wall of teliospore 2µ or less thick (78)
77. Side wall of teliospore 2.5µ or more thick (79)
78. Teliospore pedicels commonly 100µ or more
                  long.................225. schedonnardi
78. Teliospore pedicels 70µ or less long.......230. andropogonis
79. Urediniospores 20-25µ long, wall 1.5µ
                    thick........226. leptochloae-uniflorae
79. Urediniospores typically more than 25µ long,
     wall 2µ or more thick (80)
80. Teliospores mostly 30-36µ long................227. perotidis
80. Teliospores mostly 24-28µ long.................228. lepturi
81. Urediniospore wall colorless (82)
81. Urediniospore wall golden to cinnamon (83)

82. Urediniospore wall 2-3µ thick; teliospore
    pedicel thick-walled.............201. arundinellae-anomalae
82. Urediniospore wall 1-1.5µ thick, teliospore
    pedicel thin-walled.............................229. macra
83. Urediniospores typically 25µ or less long (84)
83. Urediniospores typically more than 26µ long (85)
84. Teliospore pedicels typically less than 85µ
                        long...................230. andropogonis
84. Teliospore pedicels typically more than 100µ
                        long.......................231. stipae
85. Teliospores conically attenuate, apical wall
                12-22µ thick (86)
85. Teliospores not so attenuate, apical wall usually
                12µ or less (87)
86. Urediniospores mostly 27-31µ diam..........232. changtuensis
86. Urediniospores mostly 30-36 x 27-31µ...........233. harryana
87. Teliospore pedicels typically less than 80µ long (88)
87. Teliospore pedicels typically 100µ or more (94)
88. Teliospores mostly 16-21µ wide (89)
88. Teliospores mostly more than 20µ wide (90)
89. Teliospores mostly 30-44µ long.............230. andropogonis
89. Teliospores mostly 40-50µ long............234. phaenospermae
90. Urediniospores mostly less than 30µ long (91)
90. Urediniospores mostly 30µ or more long (93)
91. Teliospore pedicels fragile, usually broken near
                the hilum..............................235. flavescens
91. Teliospore pedicels usually collapsing laterally,
                but persistent (92)
92. Teliospore wall golden brown, apical wall mostly
                4-6 thick.......................236. polypogonis
92. Teliospore wall chestnut-brown, apical wall
                mostly 7-10µ thick......................237. amphigena
93. Urediniospores mostly 30-37 x 24-28µ; telio-
                spores chestnut, mostly 40-50µ long........238. crandallii
93. Urediniospores mostly 30-34 x 26-30µ telio-
                spores pale golden, mostly 42-60µ long........239. moyanoi
94. Teliospore pedicels thin-walled, collapsing
                laterally or not (95)
94. Teliospore pedicels thick-wall, mostly remaining
                terete (98)
95. Urediniospores mostly 23-26µ long; teliospore
                pedicels to 175µ long.....................231. stipae
95. Urediniospores larger; teliospore pedicels to 115µ
                long (96)
96. Urediniospore wall 1.5-2 or -2.5µ thick (97)
96. Urediniospore wall 3-4µ thick..............240. distichlidis
97. Urediniospores mostly 26-30 x 22-26µ............217. monoica
97. Urediniospores mostly 32-39 x 29-36µ........241. durangensis
98. Teliospores mostly 50-70µ long.............242. lasiagrostis
98. Teliospores 60µ or less long (99)
99. Urediniospore wall about 2µ thick (100)
99. Urediniospore wall mostly 2.5-3.5µ thick......243. trebouxii
100. Urediniospores mostly 26-33 x 22-26µ; telio-
                spores mostly 42-54 x 25-30µ.............244. cryptandri

100. Urediniospores 26-28µ diam; teliospores
      50-58 x 20-23µ........................245. psammochloae

GROUP VII: Uredinia aparaphysate, urediniospores verrucose,
        germ pores equatorial.

1. Teliospore wall finely punctate, pedicels to
      35µ long.............................246. cagayanensis
1. Teliospore wall smooth, pedicels 50µ or more long (2)
2. Teliospores predominantly
      1-celled...............252. esclavensis var. unicellula
2. Teliospores typically 2-celled (3)
3. One-celled teliospores common but not predominant (4)
3. One-celled teliospores only occasionally produced (6)
4. Teliospores ellipsoid, mostly 20µ or less
                wide.................247. infuscans
4. Teliospores broadly ellipsoid, mostly more than
        20µ wide (5)
5. Apical wall of teliospore nearly uniformly chestnut-
      brown, mostly 6-8µ thick..............248. anthephorae
5. Apical wall of teliospore conspicuously paler
      externally, mostly 6-10µ thick.............249. miyoshiana
6. Teliospore pedicels typically less than 100µ
      long, mostly about 80µ (7)
6. Teliospore pedicels typically or commonly
      exceeding 100µ (18)
7. Teliospores broadly ellipsoid, broadly rounded
          apically (8)
7. Teliospores ellipsoid, narrowly rounded or acuminate
          apically (13)
8. Teliospore pedicel thin-walled, usually collapsing (9)
8. Teliospore pedicel thick-walled, mostly terete (11)
9. Wall of teliospore gradually thickened
                apically.............248. anthephorae
9. Wall of teliospore abruptly thickened appically (10)
10. Apical wall of teliospore uniformly brown,
        amphispores lacking............250. cymbopogonis
10. Apical wall of teliospore paler externally,
        amphispores common..................251. cesatii
11. Teliospores mostly 28-36 x 22-27µ; germ pores
      4-6, commonly 5....................252. esclavensis
11. Teliospores larger; germ pores mostly fewer (12)
12. Urediniospores 24-35µ diam, wall 3.5µ thick,
     germ pores 2 or 3............253. eragrostidis-arundinaceae
12. Urediniospores mostly 21-31 x 21-25µ, wall
     2-2.5µ thick, germ pores 3-5, mostly 4.....254. redfieldiae
13. Urediniospores usually less than 26µ long (14)
13. Urediniospore commonly to at least 30µ long (15)
14. Urediniospores mostly 19-22 x 18-20µ; teliospores
      mostly 31-45µ long.......................255. ellisiana
14. Urediniospores mostly 20-26 x 19-23µ; teliospores
      mostly 30-55µ long......................256. cynodontis
15. Urediniospore wall 1.5-2µ thick, wall of
      teliospore pedicel thin..................257. windsoriae

15. Urediniospore wall 2.5-3.5μ thick; apical wall of
    teliospore pedicel thick (16)
16. Teliospores mostly 40-50 x 18-23μ, apical wall
    mostly 6-12μ thick (17)
16. Teliospores mostly 40-56 x 19-27μ, apical wall
    mostly 10-16μ thick.....................258. crassapicalis
17. Sori in conspicuous, confluent, linear
                        series................259. daniloi
17. Sori not in such series..................260. pseudocesatii
18. Urediniospores mostly 23-26 x 20-24μ, wall near
    chestnut-brown; teliospore wall clear
                    chestnut..................261. schoenanthi
18. Urediniospores mostly more than 26μ long, golden or
    cinnamon-brown; teliospore wall deep chestnut (19)
19. Urediniospores 26-28μ diam; teliospores 37-53
                    x 18-32μ.......................262. danthoniae
19. Urediniospores attaining at least 30μ in length (20)
20. Urediniospores mostly 25-33 x 18-23μ, wall 2.5-3μ;
    teliospores mostly 40-58
                    x 20-27μ........263. aristidae var. aristidae
20. Urediniospores mostly 23-30 x 21-26μ (21)
21. Urediniospore wall 2.5-4.5μ thick; teliospore
    side wall 3-4.5μ............263. aristidae var. chaetariae
21. Urediniospore wall 2.5-3.5μ thick; teliospore
    side wall 2-3.5μ thick..................264. aeluropodis

GROUP VIII: Uredinia aparaphysate, urediniospores
             verrucose, germ pores scattered.

1. Telia covered by the epidermis, teliospore
        pedicels short...........................265. abramoviana
1. Telia exposed; teliospores pedicels long (2)
2. Urediniospore wall colorless or nearly so,
        labrynthiformly rugose; teliospore pedicels
        to 135μ long................................266. pazensis
2. Urediniospore wall at least golden, not rugose (3)
3. Urediniospore wall with rod-like papillae;
        teliospore pedicels about 60μ
                    long........267. polliniae-quadrinervis
3. Urediniospore wall finely verrucose (4)
4. Teliospore pedicels typically less than 100μ long (5)
4. Teliospores typically exceeding 100μ (9)
5. Urediniospores mostly 29-34 x 25-28μ...........268. setariae
5. Urediniospores 19-26 x 20-25μ (6)
6. Urediniospore pores 4-6, wall 1.5-2.5μ; telio-
        spores mostly 25-34 x 19-24μ, pedicels thick-
        walled....................................269. leptochloae
6. Urediniospore pores 6-8, wall mostly 2-3μ (7)
7. Teliospore pedicels thin-walled, spores mostly
                    30-36 x 19-24μ..................270. chihuahuana
7. Teliospore pedicels thick-walled (8)
8. Teliospores mostly 31-36 x 22-25μ, pedicels
        colorless................................271. pseudoatra
8. Teliospores mostly 30-46 x 21-24μ, pedicels
                    brownish.......................272. morigera

9. Urediniospores mostly 20-24μ diam, germ pores
    mostly 3, 4 or 5, equatorial, plus 1 or 2
    apical..................................273. <u>subnitens</u>
9. Urediniospores slightly or much larger, germ
    pore arrangement different (10)
10. Urediniospores mostly 20-26 x 19-26μ; telio-
    spores mostly 26-45 x 19-26μ.................274. <u>opuntiae</u>
10. Urediniospores mostly 32-45 x 18-23μ; telio-
    spores mostly 40-55 x 21-27μ...................275. <u>tarri</u>

GROUP IX: Uredinia and urediniospores unknown, or
           lacking in the life cycle (opis-forms).

1. Teliospores frequently 3- or 4-celled, wall
    uniformly 2-3μ or to 4μ at apex..........276. <u>miscanthicola</u>
1. Teliospores typically 2-celled or occasionally
    with 1-celled spores admixed (2)
2. Telia covered, teliospore pedicels very
                          short.............277. <u>lavroviana</u>
2. Telia exposed (3)
3. Telia with capitate paraphyses, spore pedicels
                   short...................278. <u>achnatheri-sibirici</u>
3. Telia without paraphyses, spore pedicels mostly long (4)
4. Telia elongate, several mm to several cm long,
    usually as wide as the leaf, deeply pulvinate (5)
4. Telia not elongate, more or less circular, may be
    deeply pulvinate (8)
5. Apex of teliospore rostroid, to 76μ
                          long...........279. <u>longirostroides</u>
5. Apex of teliospore not rostroid (6)
6. Pedicels thin-walled, fragile, broken near
    the spore.............................280. <u>avocensis</u>
6. Pedicels thick-walled, persistent, to 200μ long (7)
7. Aecia associated with telia; autoecious......281. <u>graminella</u>
7. Aecia not associated, heteroecious.........282. <u>interveniens</u>
8. Teliospore pedicels less than 100μ long
    or typically broken shorter (9)
8. Teliospore pedicels commonly much exceeding
    100μ, thick-walled, persistent (12)
9. Telia on stems and inflorescence, spores
    golden brown..............................283. <u>bewsiae</u>
9. Telia on leaves, spores chestnut-brown (10)
10. Teliospore pedicels brown....................284. <u>phaeopoda</u>
10. Teliospore pedicels colorless or yellowish (11)
11. Teliospores mostly 30-34 x 21-25μ, side wall
    mostly 2-3μ thick.....................285. <u>fushunensis</u>
11. Teliospores mostly 28-43 x 12-20μ, side wall
    mostly 1-1.5μ thick................286. <u>festucae-ovinae</u>
12. Teliospores ellipsoid or oblong-ellipsoid, the
    apex broadly rounded (13)
12. Teliospores fusiform or fusiform-ellipsoid, the
    apex narrowly round or accuminate (14)
13. Side wall of teliospore uniformly 2.5-3.5μ thick,
    uniformly pigmented.......................287. <u>oryzopsidis</u>

13. Side wall of teliospore conspicuously unilaterally
thickened, pigmentation paler externally.......288. *tenella*
14. Teliospore side wall usually obviously
unilaterally thickened, apical wall
25-75µ..................289. *flammuliformis*
14. Teliospore side wall slightly or not unilaterally
thickened, apical wall much thinner (15)
15. Apex of teliospore mostly long accuminate, the
apical wall 17-34µ thick................290. *nigroconoideae*
15. Apex of teliospore less extended, the apical wall
mostly 14-22µ thick..................291. *brachystachyicola*

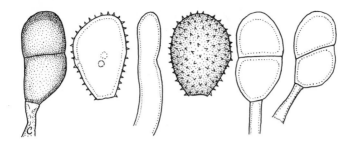

Figure 20

1. PUCCINIA CHAETOCHLOAE Arth. Bull. Torrey Bot. Club
34:585. 1907. Fig. 20.

    Uredo chaetochloae Arth. Bull. Torrey Bot. Club 33:518.
1906.

    Puccinia maublancii Rangel Arch. Mus. Nac. Rio de
Janeiro 18:159. 1916.

    Aecia unknown. Uredinia amphigenous, rather long capped
by the epidermis, cinnamon-brown, with inconspicuous, colorless
or yellowish, thin-walled paraphyses; spores (26-)30-42(-50)
x (19-)22-28(-30)μ, mostly oval or oblong and commonly
angular, wall 2μ thick, golden or cinnamon-brown, echinulate,
pores 3 or 4, equatorial. Telia blackish, covered by the
epidermis, without paraphyses; spores (28-)32-40(-45) x
(17-)20-26μ, mostly clavate or oblong-ellipsoid, usually
angular, wall 1.5μ thick at sides, 2-4μ apically, chestnut-
brown, smooth, pedicels yellowish or golden, thin-walled
and commonly collapsing, to 25μ long, persistent.

    Hosts and distribution: Ixophorus unisetus (Presl)
Sclecht., species of Paspalum, Pennisetum spicatum (L.)
Koern., Setaria geniculata Beauv., S. macrosperma (Scribn.
& Merr.) Schum: southern U.S.A. to the Dominican Republic,
Mexico, Venezuela, and Brazil.

    Type: Holway, on Setaria macrosperma, Miami, Florida
(PUR).

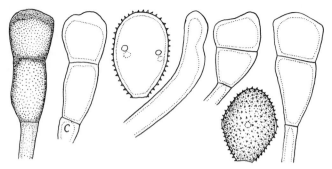

Figure 21

2. PUCCINIA STENOTAPHRI Cumm. Bull. Torrey Bot. Club
87:40. 1960. Fig. 21.

Uredo stenotaphri H. Syd. & P. Syd. Ann. Mycol. 7:544.
1909.

Aecia unknown. Uredinia amphigenous, yellowish to
cinnamon-brown, with yellowish to hyaline, usually moderately
(1.5-3µ) thick-walled, cylindric, peripheral paraphyses;
spores (28-)30-40(-46) x (22-)25-28(-30)µ, mostly oval or
ellipsoid, wall 1.5µ thick, golden to cinnamon-brown,
echinulate, pores 4 or 5, equatorial. Telia blackish,
long-covered, without paraphyses; spores (37-)44-60 x
19-26µ, mostly clavate or oblong-clavate, wall 1.5µ thick
at sides, 2.5-4(-5.5)µ apically, chestnut-brown, smooth;
pedicels brownish, thin-walled but mostly not collapsing,
to 15µ long, persistent.

Hosts and distribution: Pennisetum hordeoides (Lam.)
Steud., P. setosum (Swartz) L. Rich., Stenotaphrum dimidiatum
(L.) Brongn., S. secundatum (Walt.) O. Ktze.; Stereochlaena
cameronii (Stapf) Pilger: India, Portuguese East Africa,
Mauritius, Puerto Rico, and Florida (U.S.A.).

Type: Wiehe No. 115, on S. dimidiatum, Mauritius
(PUR; isotype IMI).

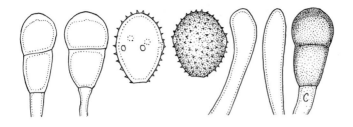

Figure 22

3. PUCCINIA OAHUENSIS Ell. & Ev. Bull. Torrey Bot. Club 22:435. 1895. Fig. 22.

Uredo digitariaecola Thuem. Myc. Univ. No. 2041. 1882.

Uredo digitariae-ciliaris Mayor Bull. Soc. Neuchâtel. Sci. Nat. 41:101. 1914.

Uredo duplicata Rangel Arch. Mus. Nac. Rio de Janeiro 18:160. 1916.

Puccinia digitariae Pole Evans Ann. Bolus Herb. 2:111. 1917.

Uredo syntherismae Speg. An. Mus. Nac. Hist. Nat. Buenos Aires 31:46. 1922.

Aecia unknown. Uredinia mostly on abaxial surface, yellowish brown, pulverulent, with hyaline, thin-walled, mostly incurved, usually clavate paraphyses; spores (23-)25-32(-40) x (18-)20-25(-28)µ, mostly oval or obovoid, wall 1.5µ thick, golden or light cinnamon-brown, echinulate, pores (3-)4 or 5(-6), equatorial or in some specimens tending to be scattered. Telia blackish, long-covered, with scant peripheral brownish paraphyses; spores (27-)35-45(-52) x (12-)16-22(-26)µ, clavate, obovoid-clavate, or oblong, wall 1-1.5(-2)µ thick at sides, 2-5(-7)µ apically, chestnut-brown, smooth; pedicels hyaline to brownish, thin-walled and collapsing or not, to 20µ, persistent.

Hosts and distribution: On species of Digitaria: circumglobal in warm regions.

Type: Heller No. 1976, on unknown grass (=D. pruriens), Oahu, Hawaii (NY; isotype PUR).

Cummins (Bull. Torrey Bot. Club 70:517-530. 1943) published a photograph of teliospores of the type.

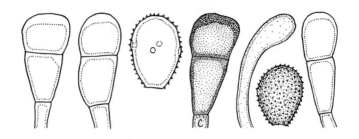

Figure 23

4. PUCCINIA CHASEANA Arth. & Fromme Torreya 15:264.
1915. Fig. 23.

Aecia unknown. Uredinia amphigenous, cinnamon-brown,
with inconspicuous, peripheral, colorless, thin-walled,
incurved, mostly clavate paraphyses; spores (24-)26-30(-33)
x (18-)20-25µ, mostly oval, wall 1.5µ thick, golden to
cinnamon-brown, echinulate, pores (3-)4, equatorial. Telia
blackish, covered by the epidermis, without paraphyses;
spores (33-)36-45(-48) x (16-)18-21(-25)µ, mostly angularly
clavate, wall 1-1.5µ thick at sides, 2.5-5µ apically,
chestnut-brown, smooth, pedicels yellowish, thin-walled and
often collapsing, to 15µ long, persistent.

Hosts and distribution: Anthephora hermaphrodita (L.)
Kuntze: Jamaica and Cuba to Guatemala and Colombia.

Type: Lloyd No. 1118, Jamaica (PUR).

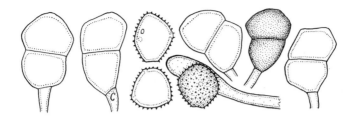

Figure 24

5. PUCCINIA DOLOSA Arth. & Fromme Torreya 15:262. 1915
var. dolosa Fig. 24.

Aecia unknown. Uredinia amphigenous or mostly on
abaxial surface, pale cinnamon-brown; paraphyses mostly
cylindrical, colorless, inconspicuous; spores (19-)24-29
x (17-)20-24µ, mostly obovoid, triangular in end-view,
wall 1-1.5µ thick, golden or pale cinnamon-brown, finely
echinulate, germ pores 3, equatorial, in the angles. Telia
covered by the epidermis, blackish brown, inconspicuous;
spores (27-)34-40(-44) x (17-)23-26µ, variable and often
angular, mostly oblong or oblong-clavate, wall 1-1.5µ
thick at sides 2-4µ apically, very brittle, chestnut-brown,
smooth; pedicels yellowish, thin-walled and mostly collapsing
to 45µ long, often broken much shorter.

Hosts and distribution: On species of Paspalum:
Southern United States southward to Puerto Rico, Panama,
Venezuela, and Brazil.

Type: E. W. D. Holway No. 3056, on P. tenellum,
Guadalajara, Mexico, 25 Sept. 1903 (PUR).

PUCCINIA DOLOSA Arth. & Fromme var. circumdata (Mains)
Ramachar & Cumm. Mycopathol. Mycol. Appl. 25:13. 1965.

Puccinia circumdata Mains Carnegie Inst. Wash. Publ.
461:101. 1935.

Urediniospores (23-)25-29(-32) x (17-)19-22(-23)µ,
oval or obovoid, triangular in end view, wall 1-1.5µ thick,
golden or light cinnamon-brown, echinulate, pores 3,
equatorial, in the angles. Teliospores (25-)27-34 x
(17-)20-24µ, variable, usually angular and mostly oblong
or oblong-ellipsoid, wall 1-1.5µ at sides, 2-3µ apically.

Hosts and distribution: Panicum fasciculatum Swartz,
P. parvifolium Lam.: Puerto Rico to Cuba, Mexico, Panama,
Brazil, and Texas (U.S.A.).

Type: Swallen No. 2592, Yucatan, Mexico (MICH; isotype
PUR).

PUCCINIA DOLOSA Arth. & Fromme var. catervaria (Cumm.) Ramachar & Cumm. Mycopathol. Mycol. Appl. 25:14. 1965.

Puccinia catervaria Cumm. Mycologia 34:679. 1942.

Urediniospores (23-)25-29(-31) x (19-)21-24µ, oval or obovoid, wall 1.5-2µ thick, cinnamon-brown, echinulate, pores 4, equatorial. Teliospores 26-33 x (18-)20-23µ, mostly ellipsoid or oblong-ellipsoid and usually angular, wall 1-1.5µ thick at sides, 2-3.5µ apically.

Hosts and distribution: Setaria geniculata (Lam.) Beauv.: Bolivia.

Type: Holway No. 348 (Reliq. Holw. No. 53), Cochabamba, Bolivia (PUR).

What appears to be the same fungus has been collected on an unknown grass in Nayarit state, Mexico.

PUCCINIA DOLOSA Arth. & Fromme var. biporula Ramachar & Cumm. Mycopathol. Mycol. Appl. 25:14. 1965.

Urediniospores (22-)23-27(-29) x (16-)17-21(-22)µ, ellipsoid or obovoid, wall 1.5-2µ thick, cinnamon-brown, echinulate, pores 2, rarely 3, equatorial. Teliospores (22-)26-32(-34) x (17-)18-22µ, ellipsoid or oblong-ellipsoid, wall 1.5-2µ thick at sides, 2-3.5µ apically.

Hosts and distribution: Setaria grisebachii Fourn.: Mexico.

Type: Cummins No. 63-174, Tamaulipas State, Mexico (PUR).

In 1942, Cummins (Mycologia 34:669-695) published photographs of teliospores of the types of the first 3 varieties (as species).

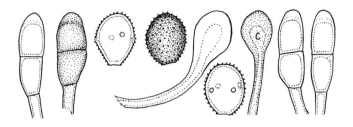

Figure 25

6. PUCCINIA AESTIVALIS Diet. Bot. Jahrb. 34:585. 1905.
Fig. 25.

Uredo ogaoensis Cumm. Mycologia 33:150. 1941.

Aecia unknown. Uredinia on abaxial leaf surface,
cinnamon-brown, with colorless to golden, capitate paraphyses, wall thickened apically to 4-8µ, the stipe commonly
thick-walled; spores 20-25(-28) x (16-)18-21(-23)µ, mostly
oval, wall 1.5µ thick, cinnamon-brown, echinulate, germ
pores 4, equatorial; amphispores (23-)26-32(-35) x
(16-)20(-23)µ, obovoid or pyriform, wall 2-3µ thick,
chestnut-brown, echinulate, pores 3(4), equatorial. Telia
cinnamon-brown, compact, exposed; spores (25-)29-40(-43) x
11-16µ, oblong or oblong-ellipsoid, wall 2µ thick at sides,
4-8µ apically, pale golden, smooth, pedicels yellowish,
thin-walled, and collapsing, to 20µ long; spores germinate
without a dormant period.

Hosts and distribution: Species of Microstegium: Japan
to Sumatra and New Guinea.

Type: Nambu, on Pollinia nuda (=Microstegium nudum
(Trin.) (A. Camus), Tokyo (S).

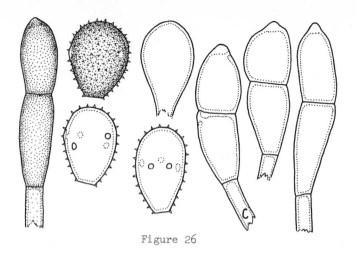

Figure 26

7. PUCCINIA GARNOTIAE T.S. Ramak. & Sund. Indian Phytopathol.
6:30-31. 1953. Fig. 26.

Aecia unknown. Uredinia on abaxial leaf surface,
cinnamon-brown or darker, with mostly obovoid paraphyses,
mostly 35-50 x 15-25μ, wall 1-1.5(-2)μ thick, colorless;
spores (25-)28-33(-40) x (19-)21-24(-26)μ, mostly obovoid
or broadly ellipsoid, wall 1.5(-2)μ thick, cinnamon-brown
or slightly darker, echinulate, germ pores (4)5(6), equatorial.
Telia on abaxial surface, erumpent but usually capped by a
loose or partially attached piece of epidermis, about
cinnamon-brown, compact; spores (38-)43-68(-75) x (16-)18-22
(-24)μ, ellipsoid or nearly cylindrical, wall 1μ thick at
sides, 3-4(-5)μ apically by a small umbo or papilla, golden
brown or pale chestnut-brown, smooth; pedicels colorless,
thin-walled, mostly 20-25μ long.

Type: Ramakrishnan and Sundaram, on Garnotia arundinacea
Hook., Burliar (Nilgiris), India (Herb. Mycol. No. 2830,
MS; isotype PUR). Known only in India.

Teliospores of the type, collected 23 Mar. 1953, are
germinating.

92

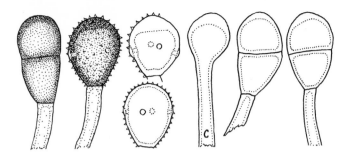

Figure 27

8. PUCCINIA ANGUSII Cumm. Mycologia 57:818. 1965.
Fig. 27.

Aecia unknown. Uredinia abaxial, yellowish brown,
paraphyses capitate, (12-)15-20(-26)μ diam apically,
wall 3-7μ thick apically, colorless or golden; spores
24-28(-32) x (19-)20-24(-26)μ, obovoid or broadly ellipsoid,
wall 1-1.5(-2)μ thick, echinulate, germ pores (3)4(5),
equatorial; amphisporic uredinia blackish brown, paraphysate;
spores (26-)28-34(-36) x (21-)24-28(-30)μ, obovoid or
globoid, wall 3-4μ thick laterally, 3-7(-9)μ apically,
verrucose-echinulate, dark chestnut-brown, germ pores
(3)4(5), equatorial. Telia abaxial, blackish brown,
compact, early erumpent; spores (27-)32-40 x (17-)20-24(-26)μ,
wall 1-2(-2.5)μ thick at sides, (2-)2.5-4(-5)μ at apex,
chestnut-brown, smooth; pedicels yellowish, persistent, to
30μ long.

Hosts and distribution: _Danthoniopsis_ _pruinosa_ C. E.
Hubb: Mt. Shimabala, N. Rhodesia, Angus No. M1144 (PUR;
isotype IMI).

P. _angusii_ is distinctive because of the amphispores.
A photograph of spores of the type was published with the
diagnosis.

9. PUCCINIA KUEHNII Butl. Ann. Mycol. 12:82. 1914.

   Uromyces kuehnii Krueger Ber. Versuchs Stat. f.
Zuckerrohr West-Java, Kagot-Tegal 1:120. 1890 (based on
uredinia).

   Uredo kuehnii (Krueger) Wakk. & Went in De Ziekten van
het suekerviet Java, Lieden, P. 144. 1898.

   Aecia unknown. Uredinia amphigenous or only hypophyllous;
cinnamon or yellowish brown, with inconspicuous, peripheral,
cylindric or capitate, thin-walled, hyaline or pale brownish
paraphyses; spores (25-)30-43(-48) x 17-26μ, mostly obovoid
or pyriform, wall 1.5-2.5μ thick at sides, often thickened
to 5μ at the apex, golden or connamon-brown, echinulate,
pores 4 or 5, equatorial. Telia small, blackish, early
exposed; spores 25-40 x 10-18μ, mostly oblong-clavate with
rounded apex, wall not thickened apically (1.5μ ?), smooth,
yellowish (immature ?); pedicel hyaline, short.

   Hosts and distribution: Saccharum arundinaceum Retz.,
S. officinarum L., S. narenga Wall., S. spontaneum L.,
Sclerostachya fusca (Roxb.) A. Camus: Africa to India,
Australia and Japan.

   Type: Butler, on Saccharum spontaneum, Bassein, Burma
(HC10).

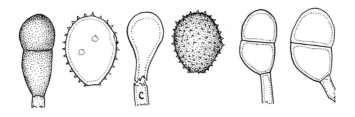

Figure 28

10. PUCCINIA LOUDETIAE Wakef. & Hansf. Linn. Soc. London, Session 161, 1948-49: 183. 1949. Fig. 28.

Puccinia trichopterygis Wakef. & Hansf. E. African Agr. J. 3: 323. 1938, nom. nud.

Aecia unknown. Uredinia mostly abaxial, paraphyses capitate, 10-17(-20)μ diam apically, wall 2-7(-12)μ thick at apex; spores (24-)26-35(-37) x (16-)20-26(-28)μ, ellipsoid or obovoid, wall 1.5-2μ thick, cinnamon- or dark cinnamon-brown, echinulate, pores 3(4), equatorial. Telia abaxial, dark brown, compact, exposed but not conspicuous; spores (28-)30-40(-42) x (14-)17-20(-23)μ, mostly ellipsoid or elongate-obovoid, wall uniformly 1-1.5μ thick or very slightly thicker apically, golden brown or clear chestnut-brown, smooth; pedicels yellowish, thin-walled, persistent, to 30μ long.

Hosts and distribution: Loudetia arundinacea (Hochst.) Steud., L. kagerensis (K. Schum.) C. E. Hubb., L. flammida (Trin.) C. E. Hubb.: Sierra Leone and Uganda.

Type: Hansford No. 1174, on Loudetia flammida (as L. phragmitoides), Uganda (K).

The uniformly thin walls of the teliospores distinguish P. loudetiae.

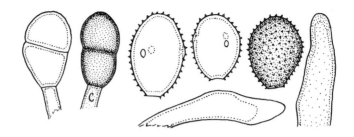

Figure 29

11. PUCCINIA CACAO McAlp. Rusts of Australia p. 117.
1906. Fig. 29.

Uredo rottboelliae Diet. Bot. Jahrb. 32:52. 1902.

Uredo mira Cumm. Bull. Torrey Bot. Club 70:528. 1943.

Aecia (Aecidium manilense Arth. & Cumm.) systemic in
species of Hygrophila; spores 24-31 x 18-26µ, wall hyaline,
2-2.5µ thick, verrucose. Uredia and telia amphigenous in
leaves. Uredinia nearly chestnut-brown, with variable,
peripheral, straight or incurved, hyaline paraphyses, wall
uniformly 2-3µ thick or often greatly thickened apically;
spores (29-)32-40(-42) x 23-29µ, mostly broadly ellipsoid
or obovoid, wall 2-2.5µ thick, dark cinnamon- or light
chestnut-brown, echinulate, pores 3, rarely 4, equatorial.
Telia not seen; spores in the uredinia 30-39 x 18-22µ,
mostly ellipsoid or obovoid, wall uniformly 2-2.5µ thick,
chestnut-brown, smooth; pedicels thin-walled, hyaline,
collapsing and deciduous.

Hosts and distribution: Hackelochloa porifera (Hack.)
Rhind, species of Hemarthria: Argentina to Africa, India,
Australia, and Japan.

Type: Robinson, on Rottboellia compressa (=Hemarthria
compressa (L. f.) R. Br., Australia (MEL).

Thirumalachar and Narasimhan (Mycologia 46:222-228.
1954) proved the life cycle by inoculation.

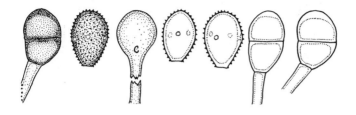

Figure 30

12. PUCCINIA SUBLESTA Cumm. Ann. Mycol. 35:99. 1937.
Fig. 30.

Aecia unknown. Uredinia on abaxial leaf surface,
cinnamon-brown, with yellowish to golden, capitate paraphyses, the wall uniformly 1.5-3µ thick, spores 19-25 x
15-19µ, ellipsoid or obovoid, wall 1.5-2µ thick, cinnamon-
brown, echinulate, pores 3 or usually 4, equatorial.
Teliospores in the uredinia 24-28 x 18-20µ, mostly oval
or oblong-obovoid, wall 2µ thick at sides, 3-3.5µ apically,
chestnut-brown, smooth; pedicels colorless, thin-walled,
collapsing, to 25µ long but usually broken near the spore.

Hosts and distribution: Isachne beneckii Hack.: the
Philippines.

Type: Clemens No. 7730, Mt. Pinatubo, Luzon (PUR).

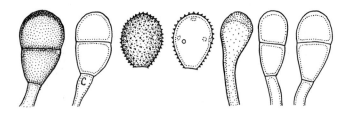

Figure 31

13. PUCCINIA BENGUETENSIS H. Syd. Ann. Mycol. 15:174.
1917. Fig. 31.

Puccinia polliniae-imberbis Hirat. f. J. Japan. Bot.
13:248. 1937.

Puccinia microstegii Saw. Taiwan Agr. Res. Inst. Rept.
86:61. 1943. Nomen nudum.

Aecia unknown. Uredinia on abaxial leaf surface,
cinnamon-brown, with yellowish to golden, capitate paraphyses, the wall thin in the stipe, 3-8µ apically; spores
(20-)23-30(-33) x (16-)19-22(-25)µ, mostly oval or obovoid,
wall 1.5-2µ thick, cinnamon-brown, echinulate, germ pores
4-6, usually 4 or 5 equatorial and 1 apical. Telia blackish
brown, exposed, compact; spores 24-33(-38) x (16-)18-21(-23)µ,
mostly obovoid, wall 1.5µ thick at sides, 2-4(-5)µ apically,
chestnut-brown, smooth; pedicels thin-walled, usually
collapsing, yellowish to brown, to 40µ long, persistent.

Hosts and distribution: Species of Microstegium: China
Taiwan, and the Philippines.

Type: Clemens No. 9272, on Pollinia sp. (=Microstegium,
probably vimineum (Trin.) A. Camus), Pauai, Luzon, Philippines.
(S.; isotype BPI).

Cummins published a photograph of teliospores of the type
(Urediniana 4: Plate III, Fig. 19. 1953).

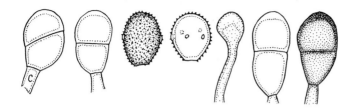

Figure 32

14. PUCCINIA MICROSPORA Diet. Bot. Jahrb. 27:101. 1905.
Fig. 32.

Aecia unknown. Sori mostly in abaxial surface of leaves,
seriate. Uredinia cinnamon-brown, with hyaline or pale
golden, capitate paraphyses, the apical wall 6-18μ thick;
spores (21-)23-27(-29) x 16-21(-23)μ, mostly oval or
obovoid, wall 1.5-2μ thick, cinnamon-brown, echinulate, pores
4(5), equatorial. Telia blackish brown, compact, early
exposed; spores (25-)28-35(-38) x (14-)16-21(-23)μ apically,
smooth, chestnut-brown; pedicels brown, thin-walled but
usually not collapsing, persistent, to 20μ long.

Hosts and distribution: Andropogon sp. (probably
Erianthus), Erianthus angustifolius Nees, E. trinii Hack.,
Hemarthria japonica (Hack.) Roshevitz, Imperata brasiliensis
Trin., I. cylindrica (L.) Beauv., I. exaltata Brogn., I.
hookeri Rupr., I. tenuis Hack., Rottboellia exaltata L. f.:
Brazil and the southwestern United States to Japan, China,
and Borneo.

Type: Nambu, on Rottboellia compressa var. japonica
(=Hemarthria japonica), Tokyo, Japan (S).

Cummins published a photograph of teliospores of the type
(Uredinia 4: Pl. I, Fig. 3. 1953).

Aecia unknown. Sori mostly in abaxial surface of leaves,
seriate. Uredinia cinnamon-brown, with hyaline or pale
golden, capitate paraphyses, the apical wall 6-18μ thick;
spores (21-)23-27(-29) x 16-21(-23)μ, mostly oval or
obovoid, wall 1.5-2μ thick, cinnamon-brown, echinulate, pores
4(5), equatorial. Telia blackish brown, compact, early
exposed; spores (25-)28-35(-38) x (14-)16-21(-23)μ, mostly
obovoid or oblong-obovoid, wall 1.5μ thick at sides, 2-3(-5)μ
apically, smooth, chestnut-brown; pedicels brown, thin-walled
but usually not collapsing, persistent, to 20μ long.

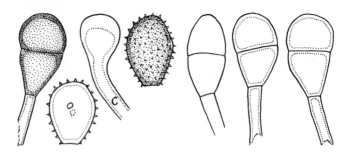

Figure 33

15. PUCCINIA THIENSIS Huguenin Bull. Trim. Soc. Mycol.
France 83(1967): 950. 1968. Fig. 33.

Aecia unknown. Uredinia on abaxial leaf surface, in
linear series, cinnamon-brown, paraphyses capitate, the
head 12-20μ wide, wall thin in the stipe, 10-15μ thick in
apex, colorless to golden; spores (24-)26-32(-34) x (19-)21-25
(-27)μ, mostly obovoid, wall 1.5-2μ thick, cinnamon-brown,
some spores chestnut-brown and with a wall 2.5-3μ thick,
echinulate, germ pores (3)4(5), equatorial. Telia on
abaxial surface, blackish brown, early exposed, compact;
spores (26-)30-40(-42) x (16-)19-22(-24)μ, mostly obovoid,
wall (1-)1.5(-2)μ thick at sides, gradually thickened
apically to 2-3(-4)μ, chestnut-brown, smooth; pedicels
brown, persistent, thick-walled, mostly 20μ or less long.

Hosts and distribution: Paspalum orbiculare G. Forst.:
New Caledonia.

Type: Collector not stated, Forêt de Thi (New Caledonia
No. 66046; isotype PUR).

The species is similar to P. microspora, but has larger
spores.

100

16. PUCCINIA ARUNDINELLAE-SETOSAE F. L. Tai Farlowia 3:114.
1947.

Aecia unknown. Uredinia amphigenous, seriately arranged,
paraphyses capitate or clavate, wall (according to Tai's
Fig. 6) thin at sides, thick (to 12µ ?) apically, brownish;
spores ovate or subglobose (obovoid as illustrated), 21-30
x 18-21µ, wall uniformly 2-2.5µ thick, echinulate, chestnut-
brown, germ pores 3-5, equatorial. Telia similar but
pulvinate, blackish; spores ellipsoid or ovate-oblong,
30-43 x 16-21µ, wall 1.5µ thick at sides, 2-4µ at apex,
chestnut-brown, smooth; pedicel short, chestnut-brown;
1-celled spores intermixed.

Hosts and distribution: Arundinella setosa Trin.:
Kunming, China, Tai No. 1939 (type presumably in Plant
Pathology Herbarium, Institute of Agricultural Research,
National Tsing Hua University).

The description is adapted from the Latin diagnosis. Tai
also listed Nos. 7575 and 7611 on A. setosa and No. 7543
on Sporobolus indicus.

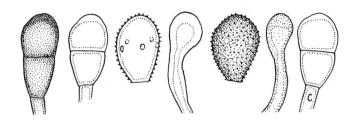

Figure 34

17. PUCCINIA MELANOCEPHALA H. Syd. & P. Syd. in Sydow &
Butler Ann. Mycol. 5:500. 1907. Fig. 34.

Puccinia erianthi Padw. & Khan Imp. Mycol. Inst. Kew
Mycol. Papers 10:32-33. 1944.

Aecia unknown. Uredinia on abaxial leaf surface,
cinnamon-brown, with capitate, colorless to golden paraphyses, the wall 1.5-3µ thick in the stipe, 3-7µ apically;
spores (25-)28-33(-36) x 18-23(-25)µ; mostly obovoid, wall
1.5µ thick, cinnamon-brown, echinulate, germ pores 4 or 5,
equatorial. Telia on abaxial surface, exposed, blackish
brown; spores (29-)30-43(-54) x (15-)17-21(-23)µ, mostly
clavate, wall 1.5-2µ thick at sides, 3-4(-6)µ apically,
chestnut-brown, smooth; pedicels thin-walled but usually
not collapsing, brown, to 12µ long.

Hosts and distribution: Erianthus ravennae (L.) Beauv.
?, E. rufipilis (Steud.) Griseb., Saccharum officinarum L.:
India and China.

Type: Butler, on Arundinaria sp. (=error for Erianthus
probably ravennae) (S).

The only specimen with telia is annotated "Uredo previously
sent (No. 512)" and date and locality are the same. The
type in S was not numbered.

An inflorescence in No. 512 was identified by John R.
Reeder as certainly Erianthus and probably E. ravennae.

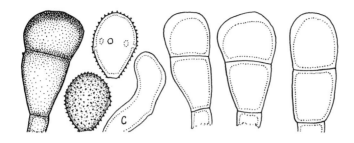

Figure 35

18. PUCCINIA ARTHRAXONIS-CILIARIS Cumm. Uredineana 4:16.
1953. Fig. 35.

Uredo arthraxonis-ciliaris P. Henn. Hedwigia 47:251.
1908.

Aecia unknown. Uredinia on abaxial leaf surface,
yellowish to yellowish brown, with cylindrical, clavate
or clavate-capitate paraphyses, usually incurved, colorless
or yellowish, the wall 2-3μ thick or thickened to 4μ
apically; spores (20-)23-30 x (16-)18-23μ, mostly oval or
obovoid, wall 1.5μ thick, yellowish brown, echinulate,
germ pores 4 or 5, equatorial. Telia on abaxial surface,
exposed, blackish brown, spores (34-)38-48 x 20-25(-27)μ,
mostly oblong-clavate, wall 1.5μ thick at sides, 3-5μ
apically, chestnut-brown, smooth; pedicels brownish, thin-walled
but usually not collapsing, persistent, to 20μ long.

Hosts and distribution: Arthraxon hispidus (Thunb.)
Merr., A. quartinianus (A. Rich.) Nash, A. mauritianus
Stapf: Uganda and Mauritius to India, New Guinea, the
Philippines, China, and Japan.

Type: Ramos No. 7021, on A. hispidus, Luzon, the Philippines
(PUR).

A photograph of teliospores of the type was published
by Cummins (Uredineana 4: Pl. II, Fig. 9. 1953).

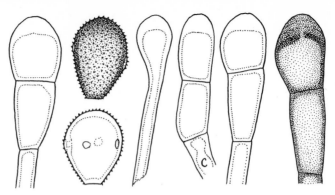

Figure 36

19. PUCCINIA VIRGATA Ell. & Ev. Proc. Acad. Philadelphia 1893:154. 1893. Fig. 36.

Caeoma andropogi Schw. Trans. Amer. Phil. Soc. II. 4:290. 1832.

Puccinia clavispora Ell. & Barth. Erythea: 4:79. 1896.

Uredo alabamensis Diet. in Atkinson Bull. Cornell Univ. 3:22. 1897.

Aecia unknown. Uredinia chestnut-brown, amphigenous, with golden brown, clavate or capitate paraphyses, wall 1.5-3µ thick in stipe, 3-9µ apically; spores 31-40(-43) x (16-)20-27(-30)µ, mostly obovoid, wall 2-3µ thick, often 3-6µ apically, chestnut-brown apically, usually paler below, echinulate, pores 4(5), equatorial. Telia blackish brown, compact, early exposed; spores (40-)45-60(-75) x 18-26µ, mostly clavate, wall 1.5-2µ thick at sides, 5-10(-12)µ apically, chestnut-brown; pedicels thick-walled, not collapsing, brown, to 20µ long.

Hosts and distribution: Erianthus (?) sp., species of Sorghastrum; northern U.S.A. to Mexico and Brazil.

Type: Bartholomew, on Panicum virgatum (error for Sorghastrum nutans (L.) Nash, Kansas, (FH; isotype PUR).

Cummins (Uredineana 4: Plate II, Fig. 13, 1953) published a photograph of teliospores of the type.

This and the following 3 species are similar in most characters and perhaps could be treated as varieties.

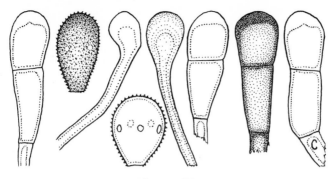

Figure 37

20. PUCCINIA MISCANTHI Miura Fl. Manchuria & E. Mongolia
Pt. 3:302. 1928. Fig. 37.

Puccinia miscanthicola Tranz. Conspectus Ured. U.S.S.R.
p. 93. 1939, not Tai & Cheo, 1937.

Aecia occur on species of Plantago; peridium short,
erose; spores (20-)22-27(-29) x (17-)20-24μ, from ellipsoid
to globoid, wall 1-1.5μ thick, colorless or pale yellowish,
verrucose. Uredinia mostly on abaxial leaf surface, cinnamon-
brown, paraphyses capitate, wall 2-3μ thick below, 6-10(-15)μ
apically, colorless or becoming brown with age; spores (25-)
29-35(-38) x 19-26μ, mostly obovoid, wall 1.5-2μ thick,
cinnamon- or dark cinnamon-brown, or the apex slightly darker,
echinulate, germ pores 4 or 5 equatorial. Telia mostly on
abaxial surface, early exposed, blackish, spores (32-)40-60
(-70) x (14-)16-23μ, mostly oblong-clavate, wall 1.5-2μ
thick at sides, 4-6μ apically, chestnut-brown, smooth;
pedicel thick-walled, not collapsing, brown to 15μ long.

Hosts and distribution: Imperata cylindrica (L.)
Beauv., species of Miscanthus, Saccharum narenga wall.:
U.R.S.S. to China, Japan and the Philippines.

Type Miura, on Miscanthus sacchariflorus (Maxim.)
Hack., Teikaton, Manchuria, not seen.

This species has been treated as Puccinia eulaliae
Barcl. in much of the literature.

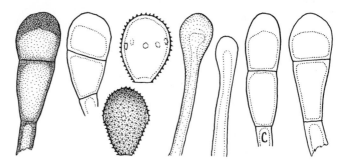

Figure 38

21. PUCCINIA POSADENSIS Sacc. & Trott. in Saccardo Syll.
Fung. 21:691. 1912. Fig. 38.

Uredo andropogonicola Speg. Anal. Mus. Nac. Buenos
Aires 19:315. 1909.

Uredo venustula Arth. Mycologia 8:21. 1916.

Puccinia andropogonicola Speg. Anal. Mus. Nac. Buenos
Aires 19:299. 1909. (Dec.) not P. andropogonicola Hariot
& Pat. 1909 (May).

Puccinia venustula Arth. Mycologia 10:128. 1918.

Puccinia kaernbachii Arth. Bull. Torrey Bot. Club
46:110. 1919.

Aecia unknown. Sori mostly in abaxial surface of leaves.
Uredinia dark cinnamon-brown, with pale golden to cinnamon-
brown, capitate paraphyses, the wall 2.5µ thick in stipe,
5-10µ thick in apex; spores (26-)28-33(-35) x 19-25µ, mostly
obovoid, wall 1.5-2µ thick, cinnamon-brown, usually darker
apically, echinulate, pores 4 or 5, equatorial. Telia
blackish brown, compact, early exposed; spores (33-)36-50(-58)
x (15-)17-20(-24)µ, mostly elongate obovoid or oblong-
obovoid, wall 1.5-2µ thick at sides, (4-)6-9µ apically,
chestnut-brown, smooth; pedicels thick-walled, not collapsing,
brown, persistent, to 20µ long, usually shorter.

Hosts and distribution: Species of Andropogon, Imperata
contracta (Kunth) Hitchc. ?: southern United States to
Panama, Trinidad, and Argentina.

Type: Spegazzini, on Andropogon condensatus, Posada,
Argentina (LPS).

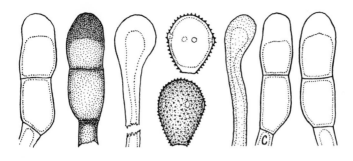

Figure 39

22. PUCCINIA DAISENENSIS Hirat. <u>f</u>. Trans. Tottori Soc. Agr.
Sci. 4:36. 1932. Fig. 39.

Aecia unknown. Uredia in abaxial surface, cinnamon-
brown, with hyaline to golden, capitate paraphyses, wall
3µ thick in the stipe, 6-9µ apically; spores (24-)26-33
x (17-)19-23(-25)µ, mostly oval or obovoid, wall 1.5µ
thick, cinnamon-brown, echinulate, pores (3)4(5), equatorial,
Telia blackish brown, compact, early exposed, spores
35-56(-66) x 15-22µ, mostly oblong-clavate or oblong, wall
1.5-2µ thick at sides, 7-13µ apically, chestnut-brown,
smooth; pedicels thick-walled, not collapsing, brown, to
15µ long.

Hosts and distribution: <u>Miscanthus oligostachyus</u> Stapf:
Japan.

Type: Hiratsuka, on <u>Miscanthus oligostachyus</u>, Japan
(Herbarium Hiratsuka; isotype PUR).

Cummins published a photograph of teliospores of the type
(Uredineana 4:Pl. II, Fig. 11, 1953).

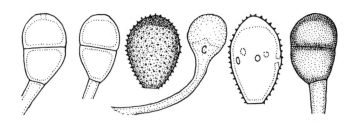

Figure 40

23. PUCCINIA FRAGOSOANA Beltrán Mem. Roy. Soc. Espan.
Hist. Nat. 50:249. 1921. Fig. 40.

Uredo schizachyrii Doidge Bothalia 2:508. 1928.

Aecia unknown. Uredinia amphigenous, paraphyses abundant,
capitate, colorless or brown, the wall usually thin in the
stipe, 7-16μ thick in the head; spores 30-43(-48) x
(16-)20-27(-30)μ, variable but mostly obovoid, wall
2(-3)μ thick at sides, 3-8μ apically, golden below to
chestnut-brown apically, echinulate, germ pores 4 or 5,
equatorial. Telia blackish brown, exposed, compact; spores
(24-)26-34 x (14-)18-23μ, mostly ellipsoid, often tending
diorchidioid, wall uniformly 2-2.5(-3)μ thick at sides and
apex or slightly thicker apically, chestnut-brown, smooth;
pedicels mostly thin-walled, collapsing or not, mostly
brownish, to 45μ long but usually broken near the spore.

Hosts and distribution: Imperata cylindrica (L.)
Beauv., Schizachyrium sanguineum (Retz.) Alst.: Spain
to Sierra Leone and South Africa; perhaps in Palestine.

Type: Beltrán, on Imperata cylindrica, Spain. Not
seen.

The isotype (PUR) of Uredo schizachyrii has teliospores.

24. PUCCINIA ARUNDINIS-DONACIS T. Hirat. Sci. Bull. Agr. Home Econ. Engin. Univ. Ryukus 5:51. 1958.

Uredo arundinis-donacis Tai Farlowia 3:133. 1947.

Aecia unknown. Uredinia amphigenous, with numerous clavate or clavate-capitate, brownish paraphyses 40-60μ long; spores 26-34 x (14-)18-20(-22)μ, broadly ellipsoid, ellipsoid, or obovoid, wall 2μ thick or to 3-4μ apically, echinulate, yellowish brown, germ pores 4, equatorial. Telia amphigenous, exposed, blackish brown, compact; spores 32-46 x 16-20μ, mostly ellipsoid, wall 1.5-2μ thick at sides, 4-8μ apically, about golden brown, smooth; pedicels thick-walled, mostly not collapsing, to 70μ long.

Hosts and distribution: Arundo donax L.: China and Japan.

Type: Tamori No. 4124, Miyako Island, Japan (herb. N. Hiratsuka).

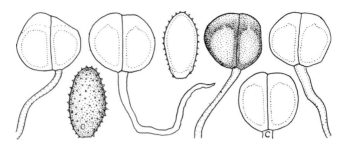

Figure 41

25. PUCCINIA ORIENTALIS (H. Syd., P. Syd. & Butl.) Arth.
& Cumm. Philippine J. Sci. 59:438. 1936. Fig. 41.

Diorchidium orientale H. Syd., P. Syd. & Butl. Ann.
Mycol. 5:500. 1907.

Puccinia ottochloae T. S. Ramak. Prod. Indian Acad.
Sci. B. 44:117. 1956.

Aecia unknown. Uredinia mostly on abaxial leaf surface,
brownish, with colorless, club-like paraphyses 13-18μ wide
and to 45μ long, the apical three-fourths solid and refractive;
spores (29-)33-44(-49) x (13-)15-19(-23)μ, ellipsoid or
obolong-ellipsoid, wall 1-1.5μ thick, golden or cinnamon-
brown, minutely and sparsely echinulate or often apparently
smooth, germ pores 2, next to the hilum. Telia on abaxial
surface, early exposed, blackish brown; spores (23-)24-26(-28)μ
high, (26)28-33(-35)μ wide, typically diorchidioid, transversely ellipsoid, wall 3-3.5μ thick at sides, (4-)5-8(-9)μ
apically, chestnut-brown but the apical wall progressively
paler externally; pedicels colorless and collapsing, long
but usually broken near the spore.

Hosts and distribution: Brachiaria ramosa (L.) Stapf,
B. reptans (L.) Gard. & C.E. Hubb., Cyrtococcum patens
(L.) A. Camus var. warburgii (Mez) Reeder, Ottochloa
nodosa (Kunth) Dandy, Panicum (?) sp.: India and Ceylon
to New Guinea and the Philippines.

Type: Sen (Butler No. 733) on Panicum prostratum
(=Brachiaria reptans), Chittagon, India (HC10).

Some Philippine collections were changed by Merrill
from Isachne miliacea to Panicum (Cyrtococcum) warburgii.
Joerstad (Nytt Mag. Bot 7:129-144. 1959) reported P.
orientalis on "Isachne sp. (very possibly error for Panicum
sp)"

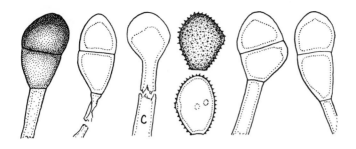

Figure 42

26. PUCCINIA LOUDETIA-SUPERBAE Cumm. Bull. Torrey Bot. Club 83:227. 1956. Fig. 42.

Aecia unknown. Uredinia amphigenous, cinnamon-brown, paraphyses capitate, 12-22μ diam apically, wall 6-9μ apically, colorless or yellowish; spores (21-)23-28 x 18-21μ, ellipsoid or obovoid, wall 1.5μ thick, dark cinnamon- or chestnut-brown, echinulate, pores 3, equatorial. Telia amphigenous, blackish brown, compact, early erumpent; spores (29-)33-42(-46) x (16-)18-21(-23)μ, mostly ellipsoid or obovoid, wall 1.5-2.5(-3)μ thick at sides, 4-6μ at apex, dark chestnut-brown, opaque apically, smooth; pedicels colorless or yellowish, thin-walled, persistent, to 50μ long.

Hosts and distribution: Tristachya superba (DeNot.) Schweinf. & Aschers.: Angola and Nyasaland.

Type: Wiehe No. 278, on Loudetia superba DeNot. (=Tristachya superba), Morshet Kasupe, Nyasaland (PUR; isotype IMI).

A photograph of teliospores of the type was published with the diagnosis.

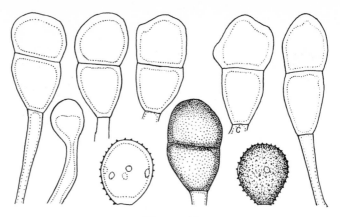

Figure 43

27. PUCCINIA TRISTACHYAE Doidge Bothalia 2:132. 1927.
Fig. 43.

Aecia (Aecidium decipiens P. Syd. & H. Syd.) on
Sphenostylis spp.: spores 26-40 x 16-20μ, oblong or
ellipsoid, wall 1μ thick, hyaline, verrucose. Uredinia
mostly abaxial, yellow, paraphyses capitate, 15-26(-36)μ
diam, wall 5-10μ thick apically, colorless; spores
(25-)27-31(-36) x (16-)19-26(-28)μ, broadly ellipsoid,
ellipsoid, or obovoid, wall (1.5-)2-2.5μ thick, cinnamon-
to chestnut-brown, echinulate, germ pores 4-6(-8), equatorial.
Telia mostly abaxial, tardily exposed, compact; spores
(37-)40-54(-60) x (16-)18-23(-25)μ, variable but mostly
oblong-ellipsoid or elongately obovoid, wall 1.5-2(-3)μ
thick at sides, 2.5-5μ at apex, golden or clear chestnut-
brown, smooth; pedicels thin-walled, yellowish brown,
persistent, to 50μ long.

Hosts and distribution: Tristachya bequaertii DeWilld.,
T. hispida (L.) K. Schm., T. rehmannii Hack.: Southern
Africa.

Type: Pole-Evans, on Tristachya rehmannii, Union of
South Africa (PRE 10039).

P. tristachyae is separable from other paraphysate
species on the Arundinelleae because of longer teliospores
and more pores in the urediniospores. A photograph of
teliospores of the type was published by Cummins and
Greene (Trans. Mycol. Soc. Japan 7:52-57 1966).

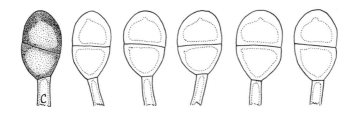

Figure 44

28. PUCCINIA EKMANII Kern, Cif., & Thurst. Ann. Mycol. 31:11. 1933. Fig. 44.

Aecia unknown. Uredinia amphigenous, yellowish brown, with hyaline or yellowish, cylindrical or clavate para- physes, the wall thin below, 3-7μ at apex; spores 23-27 x 18-23μ, broadly oval or obovoid, wall 1μ thick, yellowish to golden, echinulate, pores 4, equatorial. Telia blackish brown, compact, early exposed; spores (26-)29-36(-39) x (16-)18-23μ, mostly ellipsoid, wall 2-3.5μ thick at sides, 3-5(-6)μ apically, chestnut-brown, smooth; pedicels thin- walled and collapsing, hyaline or yellowish, persistent, to 110μ long.

Hosts and distribution: Leersia monandra Swartz: Venezuela.

Type: Ekman No. 3414, Venezuela (PAC).

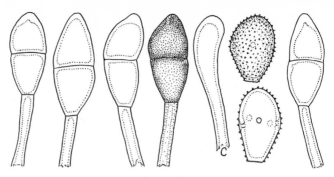

Figure 45

29. PUCCINIA INVENUSTA H. Syd. & P. Syd. in Sydow &
Butler Ann. Mycol. 5:498. 1907. Fig. 45.

Aecia unknown. Uredinia amphigenous, yellowish, long
covered by epidermis, with hyaline or yellowish, clavate
or capitate paraphyses, wall thin below, 2-4µ apically;
spores (20-)25-32(-37) x (12-)15-18µ, variable, oblong,
ellipsoid, or pyriform, wall 1.5µ thick, yellowish to
golden, minutely echinulate or verrucose-echinulate,
pores obscure, 3 or 4, equatorial. Telia blackish brown,
compact, early exposed; spores (26-)34-42(-48) x
(14-)16-20(-22)µ, mostly ellipsoid, wall 2µ thick at
sides, 3-5µ at apex, golden to chestnut-brown, smooth;
pedicels yellowish to brownish, thick-walled and not
collapsing, persistent, to 100µ long but usually shorter.

Hosts and distribution: Phragmites communis Trin.,
P. karka Trin. ex Steud., P. mauritianus Kunth: Africa,
India, the Philippine Islands, and China.

Type: Butler No. 888 on P. karka, Pusa, India (S).

Sanwal (Phytomorphology 2:35-38. 1952) reported that
Aecidium polygoni-cuspidati Diet. is the aecial stage,
but Narasimhan (Indian Phytopathol. 18:107-115. 1965)
states that the rust fungus was Puccinia phragmitis.

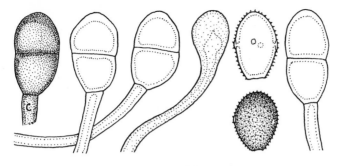

Figure 46

30. PUCCINIA RUFIPES Diet. Bot. Jahrb. 32:48. 1902.
Fig. 46.

Puccinia stichosora Diet. Bot. Jahrb. 37:100. 1905.

Aecia on Thunbergia; spores 19-28 x 16-25μ, wall thin
(1μ ?), hyaline, finely verrucose. Uredinia and telia in
leaves, usually amphigenous. Uredinia cinnamon-brown,
with capitate, yellowish to golden paraphyses, the wall
usually thin below, 11-19μ in the apex; spores (24-)27-33
(-37) x 18-25μ, mostly oval or obovoid, wall 2-2.5μ thick
and dark cinnamon-brown at sides, chestnut-brown and occasionally slighter thicker apically, echinulate, pores 4,
equatorial. Telia blackish brown, compact, early exposed;
spores (28-)30-36(-38) x 18-23(-25)μ, ellipsoid, wall
uniformly 2.5-3.0μ thick or very slightly thicker apically,
chestnut-brown, smooth; pedicels brown, thick-walled and
not collapsing, to 90μ long, persistent.

Hosts and distribution: Imperata cylindrica (L.) P.
Beauv.: Gold Coast and South Africa to India, Australia,
the Philippines, Japan, and U.R.S.S.

Type: Kusano, on Imperata arundinacea var. koenigii
(=I. cylindrica var. koenigii), Tokyo (S).

Sundaram (Indian Phytopathol. 9:133-137. 1956) proved
the life cycle by inoculation.

The fungus reported by Teng and Ou (Sinensia 8:255.
1937) as P. pachypes Syd. on Spodiopogon sp. probably
belongs here. The host plant of P. stichosora was
reported as Calamagrostis sciuroides but undoubtedly
is some species of Imperata.

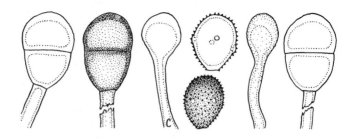

Figure 47

31. PUCCINIA PUSILLA H. Syd. & P. Syd. in Sydow & Butler
Ann. Mycol. 4:435. 1906. Fig. 47.

Puccinia andropogonis-micranthi Diet. Ann. Mycol.
7:354. 1909.

Aecia unknown. Uredinia in abaxial surface, cinnamon-
brown, with capitate, hyaline to golden paraphyses, the
wall usually thin below, 4-8(-12)μ apically; spores
(17-)20-28(-30) x (14-)16-22(-24)μ, mostly oval or
obovoid, wall 1.5-2μ thick and cinnamon-brown at sides,
darker and sometimes slightly thicker apically, echinulate,
pores (3-)4(-5), equatorial. Telia blackish brown, compact,
early exposed; spores (25-)29-36(-38) x (18-)20-25(-27)μ,
mostly broadly ellipsoid or oval, wall 1.5-2.5μ thick at
sides, 2.5-5μ apically, chestnut-brown, smooth; pedicels
hyaline or pale yellowish, thin-walled, usually collapsing,
to 65μ long, persistent.

Hosts and distribution: Capillipedium glaucopsis
(Steud.) Stapf, C. parviflorum (R. Br.) Stapf (Andropogon
micranthus): India and Burma to Sumatra, the Philippines,
China and Japan.

Type: Butler No. 541, on Andropogon assimilis Steud.
(=Capillipedium glaucopsis) (S).

Cummins published a photograph of teliospores of the
type (Uredineana 4:Pl. III, Fig. 16. 1953).

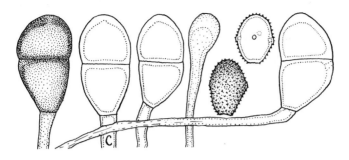

Figure 48

32. PUCCINIA APLUDAE H. Syd. & P. Syd. in Sydow & Butler Ann. Mycol. 4:436. 1906. Fig. 48.

Aecia unknown. Uredinia not seen; paraphyses capitate, hyaline or yellowish, the wall thin below, 6-9µ apically; urediniospores 19-26(-29) x 15-19µ, obovoid or oval, wall 1.5µ thick, pale at hilum to nearly chestnut-brown apically, echinulate, pores 4, equatorial. Telia on abaxial surface, blackish brown, compact, early exposed; spores (31-)36-43 (-46) x (17-)20-26µ, mostly oblong-ellipsoid or ellipsoid, wall 2-2.5µ thick at sides, 3-5µ at apex, chestnut-brown; pedicels hyaline or pale yellowish brown, thin-walled and usually collapsing, to 80µ long, persistent.

Hosts and distribution: Apluda mutica L. var. aristata (L.) Pilger: India.

Type: Butler No. 536 on Apluda aristata (=A. mutica var. aristata), Dehra Dun (S).

Cummins published a photograph of teliospores of the type (Uredineana 4:Pl. II, Fig. 15. 1953).

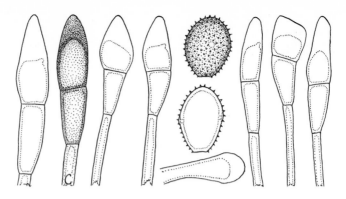

Figure 49

33. PUCCINIA KIUSIANA Hirat. f. in Ito & Murayama Trans.
Sapporo Nat. Hist. Soc. 17:167. 1943. Fig. 49.

Aecia unknown. Uredinia on adaxial leaf surface, cinnamon-
brown, paraphyses clavate or clavate-capitate, often curved or
geniculate, 10-18μ wide, the wall thin below, to 16μ in the
apex, colorless or yellowish; spores 22-28(-30) x (17-)18-21(-22)μ
mostly ellipsoid or obovoid, wall 1.5μ thick, cinnamon-brown,
echinulate, germ pores 2, equatorial. Telia on adaxial surface
and on sheaths, early exposed, compact, chocolate-brown; spores
(30-)40-56(-58) x (12-)13-19(-21)μ, mostly fusiform or elongately obovoid, wall 1(-1.5)μ thick at sides, (8-)11-20μ
apically, golden to clear chestnut-brown, the pigmentation
apparently developing slowly, smooth; pedicels persistent,
colorless, narrow, collapsing or not, to 60μ long.

Hosts and distribution: Hystrix japonica (Hack.) Ohwi: Japan.

Type: Tobinaga, Mt. Hikosan, Prov. Busen, Kiushu Japan
(Herb. Hirat.; isotype PUR).

Uredinia as such were not seen but urediniospores and
paraphyses were seen in telia of the isotype. Spores with
only 2 germ pores are uncommon in species on grasses.

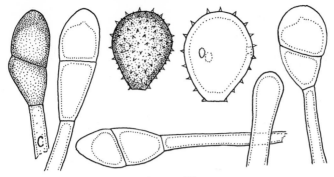

Figure 50

34. PUCCINIA OBLIQUO-SEPTATA V.-Bourgin Uredineana 5:219.
1958. Fig. 50.

Uredo bambusarum P. Henn. Hedwigia 35:255. 1896.

Uredo detenta Mains Bull. Torrey Bot. Club 66:621. 1939.

Aecia unknown. Uredinia amphigenous or mostly on abaxial
leaf surface, cinnamon-brown, with inconspicuous periperal,
cylindrical to capitate, paraphyses, the wall usually thin,
yellowish or pale brownish, spores (27-)30-36(-39) x (22-)24-31μ,
broadly ellipsoid or obovoid, wall 2-2.5(-3.5)μ thick,
cinnamon-brown, strongly echinulate, germ pores 3 or 4,
equatorial. Telia mostly on abaxial surface, chocolate-brown,
early exposed, compact; spores (25-)28-40(-44) x (12-)15-20(-23)μ,
mostly ellipsoid or narrowly obovoid, the septum commonly
oblique but diorchidioid spores rare, wall 1-1.5μ thick at
sides, 4-8(10)μ apically, yellowish to golden brown, smooth;
pedicels colorless or yellowish, moderately thick-walled but
usually collapsing, to 60μ long but usually shorter.

Hosts and distribution: Olyra micrantha H.B.K.: Brazil,
Paraguay.

Type: Maublanc, on Olyra sp. (probably O. micrantha)
Corcovado, Brazil (PC).

This fungus has long been confused with Puccinia bambusarum
but is distinct.

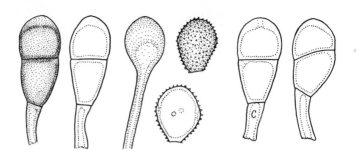

Figure 51

35. PUCCINIA POLLINIAE Barcl. J. Asiatic Soc. Bengal
58:243. 1889. Fig. 51.

Puccinia oplismeni H. Syd. & P. Syd. in Sydow & Butler Ann.
Mycol. 4:436. 1906.

Aecia, Aecidium strobilanthis Barcl., occur on species of
Strobilanthes; spores nearly globoid, 16-18μ diam, wall
yellowish, verrucose. Uredinia mostly on abaxial leaf
surface, yellowish brown, paraphyses yellowish to nearly
chestnut-brown, capitate, the wall usually thin below, 5-12μ
apically, spores 23-27 x 18-22μ, mostly oval, wall 1-1.5μ thick,
pale cinnamon-brown, echinulate, pores 3 or 4 equatorial.
Telia blackish brown, compact, early exposed; spores
(27-)33-43(-45) x 15-23μ, mostly ellipsoid or oblong-ellipsoid,
wall 1.5-2μ thick at sides, 4-7(-9)μ apically, golden to
chestnut-brown, the apex progressively paler externally,
smooth; pedicels golden, moderately thick-walled and mostly
not collapsing, to 70μ long, persistent; germinating, at
least in part, without a dormant period.

Hosts and distribution: Microstegium nudum (Trin.) A.
Camus, M. vimineum (Trin.) A. Camus: India, China and Japan
(aecial).

Type: Barclay, on Pollinia nuda (=Microstegium nudum),
Simla, India (S).

Ramachar and Cummins (Mycopathol. Mycol. Appl. 25:59.
1965) reported that the host of P. oplismeni is Microstegium,
not Oplismenus.

Cummins published a photograph of teliospores of the type
(Uredineana 4:Pl. III, Fig. 20. 1953).

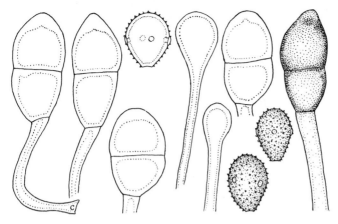

Figure 52

36. PUCCINIA ISACHNES Petch Ann. Roy. Bot. Gard. Peradeniya
7:293. 1922. Fig. 52.

Uromyces isachnes Petch Ann. Roy. Bot. Gard. Peradeniya
6:209. 1917. Based on uredinia.

Puccinia kunthiana Ramak., Srin. & Sund. Proc. Indian
Acad. Sci. B. 37:88. 1953.

Aecia unknown. Uredinia on abaxial leaf surface, dark
cinnamon-brown, with mostly golden, capitate paraphyses, the
wall uniformly 1.5-2µ thick; spores (20-)23-27(-29) x
(15-)16-19µ, mostly obovoid, wall 1.5µ thick, dull cinnamon-
brown or with an olivaceous tint, echinulate, germ pores
4, equatorial. Telia exposed, blackish brown, compact;
spores 35-42(-47) x (18-)20-25(-27)µ, mostly oblong-ellipsoid,
wall 1.5-2.5µ thick at sides, 4-7µ apically, chestnut-brown,
smooth; pedicels brownish, rather thin-walled, collapsing or
not, to 65µ long, persistent.

Hosts and distribution: Isachne gardneri Benth., I.
kunthiana Wight & Arn.: Ceylon and India.

Type: Petch, on I. kunthiana, Hakgala, Ceylon (K).

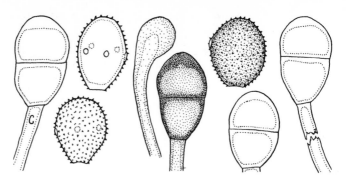

Figure 53

37. PUCCINIA NAKANISHIKII Diet. Bot. Jahrb. 34:585. 1905.
Fig. 53.

Uredo tonkinensis P. Henn. Hedwigia 34:11. 1895.

Uredo andropogonis-schoenanthi P. Henn. Bot. Jahrb. 25:496.
1898.

Puccinia citrata H. Syd. & P. Syd. Ann. Mycol. 10:78. 1912.

Uredo cymbopogonis-polyneuri Petch. Ann. Roy. Bot. Gard.
Peradeniya 6:216. 1917.

Puccinia cymbopogonicola Saw. J. Taihoku Soc. Agr. For.
7:23. 1943.

Aecia unknown. Uredinia amphigenous or on abaxial surface,
cinnamon-brown, or dark cinnamon-brown, paraphyses yellowish
to golden, capitate or clavate-capitate, wall thin below,
6-10μ thick apically; spores 26-36(-38) x (17-)19-24(-26)μ,
oval or obovoid, wall 1.5-2.5μ thick at sides, dark cinnamon-
brown, or often chestnut-brown apically, echinulate, pores
4 or 5, equatorial. Telia blackish brown, compact, early
exposed; spores (29-)33-44(-48) x (16-)20-25(-28)μ, mostly
ellipsoid, wall 2-3(-3.5)μ thick at sides, 4-8μ apically,
chestnut-brown, smooth; pedicels brown, thick-walled and not
collapsing, to 65μ long.

Hosts and distribution: Andropogon kwashotensis Hayata,
Bothriochloa intermedia (R. Br.) A. Camus, Capillipedium
parviflorum (R. Br.) Stapf ?., species of Cymbopogon, Sorghum
nitidum (Vahl.) Pers. ?: Africa to India, Ceylon, New Guinea,
the Philippines, China and Japan.

Type: Nakanishiki on Andropogon nardus var. goeringii
(=Cymbopogon nardus L. var. g.), Tosa, Japan (S).

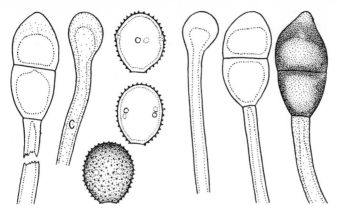

Figure 54

38. PUCCINIA PAPPIANA H. Syd. & P. Syd. Ann. Mycol. 9:142.
1911. Fig. 54.

Aecia unknown, Uredinia not seen; paraphyses yellowish to
golden, capitate, the wall 2μ thick below, 5-8μ apically;
urediospores 25-29 x (18-)20-25μ, mostly oval, wall 1.5-2μ
thick, cinnamon-brown or slightly darker apically, echinulate,
pores equatorial, probably 4. Telia on abaxial surface,
blackish brown, compact, early exposed; spores (35-)40-55
x 17-24(-27)μ, ellipsoid, or oblong-ellipsoid, wall 2.5-3.5μ
thick at sides, 5-10(-13)μ apically, chestnut-brown, the
apical thickening progressively paler externally, smooth;
pedicels yellowish, thick-walled and not collapsing, to 85μ
long, persistent.

Hosts and distribution: Hackelochloa granularis (L.)
O. Ktze.: Eritrea.

Type: Pappi, on Manisuris granularis (=Hackelochloa
granularis), Dongollo (S). Not otherwise known.

Cummins (Uredineana 4: Pl. IV, Fig. 22. 1953) published a
photograph of teliospores of the type.

123

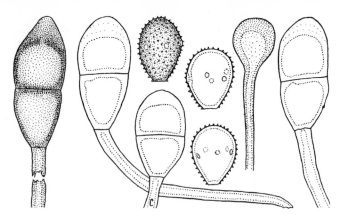

Figure 55

39. PUCCINIA POGONATHERI Petch Ann. Roy. Bot. Gard. Peradeniya
5:235. 1912. Fig. 55.

Aecia unknown. Uredinia on abaxial surface, dark cinnamon-
brown, pulverulent; paraphyses deep golden, capitate, wall
usually thin below, 3-8µ apically; spores 25-30 x (17-)19-23µ,
oval or obovoid, wall 1.5-2µ thick, dark cinnamon-brown,
echinulate, pores 5, equatorial. Telia blackish brown, compact,
early exposed; spores (34-)40-48(-50) x (18-)20-24(-26)µ,
ellipsoid or clavate-ellipsoid, wall 2-2.5µ thick at sides,
(5-)7-10(-12)µ apically, chestnut-brown, smooth; pedicels
brown, thick-walled and usually not collapsing, to 85µ long,
persistent.

Hosts and distribution: Pogonatherum paniceum (Lam.) Hack.,
and varieties: India and Ceylon to New Guinea, Formosa and
the Philippines.

Type: Petch No. 3132 on Pogonatherum crinitum (=P. paniceum
var. monandrum), Hakgala, Ceylon (K).

Cummins published a photograph of teliospores of the type
(Uredineana 4: Pl. IV, Fig. 23. 1953).

This species is remarkably similar to P. pappiana but
because the latter is so poorly known, both are retained.

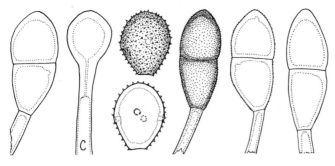

Figure 56

40. PUCCINIA PHYLLOSTACHYDIS S. Kusano Bull. Coll. Agr.
Tokyo Imp. Univ. 8:2. 1908. Fig. 56.

Aecia unknown. Uredinia in leaves, on abaxial surface,
cinnamon-brown, with mostly hyaline capitate paraphyses which
are septate, the septum usually near the head, and borne in
groups on basal cells which also produce urediniospores (and
probably teliospores), wall mostly uniformly 1.5-3.5µ thick
but the paraphyses sometimes becoming thicker-walled, pigmented,
and teliospore-like, even to being minutely verrucose; spores
(24-)28-34(-37) x (20-)22-26µ, mostly broadly oval or broadly
obovate, wall 2-3µ thick, cinnamon-brown, echinulate, pores
4 or 5, equatorial. Telia blackish brown, moderately compact,
early erumpent; spores (35-)40-50(-55) x (17-)19-22µ, mostly
clavate-ellipsoid or ellipsoid, wall 2-3µ thick at sides,
3.5-5(-7)µ at apex, golden to chestnut-brown, minutely
verruculose; pedicels hyaline, mostly thin-walled and not
collapsing, slender and tapering downward, to 150µ long but
usually broken short; germinating without dormancy.

Hosts and distribution: On species of Phyllostachys:
southeastern United States, Hawaii, Japan, and China.

Lectotype: Kusano, on Phyllostachys bambusoides, 2 Mar.
1903, Tokyo, Japan (TNS). Kusano listed several collections.
The lectotype designated here is one of 4 numbered 387; the
others are dated 5. III, 13. III, and 22. III 1903.

It is certain that the rust of the southeastern United
States, occurring only on introduced bamboos, is identical
with P. phyllostachydis rather than P. melanocephala whose
host plant is of the genus Erianthus (Andropogoneae).

Katumoto (Bull. Fac. Agr. Yamaguti Univ. 19:1135-1158. 1968)
lists 10 species of Phyllostachys and states the opinion that
only Phyllostachys spp. are hosts.

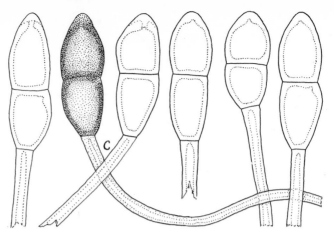

Figure 57

41. PUCCINIA TEPPERI F. Ludwig Z. Pflanzenkr. 2:132. 1892.
Fig. 57.

Aecia unknown. Uredinia not seen; paraphyses (McAlpine,
The Rusts of Australia, p. 131)" in clusters, hyaline or pale
yellow, capitate, thickened at apex, up to 75μ long"; urediniospores (Ludwig) "27-30 x 20-23μ, elliptisch oder birnförmig."
Telia on the adaxial surface, early exposed, compact, chocolate-
brown, confluent in large groups to 3 cm long; spores (40-)47-64
(-70) x (18-)20-23(-25)μ, mostly ellipsoid, wall 2.5-3.5(-4.5)μ
thick at sides, (4-)5-7(-8)μ at apex, clear chestnut-brown or
dark golden brown, the apex usually paler externally, smooth;
pedicels persistent, mostly thick-walled and not collapsing,
to 180μ.

Hosts and distribution: Phragmites communis Trin: Australia.

Type: Tepper, Grange, S. Australia (S).

McAlpine (loc. cit.) described the urediniospores as
"ellipsoid or pear-shaped, echinulate, pale yellowish, with
as many as 9 scattered germ-pores on one face,...." He
listed the type only. In the type material which I examined
I saw only 2 urediniospores. The wall was 3-3.5μ thick and
the pores equatorial and probably 5.

The status of the species is uncertain.

126

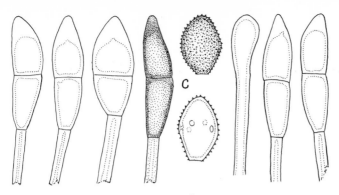

Figure 58

42. PUCCINIA MORIOKAENSIS S. Ito J. Coll. Agr. Tohoku Imp. Univ.
3:224. 1909 var. moriokaensis. Fig. 58.

Puccinia akiyoshidanensis Y. Morim. Japan. J. Bot. 34:187.
1959.

Aecia unknown. Uredinia not seen or previously described;
paraphyses and some urediniospores occasionally in telia,
paraphyses 12-20µ wide, to 80µ long, yellowish, urediniospores
24-30 x (14-)16-20µ, mostly ellipsoid or obovoid, wall about
1.5µ thick, yellowish brown or paler, echinulate, pores 4-6,
equatorial. Telia amphigenous or mostly on abaxial surface,
mostly discrete, early exposed, blackish brown, compact; spores
(36-)46-70(-75) x (11-)14-21(-25)µ, narrowly ellipsoid, shorter
spores narrowly ellipsoid or obovoid, wall at sides 1.5-2.5(-4)µ,
usually in the thicker range in short robust spores, (4-)8-12(-14)µ
at apex, chestnut-brown, the long spores usually paler than the
shorter spores, smooth; pedicels persistent, thick-walled,
not collapsing, yellowish, to 150µ long.

Hosts and distribution: Phragmites communis Trin., P.
longivalvis Steud., P. prostratus Makino, P. sp.: Japan,
China and easternmost U.S.S.R.

Type: Yamada and Sawada on Phragmites longivalvis (originally
reported as P. communis), Morioka, Prov. Rikuchu, Japan (SAPA;
isotype PUR).

Morimoto (loc. cit. p. 185) reported Phalaris arundinacea
var. genuina as a host.

The species differs from Puccinia magnusiana particularly
because of equatorial pores in the urediospores, although
the paraphyses are similar, and because the teliospores are
longer and have longer pedicels.

The following variety differs in habit only and is maintained
as a unit only because no urediniospores are known.

PUCCINIA MORIOKAENSIS S. Ito var. okatamaensis (S. Ito) comb. nov.

Puccinia okatamaensis S. Ito. J. Coll. Agr. Tohoku Imp. Univ.

127

3:226. 1909.

Aecia, uredinia and urediniospores unknown. Telia predominantly on the sheaths, confluent in closely parallel lines
several (at least 12) cm long, early exposed, blackish brown,
compact; spores (40-)50-70(-80) x (15-)17-22(-26)μ, mostly
narrowly ellipsoid; wall at sides 1.5-3(-4)μ, at apex
(6-)8-12(-15)μ, chestnut-brown; pedicel persistent, thick-
walled, not collapsing, golden brown or paler, to 150μ long,
usually near 100μ.

Hosts and distribution: <u>Phragmites</u> <u>communis</u> Trin., Steud,
<u>P</u>. <u>prostratus</u> Makino: China and Japan.

Type: K. Miyabe, on <u>P</u>. <u>communis</u> (as <u>P</u>. <u>vulgaris</u>), Okatama,
Prov. Ishikari, Japan (SAPA; isotype PUR - received from Ito
designated as "Type collection."

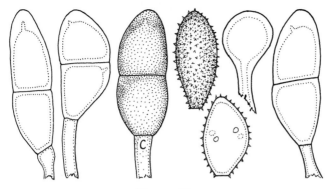

Figure 59

43. PUCCINIA XANTHOSPERMA H. Syd. & P. Syd. in Sydow and Butler
Ann. Mycol. 4:437. 1906. Fig. 59.

Aecia unknown. Uredinia on the abaxial leaf surface, pale
yellowish dry, with yellowish capitate, thick-walled paraphyses, the wall mostly uniformly 3-5(-6)µ thick in the head;
spores (24-)27-37(-42) x (14-)17-21µ, mostly ellipsoid, wall
1.5-2µ thick or the apex 3-3.5µ thick, yellowish, echinulate,
germ pores (4?) 5, equatorial, obscure. Telia on the abaxial
surface, early exposed, pulvinate but rather loose, yellowish
brown or near cinnamon-brown; spores (38-)45-60(-66) x (16-)
18-22(-25)µ, mostly oblong-ellipsoid, wall 1.5-2(-2.5)µ thick
at sides, (4-)5-10(-12)µ apically, yellowish or pale golden brown,
smooth; pedicels nearly colorless, thin-walled and mostly
collapsing, to at least 100µ long but usually broken shorter.

Type: Butler No. 539, on Bambusa sp., Mussoorie, India, 9
May 1903 (S). Not otherwise known.

44. PUCCINIA PUGIENSIS Tai in Wang Acta Phytotax. Sinica
10:294. 1965.

Aecia unknown. Uredinia hypophyllous, sometimes on sheaths,
paraphyses clavate or cylindrical, brownish or hyaline, 10-18µ
wide apically, wall 1.5-2µ thick below, 3-6µ apically; spores
27-39 x 17-24µ, subglobose, ellipsoid, or pyriform, wall
1.5-2µ thick, subhyaline or yellowish, echinulate, germ pores 4,
equatorial. Telia hypophyllous, exposed, pulvinate, blackish
brown; spores 36-56 x 14-21µ, ellipsoid, broadly ovoid, or
oblong, wall 1.5-2µ thick at sides, 4-7µ apically, cinnamon-
brown, smooth; pedicels brown, persistent, to 140µ long.

Type: Tai, on <u>Saccharum</u> <u>spontaneum</u> L., Pugi, Kunming,
Yunnan, China(Plant Pathol. Herb. No. 8377, Tsing Hua Univ. =
Inst. Microbiol., Peking No. 4377; not seen).

The description is adapted from the original. Tai (Farlowia
3:112. 1949) reported this fungus as <u>P</u>. <u>kuehnii</u>. Wang (loc.
cit.) published a photograph of the teliospores.

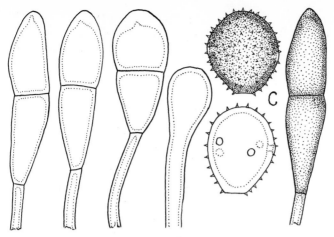

Figure 60

45. PUCCINIA HIKAWAENSIS Hirat. f. & S. Uchida in Uchida Mem.
Mejiro Gakuen Woman's Junior Coll. 2:24. 1965. Fig. 60.

Aecia unknown. Uredinia on abaxial leaf surface, dark brown,
with hyaline capitate paraphyses; spores 32-38 x (26-)28-32
(-35)μ, mostly broadly obovoid, wall 3-4μ thick, nearly
chestnut-brown, echinulate, germ pores (4)5,equatorial. Telia
on abaxial surface, early exposed, deeply cushion-shaped,
cinnamon-brown; spores (44-)52-70(-80) x 16-24(-28)μ, narrowly
obovoid or more or less fusiform, wall 1.5-2μ thick at sides,
(3-)4-9(-10)μ apically, yellowish or golden brown, or the
shorter broader spores nearly chestnut-brown, smooth; pedicels
colorless, thick-walled, not collapsing, to 200μ long.

Hosts and distribution: Sasa kesuzu Muroi & Okam.: Japan.

Type: Uchida No. 1527, Sasamorpha mollis Nakai (=Sasa
kesuzu), Hikawa-mura, Pref. Tokyo (Herb. Hiratsuka).

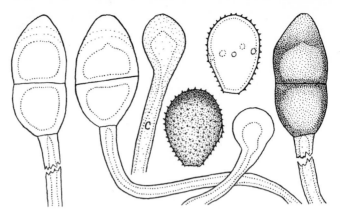

Figure 61

46. PUCCINIA ANDROPOGONICOLA Hariot & Pat. Bull. Mus. Hist. Nat. Paris 1909: 199. 1909. Fig. 61.

Aecia unknown. Uredinia mostly on abaxial surface, dark cinnamon-brown; paraphyses hyaline to golden, straight or curved, capitate or clavate-capitate, wall moderately thick below, 5-10µ apically; spores (26-)29-35 x (18-)20-26(-28)µ, oval or obovoid, wall 1.5-2(-2.5)µ thick, cinnamon- or dark cinnamon-brown at sides, usually chestnut-brown apically, echinulate, pores (4)5, equatorial. Telia blackish brown, compact, early exposed; spores (38-)40-56(-65) x (18-)22-27µ, ellipsoid or oblong-ellipsoid, wall 2.5-3.5µ thick at sides, chestnut-brown, (6-)9-12(-15)µ apically, chestnut-brown becoming progressively paler externally, smooth; pedicels hyaline, thick-walled and mostly not collapsing, to 150µ long, persistent.

Hosts and distribution: Andropogon (Cymbopogon?) sp., Cymbopogon giganteus (Hochst.) Chiov., C. proximus (Hochst.) Stapf, Hyparrhenia dissoluta (Nees) C. E. Hubb., H. rufa (Nees) Stapf: Ethiopia to Mauritius, French Congo and Gold Coast.

Lectotype: Chevalier or Andropogon sp. (=Cymbopogon?), Cubangui, French Congo (Vestergren, Micromycetes rar. sel. No. 1563).

Cummins (Uredineana 4:Pl. IV, Fig. 25. 1953) published a photograph of teliospores of the type.

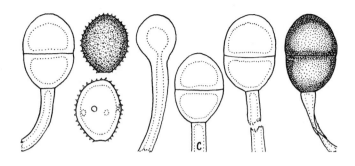

Figure 62

47. PUCCINIA SONORICA Cumm. & Husain Bull. Torrey Bot. Club
93:57. 1966 var. sonorica. Fig. 62.

Aecia unknown. Uredinia adaxial, cinnamon-brown; paraphyses capitate, to 90μ long, 24μ wide, wall to 6μ thick
apically, colorless or golden brown; spores (23-)25-30(-33)
x (18-)20-24(-25)μ, mostly ellipsoid or obovoid, wall
(2-)2.5-3.5(-4)μ thick, dark cinnamon- or nearly chestnut-
brown, echinulate, pores 3 or 4(5), equatorial. Telia
adaxial and often on the stems and inflorescence, blackish
brown, compact, early exposed, often confluent; spores
(29-)32-37(-40) x (21-)23-26(-28)μ, mostly broadly ellipsoid,
wall (2-)2.5-3.5(-4)μ thick at sides, 3.5-5μ at apex, uniformly chestnut-brown, smooth; pedicels colorless persistent,
to 175μ long.

Hosts and distribution: Aristida hamulosa Henrard, A.
ternipes Cav.: Arizona, U.S.A. and Sonora, Mexico.

Type: Cummins No. 62-65 (PUR 59369), on Aristida hamulosa,
Arizona, U.S.A.

The species is similar to P. unica var. unica except for
smaller spores and urediospores with equatorial pores. A
photograph of teliospores of the type was published with the
diagnosis.

PUCCINIA SONORICA Cumm. & Husain var. minor Cumm. & Hussain Bull. Torrey Bot. Club 93:57. 1966.

Aecia unknown. Uredinia and paraphyses as in var. sonorica; spores (20-)21-24(-26) x (16-)17-20(-22)μ, wall 1.5-2(-3)μ thick, cinnamon-brown, echinulate, pores (3)4(5), equatorial, rarely 4 equatorial and 1 apical; teliospores (26-)28-32 x 21-25(-27)μ, wall (1.5-)2-2.5(-3.5)μ thick at sides, (3.5-)4-5μ at apex, uniformly chestnut-brown, smooth; pedicels colorless, persistent, to 140μ long.

Hosts and distribution: Aristida ternipes Cav.: Guerrero, Sinaloa, and Zacatecas, Mexico.

Type: Cummins No. 63-673 (PUR 59378), Sinaloa.

A photograph of spores of the type was published with the diagnosis.

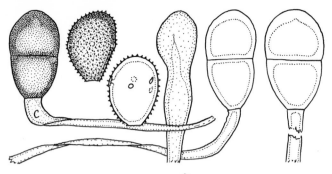

Figure 63

48. PUCCINIA OPERTA Mund. & Thirum. Imp. Mycol. Inst. Kew
Papers 16:10. 1946. Fig. 63.

Uredo operta H. Syd., P. Syd. & Butl. Ann. Mycol. 5:509.
1907.

Aecia unknown. Uredinia amphigenous, cinnamon-brown but
long capped by the epidermis, paraphyses hyaline or yellowish,
capitate or clavate, often incurved, wall thick throughout but
thicker (4-8µ) apically; spores (23-)27-34(-36) x (17-)20-25(-27)µ,
mostly obovoid, wall 1.5µ thick at sides, 2-3(-4)µ apically,
cinnamon-brown but darker apically, germ pores 4-6, equatorial.
Telia blackish brown, compact, early exposed; spores (35-)39-46
(-55) x (20-)23-30(-33)µ, variable but mostly ellipsoid or
oblong-ellipsoid, wall 2.5-3.5µ thick at sides, 3.5-5.5µ
apically, chestnut-brown, smooth; pedicels yellowish to
brownish, thin-walled and mostly collapsing, to 110µ long but
usually broken shorter.

Hosts and distribution: Coix lachryma-jobi L.: India
and Ceylon to New Guinea and the Philippines.

Type: Ajrekar, Girnar Hills, India (HC10; isotype IMI, PUR).

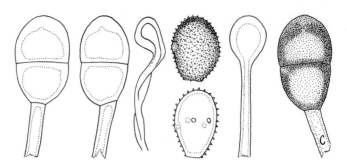

Figure 64

49. PUCCINIA ERAGROSTIDIS-SUPERBAE Doidge Bothalia 3:500. 1939.
Fig. 64.

Aecia unknown. Uredinia amphigenous, cinnamon-brown,
paraphyses hyaline to golden, clavate or clavate-capitate,
often somewhat curved, the wall mostly thin below, 4-9(-12)μ
apically; spores 27-32 x (17-)20-25μ, mostly oval or obovoid,
wall 1.5-2μ thick, cinnamon-brown, the apex darker and often
slightly thicker, echinulate, pores 4-6, equatorial. Telia
blackish brown, compact, early exposed; spores (35-)38-42(-45)
x (22-)24-27(-30)μ, mostly ellipsoid or oblong-ellipsoid, wall
3-4μ thick at sides, 5-8(-10)μ apically, chestnut-brown, smooth;
pedicels hyaline to golden, moderately thin-walled and mostly
collapsing, to 112μ long, persistent.

Hosts and distribution: Eragrostis happula Nees var.
divaricata Stapf, E. superba Peyr.: South Africa.

Type: Doidge and Bottomley on Eragrostis superba, Derdepoort
(PRE 29811).

136

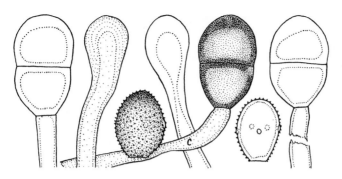

Figure 65

50. PUCCINIA DUTHIAE Ell. & Tracy in Ellis and Everhart Bull.
Torrey Bot. Club 24:283. 1897. Fig. 65.

Puccinia amphilophidis Doidge Bothalia 3:496. 1939.
    Aecia, Aecidium barleriae Doidge, occur on Barleria cuspidata
Heyne, systemic; spores 17-22μ diam, more or less globoid, wall
1-1.5μ thick, hyaline, verrucose. Uredinia mostly on abaxial
leaf surface, cinnamon-brown; paraphyses capitate, mostly
lemon yellow, the wall usually thick in the stipe, 4-8(-10)μ,
apically; spores (24-)26-32(-35) x 18-23(-25)μ, mostly oval,
wall 1.5-2.5μ thick, cinnamon-brown, the apex usually darker,
echinulate, germ pores (4)5(6), equatorial. Telia mostly on
abaxial surface, exposed, blackish brown, compact; spores
(30-)25-42(-49) x 22-27(-30)μ, mostly broadly ellipsoid, wall
(2-)2.5-3(-3.5)μ thick at sides, 4-8μ apically, chestnut-brown,
smooth; pedicels yellowish to brownish, moderately thin-walled,
collapsing or not, to 120μ long, often broad.

    Hosts and distribution: Andropogon (Bothriochloa ?) sp.,
species of Bothriochloa, Dichanthium annulatum (Forssk.)
Stapf: South Africa and Tanzania to India, Australia, and
China.

    Type: Duthie, on Andropogon pertusus (=Bothriochloa pertusa
(L.) A. Camus), Saharanpur, India (NY; isotype PUR).

    Narasimhan (Indian Phytopathol. 18:107-115. 1965) proved
the life cycle by inoculation. Cummins (Uredineana 4: Plate
IV, Fig. 24, 1953) published a photograph of teliospores of the
type.

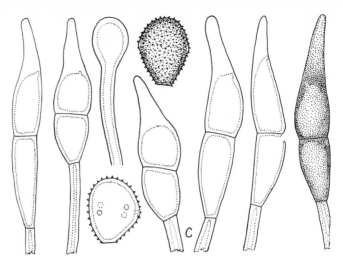

Figure 66

51. PUCCINIA LONGICORNIS Pat. & Hariot Bull. Soc. Mycol. France
7:143. 1891. Fig. 66.

Aecia unknown. Uredinia in abaxial surface of leaves,
yellowish brown, paraphyses hyaline or yellowish, capitate
or clavate-capitate, the wall uniformly 2-3µ thick or only
slightly thicker apically; spores (24-)28-34(-36) x (21-)24-30
(-32)µ, mostly broadly obovoid or nearly globose, wall
2.5-3.5(-4)µ thick, golden or cinnamon-brown, echinulate,
pores 4 or 5, equatorial. Telia blackish brown, compact,
early exposed; spores (50-)65-100(-110) x (12-)14-19(-21)µ,
fusiform or cylindrical-fusiform, wall 2µ at sides, 14-33µ
at apex, golden to light chestnut-brown, the apical wall
progressively paler externally, smooth or minutely verruculose;
pedicels hyaline, thick-walled and not collapsing, tapering,
to 200µ long, persistent.

Hosts and distribution: On species of <u>Nipponobambusa</u>,
<u>Phyllostachys</u> (?), <u>Pleioblastus</u>, <u>Pseudosasa</u>, <u>Sasa</u>, and
<u>Sasaella</u>. Japan and China.

Type: Faurie, on Bambuseae, Japan (PC).

The latest detailed list of hosts is Katumoto's (Bull. Fac.
Agr. Yamaguti Univ. 19:1135-1158. 1968).

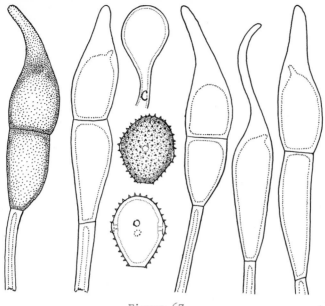

Figure 67

52. PUCCINIA SASICOLA Hara ex Hino & Katumoto Bull. Fac. Agr.
Yamaguti Univ. 6:68. 1955. Fig. 67.

Aecia on <u>Corylopsis</u>, few in small wart-like galls, mostly
on abaxial leaf surface; spores (23-)25-30(-34) x (17-)20-24(-26)μ,
mostly ellipsoid, oblong-ellipsoid or angularly globoid, wall
1-1.5μ thick at sides, 4-10(-12)μ apically, often also thickened
elsewhere, echinulate-verrucose, hyaline. Uredinia on abaxial
surface of leaf, pale brownish, with short capitate, thin-
walled, colorless paraphyses; spores (24-)26-30(-33) x
(18-)20-24μ, mostly broadly obovoid, wall 1.5-2(-2.5)μ thick,
yellowish or pale cinnamon-brown, occasional spores with
thicker darker brown walls, echinulate, pores 4 or 5, equatorial.
Telia brown, compact, early exposed; spores (70-)90-125(-135)
x (12-)16-22(-24)μ, ellipsoid-cylindrical with a greatly
elongate tapering apex, wall 1.5-2.5μ thick and golden at
sides, (20-)30-50(-75)μ apically and nearly colorless, smooth;
pedicels hyaline, thick-walled and not collapsing, slender
and tapering, to 200μ long.

Hosts and distribution: <u>Sasa borealis</u> Makino, <u>S. kesuzu</u>
Muroi & Okam.: Japan.

Type: K. Hara, on <u>Sasamorpha purpurascens</u> (=<u>Sasa borealis</u>)
Prov. Mino, Japan (YAM).

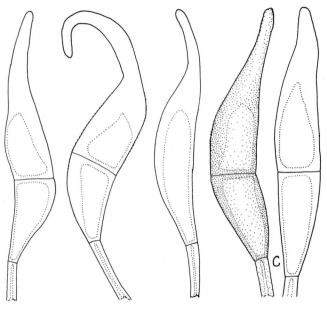

Figure 68

53. PUCCINIA MITRIFORMIS S. Ito. J. Coll. Agr. Tohoku Imp.
Univ. 3:233. 1909. Fig. 68.

Aecia unknown. Uredinia on abaxial leaf surface, cinnamon-
brown, rather compact, paraphyses capitate, colorless to
brownish, the wall uniformly 1.5-2.5μ thick in the head,
usually thicker in the stipe; spores (29-)31-36 x 26-30(-33)μ,
broadly obovoid or nearly globoid, wall (2-)3-4μ thick,
cinnamon- or dark cinnamon-brown, echinulate, pores 4 or 5,
equatorial. Telia on abaxial surface, early exposed,
chocolate-brown; spores (75-)95-130(-145) x (15-)18-22(-26)μ,
wall usually unilaterally thickened, 2-3μ on the thin side,
somewhat to much thicker on opposite side, (30-)40-70(-80)μ
apically, golden brown but the apical thickening much paler,
smooth or perhaps minutely rugose on the broad part of the
spore; pedicels mostly not collapsing, colorless, long and
tapering, to 250μ long; 1-celled spores sometimes rare, sometimes predominating.

Hosts and distribution: Species of Sasa; China and Japan.

Type: Yamada, on Sasa paniculata (=Sasa borealis (Hack.)
Makino, Prov. Rikucku, Japan (SAPA).

140

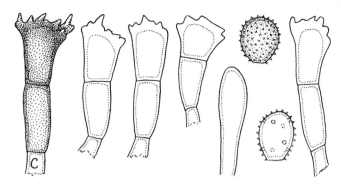

Figure 69

54. PUCCINIA CORONATA Corda Icon. Fung. 1:6.　1837 var.
coronata.　Fig. 69.

Puccinia sertata Preuss in Sturm Deutschl. Flora Abt. 6:25.
1848.

Puccinia lolii Niels. Ungeskr. Landm. IV.　9:549.　1875.

Puccinia coronifera Kleb. Z. Pflanzenkr.　4:135.　1894.

Puccinia paniculariae Arth. Bull. Torrey Bot. Club 28:663.
1901.

Puccinia beckmanniae McAlp. Rusts Australia p. 116.　1906.

Puccinia hierochloae Ito　J. Coll. Agr. Tohoku Imp. Univ.
3:193.　1909.

Puccinia pertenuis Ito J. Coll. Agr. Tohoku Imp. Univ. 3:193.
1909.

Puccinia mediterranea Trott. Ann. Mycol. 10:510.　1912.

Puccinia pumilae-coronatae Paul in Poeverlein & Schoenau
Kryptog.　Forsch. Bayern 2:95-96.　1929.

Puccinia coronata Corda var. calamagrosteos Fraser &
Ledingham Sci. Agr.　13:322.　1933.

Puccinia coronata Corda var. bromi Fraser & Ledingham Sci.
Agr.　13:322.　1933.

Puccinia coronata Corda var. elaeagni Fraser & Ledingham
Sci. Agr.　13:322.　1933.

Puccinia deyeuxiae Tai & Cheo Bull. Chinese Bot. Soc. 3:65.
1937.

Puccinia corniculata Mayor & V. -Bourgin Rev. Mycol. 15:103.
1950.

Puccinia coronata Corda var. intermedia Urban Ceska Mycol.
21:13.　1967.

Aecia (Aecidium rhamni Pers.) occur on species of Berchemia,
Rhamnus, and Elaeagnus (incl. Shepherdia); spores 16-24 x

141

15-19μ, wall 1-1.5μ thick, verrucose. Uredinia amphigenous or mainly on adaxial surface, brownish yellow to yellow (fresh), with few colorless, more or less cylindrical, mostly thin-walled paraphyses marginally, these rarely abundant; spores (17-)19-25(-28;-30) x (14-)17-21(-25)μ, mostly ellipsoid or broadly ellipsoid, wall 1.5-2μ thick, pale yellowish to nearly colorless, echinulate, germ pores 8-10, obscure. Telia amphigenous, long covered by the epidermis or only tardily exposed, blackish, with brownish paraphyses present but seldom abundant and the sori scarcely loculate; spores (30-)36-65(-70;-80) x (12-)14-19(-22)μ excluding digitations, wall 1-1.5(-2)μ thick at sides, about 2-4μ apically excluding digitations, golden to chestnut-brown, apex coronate with digitations (0-)3-10(-14)μ long; pedicels short, yellowish to brownish.

Hosts and distribution: On species of Agropyron, Agrostis, Alopecurus, Ammophila, Anthoxanthum, Apera, Arrhenatherum, Arundinella, Avenochloa, Beckmannia, Bothriochloa (?), Brachypodium, Briza, Bromus, Calamagrostis, Catabrosa, Chrysopogon (?), Cinna, Cynosurus, Dactylis, Deschampsia, Desmazeria, Deyeuxia, Elymus, Festuca, Glyceria, Helictotrichon, Hierochloë, Holcus, Hordeum, Hystrix, Koeleria, Lagurus, Lamarckia, Lolium, Melica, Milium, Molinia, Paspalum, Phalaris, Phleum, Poa, Polypogon, Puccinellia, Scolochloa, Sesleria, Trisetum, Vulpia: circumglobal.

Type: Corda, on Luzula albida (= error for Calamagrostis arundinacea (L.) Roth or C. villosa (Chaix) J. F. Gmel. -det. M. Deyl), Liberec, Reichenberg (PR 155608). This correction of the identity of the host was reported by Urban (Ceska Mykol. 21:12-16. 1967).

The first inoculations to prove the life cycle were done by de Bary (Monatsber. K. Preuss. Adak. Wiss. Berlin 1866:205-215) using telia from an undesignated grass to inoculate Rhamnus frangula.

In addition to var. avenae, I recognize vars. gibberosa, himalensis, and rangiferina but the varieties are not very distinct.

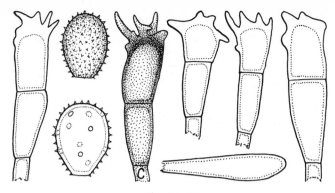

Figure 70

PUCCINIA CORONATA Corda var. avenae Fraser & Ledingham Sci. Agr. 13:322. 1933. Fig. 70.

Aecia occur on Rhamnus. Uredinia amphigenous, with few or no paraphyses; spores (21-)25-30(-34) x (17-)20-24(-26)μ, germ pores scattered, obscure, with slight or no invagination of the wall and slight or no cuticular caps, mostly 9-11. Telia covered, tending to be loculate with brownish stromatic paraphyses; spores variable, (33-)40-60(-75) x (12-)14-19(-23)μ excluding digitations, digitation 2-several, 4-16μ long, mostly 5-10μ long.

Hosts and distribution: On species of Avena and occasional other grasses: common where oats (A. sativa L.) are grown.

Lectotype: Fraser, on Avena sativa, Saskatoon, Sask., Canada, 25 July 1923 (SASK; isotype PUR). Lectotype designated here.

The first successful inoculations were by Cornu (Bull. Soc. Bot. France 27:209-210. 1880) using aeciospores to infect oats.

The size of the urediniospores, and the number of germ pores is greater than in most variants of P. coronata.

Variety avenae has been proved capable of infecting various grasses and occasional collections in the "wild" have spores of the proper size.

143

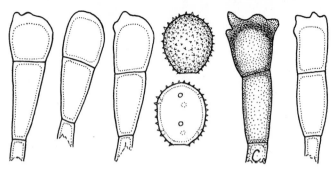

Figure 71

PUCCINIA CORONATA Corda var. gibberosa (Lagerh.) Joerst.
Avh. Norske Videnskaps-Akad. Oslo I. 1948:9. 1949. Fig. 71.

Puccinia gibberosa Lagerh. Ber. Deut. Bot. Ges. 6:124. 1888.

Aecia unknown. Urediniospores (22-)25-32(-35) x (20-)21-24
(-27)μ, wall (1.5-)2μ thick, germ pores 8 or 9, obscure, with
slight or no cuticular caps and no invagination of the wall.
Telia covered, loculate with brown stromatic paraphyses; spores
(33-)40-60(-65) x (12-)14-19μ, excluding digitations, digitations from none to 5, short, 2-6μ long, often only tubercle-
like.

Hosts and distribution: Festuca altissima All.: Europe.

Type: Lagerheim, near Frieburg in Baden, Germany (S).

The "gibberose" character of the teliospore apex is not
uncommon, e. g. P. brevicornis = himalensis, and hence is of
doubtful value. The urediniospores are similar to those of
var. avenae but with fewer and usually more obvious germ pores.
It remains in doubt whether other fescues serve as hosts, but
some, e. g. F. montana, have been reported.

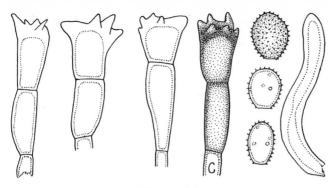

Figure 72

PUCCINIA CORONATA Corda var. himalensis Barcl. Trans. Linn. Soc. London 3:227. 1891. Fig. 72.

Puccinia himalayensis (Barcl.) Diet. in Engler-Prantl Natur. Pflanzenfam. 1 (1**): 63. 1900.

Puccinia melicae P. Syd. & H. Syd. Monogr. Ured. 1:760. 1903.

Puccinia erikssoni Bub. Pilze Boehm. p. 107. 1908.

Puccinia brevicornis Ito J. Coll. Agr. Tohoku Imp. Univ. 3:191. 1909.

Uredo jozankensis Ito J. Coll. Agr. Tohoku Imp. Univ. 3:245. 1909.

Puccinia subdigitata Arth. & Holw. Amer. J. Bot. 5:468. 1918.

Puccinia poae-pratensis M. Miura Fl. Manchuria & E. Mongolia III:280. 1928.

Puccinia coronata Cda. var. melicae (Syd.) Joerst. Avh. Norske Vidensk.- Akad. Oslo, I. 1948:7. 1949.

Uredinia mostly on adaxial leaf surface, with scanty hyaline or golden, cylindrical or clavate paraphyses whose wall varies from thin to moderately and uniformly thick; spores (12-)14-20 (-21) x (11-)13-16(-18)μ, mostly oval or ellipsoid, wall 0.5-1.0μ thick, hyaline or yellowish, echinulate, pores probably 4-8 obscure, scattered or tending equatorial. Telia exposed, without paraphyses; spores (26-)35-55(-65) x (10-)12-18(-22)μ (excluding crown), mostly oblong or clavate, wall 1-1.5μ thick at sides, 2.5-5μ apically (excluding crown), apex coronate with digitations 2-10μ long; pedicels hyaline to brownish, thin-walled, persistent, to 15μ long.

Hosts and distribution: Agrostis gigantea Roth, Arundinella sp., Brachypodium formosanum Hayata, B. japonicum Miq., B. mexicanum (Roem. & Schult.) Link, B. sylvaticum (Huds.) Beauv., Calamagrostis arundinaceum (L.) Roth, C. langsdorfii Trin., Melica ciliata L., M. nutans L., Phalaris arundinacea L., Poa

pratensis L., Schizachne purpurascens (Torr.) Swallen: Europe
to India, Japan, and North and South America.

Type: Barclay, on Brachypodium sylvaticum, Simla, India (K).

The aecial hosts include Rhamnus dahurica Pall. in India and
Asia and probably R. japonica Maxim. in Asia, and R. serrata
Roem. & Schult. in Mexico. Tranzschel (Trudy Bot. Inst. Akad.
Nauk 4:327. 1940), using teliospores on Melica from the Far
East, obtained spermogonia on Rhamnus dahurica but the excised
branches died before aecia developed. Barclay noted the
association of the rusted grass and Rhamnus and conducted
successful reciprocal inoculations using Brachypodium sylvaticum
and Rhamnus dahurica.

The variety is constant in having small thin-walled urediniospores and exposed aparaphysate telia. It is variable in the
abundance of uredial paraphyses, the thickness of the paraphysis
wall, the size of the teliospores, and the length of the digitation of the "crown".

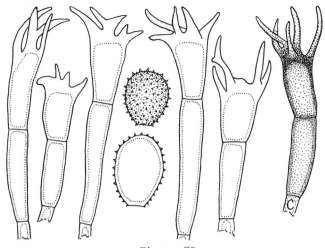

Figure 73

PUCCINIA CORONATA Corda var. rangiferina (Ito) Cumm. comb. nov. Fig. 73.

Puccinia rangiferina Ito J. Coll. Agr. Tohoku Imp. Univ. 3:194. 1909.

Puccinia epigeios Ito J. Coll. Agr. Tohoku Imp. Univ. 3:192. 1909.

Aecia unknown. Uredinia on the adaxial surface, with few short, club-shaped, inconspicuous paraphyses or these sometimes prominent; spores (22-)24-30(-35) x (17-)19-24(-26)μ, mostly broadly ellipsoid, wall 1.5μ thick, yellowish, echinulate, germ pores obscure, scattered. Telia early exposed, blackish, on the abaxial leaf surface and on sheaths; spores (55-)65-95(-105) x (12-)14-17(-19)μ wide, narrowly clavate or nearly cylindrical, wall 1.5μ thick at sides, 3-5μ apically, excluding digitations, yellowish basally to chestnut-brown apically, digitations (3-)10-20(-30)μ long; pedicels brownish, 12μ or less long.

Hosts and distribution: Calamagrostis arundinacea Roth, C. epigeios (L.) Roth, C. sp.: China and Japan.

Type: Yamada, on Calamagrostis arundinacea, Morioka, Prov. Rikuchu (SAPA; isotype PUR).

The species is characterized by large urediniospores, longer teliospores, and longer digitations than other "coronate" species and, as in the others, uredinial paraphyses seem not to provide a dependable character. Ito described paraphyses of P. epigeios but did not describe uredinia of P. rangiferina. A few urediniospores were found in a portion of the type and they are large as in P. epigeios. It is doubtful if the two are distinct.

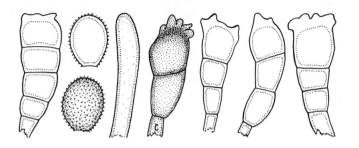

Figure 74

55. PUCCINIA ADDITA H. Syd. Ann. Mycol. 35:245. 1937. Fig. 74.

Rostrupia addita (Syd.) V. -Bourgin Ann. Ecole Agric.
Grignon Ser. 3. 1:124. 1938-1939. 1939.

Aecia unknown. Uredinia amphigenous, golden, with scant
and inconspicuous, hyaline, uniformly thin-walled paraphyses;
spores (16-)19-25(-27) x (14-)15-19(-21)μ, ellipsoid, oval
or globoid, wall 1-1.5μ thick, hyaline or yellowish, minutely
echinulate, pores probably 7-10, very obscure. Telia blackish
brown, pulvinate, compact, early exposed, with no or scanty
marginal paraphyses; spores (25-)36-50(-56) x (12-)14-18(-21)μ
(without digitations), 2-4(-5)-celled, cylindrical, oblong-
clavate or clavate, wall 1-1.5μ thick at sides, 3-4.5μ apically
(without digitations), apex coronate, the digitations 3-8(-10)μ
long, chestnut-brown; pedicels thin-walled, brownish, persistent,
to 10μ long.

Hosts and distribution: Phalaris brachystachys Link:
Madeira.

Isotype: Viennot-Bourgin No. 22, Funchal, Madeira (Herb.
Viennot-Bourgin).

The aecial stage is not known but the species doubtless is
related to P. coronata from which it differs primarily because
of the abundance of teliospores having more than one septum.

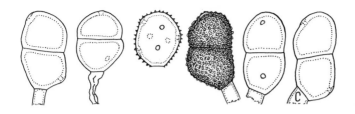

Figure 75

56. PUCCINIA PARADOXICA Ricker J. Mycol. 11:114. 1905. Fig. 75.

Aecia unknown. Uredinia on abaxial leaf surface, orange to brownish yellow, with peripheral mostly broadly capitate, hyaline, thin-walled paraphyses: spores 20-25(-28) x 17-20μ, broadly ellipsoid or globoid, wall 2-3μ thick, yellowish, echinulate, pores 7-9, scattered, obscure. Telia on abaxial surface, cinnamon-brown, pulverulent, early exposed; spores (27-)30-38(-42) x (16-)19-23(-26)μ, variable but mostly ellipsoid, wall uniformly 1.5-2.5μ thick, verrucose or labyrinthiformly verrucose, dark cinnamon- to chestnut-brown, pore in each cell usually depressed, under a small hyaline papilla; pedicels thin-walled, hyaline, to 25μ but breaking near the spore, sometimes displaced laterally.

Hosts and distribution: Melica smithii (Porter) Vasey: U.S.A. (Michigan).

Type: C.H. Wheeler, Chatham Station, Michigan (Wis; isotype PUR).

57. PUCCINIA NAUMOVII Kazenas Akad. Nauk Bot. Odt. Sporov.
Rast. Bot. Mater. 12:232. 1959.

Aecia unknown. Uredinia epiphyllous, arranged in lines,
paraphyses clavate; spores 20-30 x 16-19µ, mostly broadly
ellipsoid, or obovoid, wall 3µ thick, orange color, echinulate, germ pores not described but undoubtedly several and
scattered. Telia epiphyllous, covered by the epidermis,
blackish, loculate with brown paraphyses; spores 2-4-celled,
apparently (from Fig. 3) mostly 3- or 4-celled, 73-86 x
16-24µ when 4-celled, 57-73 x 13-19µ when 3-celled, 39-62 x
13-16µ when 2-celled, mostly more or less cylindrical, wall
thin (1-1.5µ ?) at sides, 2-4, mostly 4µ apically, chestnut-
brown (?), smooth; pedicel very short.

Type: Kazenas, on Agropyron ramosum Richt., near Lake
Temir-Tau, Karaganda region, Kazachstan SSR. (LE?; not seen).

The description is adapted from the original diagnosis and
Figs. 2 and 3. Kazenas described uredinial paraphyses as
filiform or clavate. From the illustration it is obvious
that the filiform structures are spore pedicels but the
clavate structures appear to be peripheral, thick-walled
(3µ ?) paraphyses.

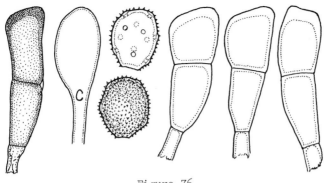

Figure 76

58. PUCCINIA STRIIFORMIS Westend. Bull. Roy. Acad. Belg., Cl. Sci., 21:235. 1854 var. striiformis. Fig. 76.

Uredo glumarum J.K. Schmidt, Allgem. Oekon.-tech. Fl. 1:27. 1827.

Puccinia straminis Fuckel Jahrb. Nass. Ver. Nat. 15:9. 1860 (in part).

Puccinia neglecta Westend. Bull. Soc. Bot. Belg. 2:248. 1863.

Puccinia glumarum Eriks. & Henn. Z. Pflanzenkr. 4:197. 1894.

Puccinia lineatula Bub. Ann. Nat. Hofmus. Wien 28:193. 1914.

Puccinia stapfiolae Mundk. & Thirum. Imp. Mycol. Inst. Kew Mycol. Papers 16:14. 1946.

Aecia unknown. Uredinia amphigenous or mostly on adaxial leaf surface, in linear series in chlorotic streaks, bright orange-yellow, with peripheral cylindrical or mostly saccate or saccate-capitate paraphyses (12-25(-30)μ diam) that collapse readily, wall colorless, 0.5μ thick; spores (20-)25-30(-34) x (15-)20-24(-26)μ, mostly broadly ellipsoid or broadly obovoid, wall 1.5(-2)μ thick, pale yellowish or nearly colorless, closely echinulate, germ pores (9)10-14(15), scattered. Telia mostly on abaxial leaf surface and sheaths in linear series, covered by the epidermis, with brown paraphyses peripherally or forming locules; spores (30-)40-60(-70) x (12-)17-23(-26)μ, variable but mostly oblong-clavate or oblong, wall 1.5-2(-2.5)μ at sides, (3-)4-6(-10)μ apically, deep golden brown or chestnut-brown, smooth; pedicels colorless to brownish, thin-walled, collapsing, less than 20μ long; 1-celled spores sometimes common.

Hosts and distribution: On species of Aegilops, Agropyron (incl. Elytrigia; Roegneria), Agrostis, Aira, Alopecurus, Arrhenatherum, Avena, Beckmannia, Boissiera, Brachypodium, Briza, Bromus, Calamagrostis, Catabrosa, Chloris (?), Dactylis

151

(?), <u>Desmostachya</u>, <u>Elymus</u> (incl. <u>Hordelymus</u>; <u>Clinelymus</u>),
<u>Festuca</u>, <u>Gaudinia</u>, <u>Glyceria</u>, <u>Haynaldia</u>, <u>Hesperochloa</u>,
<u>Heteranthelium</u>, <u>Holcus</u>, <u>Hordeum</u>, <u>Hystrix</u>, <u>Koeleria</u>, <u>Lamarckia</u>,
<u>Leersia</u>, <u>Lolium</u>, <u>Milium</u>, <u>Muhlenbergia</u>, <u>Phalaris</u>, <u>Phleum</u>, <u>Poa</u>,
<u>Puccinellia</u>, <u>Secale</u>, <u>Sitanion</u>, <u>Stipa</u>, <u>Taeniatherum</u>, <u>Trisetum</u>,
<u>Triticum</u>, and <u>Vulpia</u> (?): circumglobal, especially in the
northern hemisphere.

Lectotype: on "chaumes des cereales " (=<u>Triticum aestivum</u>),
environs de Courtray, Belgium (isotypes Westendorp et Wallays
Herb. Crypt. Belg. No. 1077; lectotype designated by Hylander,
Jørstad, and Nannfeldt (Opera Bot. 1:75. 1953).

Most of the host genera are those listed by Hassebrauk (Mitt.
Biol. Bundesanstalt Land - u. Forstwirschaft Berlin-Dahlem
116:1-75. 1965) as naturally infected. <u>Puccinia</u> <u>hordei</u> is
common on <u>Gaudinia</u>, <u>Holcus</u>, <u>Lolium</u>, <u>Koeleria</u>, and <u>Vulpia</u>, and
the two rusts have sometimes been confused. The telia of <u>P</u>.
<u>montanensis</u> and <u>P</u>. <u>brachypodii</u> usually are conspicuously linear
in arrangement but the teliospores are distinctive.

The numerous pores of the urediniospores are readily observed
in chloral hydrate solution despite the nearly colorless wall.
Uredinia have been described as having and as lacking paraphyses.
They are constantly present at the edge of the sorus.

The following variety has smaller spores than var. <u>striiformis</u>
and tolerates higher temperatures.

PUCCINIA STRIIFORMIS Westend. var. dactylidis Manners
Trans. Brit. Mycol. Soc. 43:65. 1960.
Aecia unknown. Urediniospores 18.5-25 x 15-20.5; teliospores
30-49 x 12-22.5μ.

Hosts and distribution: <u>Dactylis</u> <u>glomerata</u> L.: the British
Isles to Russia, Iran, and India.

Type: Viennot-Bourgin, on <u>Dactylis</u> <u>glomerata</u> var. <u>hispanica</u>,
Facham near Teheran, Iran (IMI 76632).

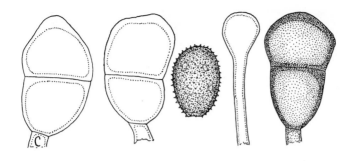

Figure 77

59. PUCCINIA MONTANENSIS Ellis J. Mycol. 7:274. 1893. Fig. 77.

Aecia (Aecidium fendleri Tracy & Earle) localized, on Berberis; spores (20-)22-27(-29) x (17-)19-23µ thick, verrucose. Uredinia on adaxial leaf surface, pale cinnamon-brown, paraphyses abundant, capitate, mostly 16-24µ wide in the head, hyaline, uniformly thin-walled, spores (25-)27-33(-36) x (19-)21-25(-27)µ, ellipsoid or broadly ellipsoid, wall 1.5-2(-2.5)µ thick, cinnamon-brown, echinulate, pores 8-10, scattered. Telia mostly on abaxial surface, blackish, covered by epidermis, weakly loculate with brownish paraphyses; spores (36-)40-55(-60) x (18-)22-32 (-35)µ, mostly obovoid or broadly obovoid, occasionally 3-celled, wall 1-2(-3)µ thick at the sides, (2.5-)3.5-7(-10)µ at apex, chestnut-brown, smooth; pedicels brownish, 15µ or less long.

Hosts and distribution: species of Agropyron, Elymus, Hordeum, Hystrix, and Sitanion: The United States and Canada, mostly the western half.

Type: Kelsey, on Elymus condensatus (=error for E. cinereus), Montana (NY: isotypes distributed as No. 2892 Ellis & Ev. N. Amer. Fungi).

The southernmost record is the summit of the Chisos Mts., Big Bend National Park, Texas. Reports of the species in South America are erroneous.

Mains (Mycologia 13:315-322. 1921) proved, by inoculation, that the aecia occurred on Berberis fendleri.

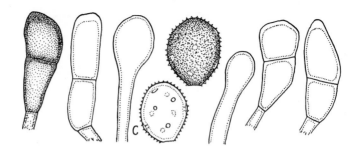

Figure 78

60. PUCCINIA PYGMAEA Eriks. Bot. Centralbl. 64:381. 1895 var.
pygmaea. Fig. 78.

Puccinia ishikawai Ito. J. Coll. Agr. Tohoku Imp. Univ.
3(2):210. 1909.

Uredo agrostidis Arth. & Cumm. Philippine J. Sci. 59:443.
1936.

Aecia on species of Berberis; spores (17-)19-24(-26) x
(14-)16-20(-22)µ, wall 1-1.5µ thick, hyaline, verrucose. Uredinia mostly on adaxial leaf surface, pale cinnamon-brown, with
abundant peripheral and some intermixed, capitate or clavate-
capitate paraphyses, 10-20µ wide, to 80µ long, the wall thin
in the head (1-1.5µ), usually thin in the stipe, the head often
collapsing, hyaline; spores (24-)26-32(-35) x (18-)21-24(-26)µ,
ellipsoid, broadly ellipsoid, or obovoid, wall (1-)1.5-2µ thick,
thicker when immature, yellowish to cinnamon-brown, finely
echinulate, pores usually obscure, (6-)8-10, scattered. Telia
mostly on abaxial surface, blackish, covered by epidermis,
weakly loculate with scanty brownish paraphyses; spores
(32-)36-48(-58) x (14-)17-22(-26)µ, mostly oblong-obovoid or
oblong, wall 1-1.5µ thick at sides, (2-)3-5(-6)µ at apex,
chestnut-brown, smooth; pedicels 15µ or less long, yellowish.

Hosts and distribution: species of Agrostis, Ammophila,
Calamagrostis, Deschampsia klossii Ridl., Festuca idahoensis
Elmer, F. subuliflora Scribn.: circumglobal, mostly in
temperate zones and at higher elevations.

Neotype: Sydow, on Calamagrostis epigeios, Germany (Sydow
Mycoth. Germ. 1480 = PUR F15967); designated, with reasons, by
Cummins & Greene (Mycologia:58: 713. 1966.)

P. pygmaea differs from the P. brachypodii complex in
having thin-walled, collapsing paraphyses and narrow teliospores.

Tranzschel (Compt. Rend. Acad. Sci. URSS. 1931: 45-48.
1931) first proved the life cycle by inoculation.

Four varieties have been recognised; var. ammophilina differs

154

because of longer teliospores; var. minor because of smaller urediospores; var. angusta because of small urediniopsores and narrow teliospores; and var. major because of larger urediospores and long teliospores.

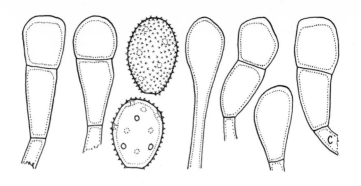

Figure 79

PUCCINIA PYGMAEA Eriks. var. ammophilina (Mains) Cumm. &
H. C. Greene Mycologia 58: 714. 1966. Fig. 79.

Uredo ammophilina Kleb. Kryptog.-Fl. Mark Brandenb. 5a:882.
1914.

Puccinia ammophilina Mains ex Cumm. Mycologia 48: 604. 1956.

Aecia unknown. Uredinial paraphyses less abundant than in
var. pygmaea; spores (26-)28-35(-40) x 20-25(-28)µ. Telial
paraphyses scant to none; spores (38-)43-63(-70) x (14-)16-22(-26)µ.

Hosts and distribution: Ammophila arenaria (L.) Link:
Europe and the Pacific Coast, U.S.A.

Type: Sprague, Ore. 10-733, on Ammophila arenaria (MICH;
isotypes BPI, PUR).

PUCCINIA PYGMAEA Eriks. var. major Cumm. & H. C. Greene
Mycologia 58:716. 1966.

Aecia unknown. Uredinial paraphyses as in var. pygmaea but spores large, (26-)30-40(-44) x 22-26(-29)μ. Telial paraphyses none or scant, sori not loculate; spores (40-)46-64(-70) x (14-)16-21(-24)μ.

Hosts and distribution: Calamagrostis ? sp., Festuca (tolucensis H.B.K.?): mountains of Mexico; 3 collections.

Type: Cummins No. 63-554 (=PUR 60267), on Calamagrostis ? sp., Durango, Mexico.

The leaves of the Calamagrostis are similar to those of C. tolucensis (H.B.K.) Trin. but some hard-leafed species of Muhlenbergia also are similar. A portion of an old inflorescence indicates that the fescue is probably correct.

PUCCINIA PYGMAEA Eriks. var. minor Cumm. & H. C. Greene
Mycologia 58:714. 1966.

Aecia unknown. Uredinial paraphyses as in var. pygmaea;
spores (20-)21-26(-28) x (17-)19-22(-24)μ. Telial paraphyses
scant, sori not loculate; spores (32-)36-43(-49) x (12-)15-21
(-23)μ.

Hosts and distribution: species of Calamagrostis: mountains
of western Europe and in Japan.

Type: Wagner (Sydow Ured. No. 1603 as P. pygmaea), on
Calamagrostis villosa (Chaix) Mutel (as C. halleriana),
Switzerland (holotype PUR F4634).

PUCCINIA PYGMAEA var. angusta Cumm. & H. C. Greene Mycologia
58: 715. 1966.

Aecia unknown. Uredinial paraphyses typical, the few spore
seen as in var. minor. Telia with few paraphyses; spores
(34-)39-54(-60) x (10-)12-16(-18)µ.

Hosts and distribution: Calamagrostis arundinacea (L.) Roth,
C. sachalinensis Schmidt: China, Japan and the Philippines.

Type: Cheo No. 655 = PUR F14403, on Calamagrostis
arundinacea, China.

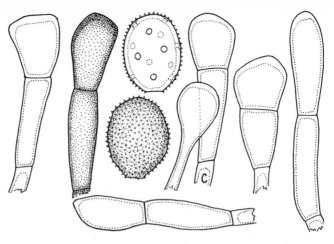

Figure 80

61. PUCCINIA CRINITAE McNabb Trans. Roy. Soc. N. Zealand 1:241.
1962. Fig. 80.

Uredo crinitae Cunn. Trans. N. Zealand Inst. 55:41. 1924.

Aecia unknown. Uredinia amphigenous or mostly on the adaxial
leaf surface, with colorless, mostly capitate paraphyses, to
25µ wide in the head, wall 1µ thick or to 3.5µ apically; spores
(27-)30-35(-38) x (24-)26-30(-32)µ, mostly broadly ellipsoid,
wall (1.5-)2(2.5)µ thick, closely echinulate, pale yellowish,
germ pores 10-14, scattered, rather obscure. Telia on adaxial
surface, blackish, covered by the epidermis, with few or no
brown paraphyses, not loculate; spores variable in size,
(36-)50-80(-88) x (11-)15-20(-24)µ, oblong, oblong-clavate,
or cylindrical, wall 1-1.5(-2)µ thick at sides, to 4µ apically,
chestnut-brown, smooth; pedicels brown adjacent to the hilum,
15µ or less long; 1-celled spores occur.

Hosts and distribution: Dichelachne crinita (Forst. f.)
Hook. f.: New Zealand.

Type: McNabb, Mt. Victoria, Wellington (PDD 19636). McNabb
lists 5 other specimens.

Telia are very rare on the type, which probably accounts for
the discrepancy between the size of the teliospore (but not
urediniospore) as published by McNabb and those above.

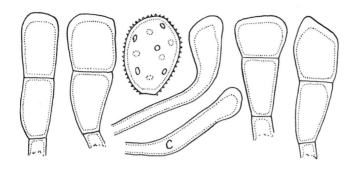

Figure 81

62. PUCCINIA BRACHYPODII-PHOENICOIDIS Guyot & Malen. Trav.
Inst. Sci. Cherifien, Ser. Bot. No. 28: 37. 1963 var.
brachypodii-phoenicoides. Fig. 81.

Aecia unknown, presumably on Berberis. Uredinia on adaxial
leaf surface, about cinnamon-brown, with abundant, mostly
peripheral paraphyses, mostly clavate-capitate, (7-)10-15(-18)μ
wide apically, wall 1μ thick basally, (1-)3-7μ apically; spores
(24-)27-32(-34) x (20-)22-25(-26)μ, mostly ellipsoid or broadly
ellipsoid, wall 2-2.5(-3)μ thick, yellowish to pale chestnut-
brown, echinulate, spines mostly spaced 1.5-2μ on centers, germ
pores obscure, 8-10, scattered. Telia on abaxial surface,
blackish, covered by epidermis, with scant marginal brownish
paraphyses, sori not loculate; spores (34-)40-56(-62) x
(14-)19-24(-26)μ, wall 1-1.5(-2)μ thick at sides, 2.5-4(-6)μ
apically, chestnut-brown or paler basally; pedicels brownish,
10μ or less long.

Type: Guyot & Malençon, on Brachypodium phoenicoidis (L.)
Roem. & Schult., Morocco (Herb. Guyot; isotype PUR F16920); not
otherwise known.

This species and its varieties differ from the P. brachypodii
complex because the paraphyses wall is thickened apically and
thin below and from the P. pygmaea complex because of the
apically thick-walled paraphyses.

PUCCINIA BRACHPODII-PHOENICOIDIS Guyot & Malen. var. davisii
Cumm. & H. C. Greene Mycologia 58: 719. 1966.

Aecia unknown. Uredinia on adaxial leaf surface, paraphyses
mostly capitate, to 24µ wide apically, wall 1µ thick at sides,
2-7(-10)µ thick at apex, mostly golden brown; spores (25-)28-32
(-35) x (20-)22-26(-28)µ, obovoid or broadly ellipsoid, wall
(1-)1.5(-2.5)µ thick, yellowish to cinnamon-brown, echinulate,
germ pores obscure, 8(-10?), scattered. Telia as in the species;
spores (30-)36-46(-52) x (12-)14-18(-20)µ, wall 1µ thick at
sides, 3-5(-7)µ at apex.

Hosts and distribution: Oryzopsis asperifolia Michx.: the
Great Lakes region, U.S.A.

Type: Davis, on Oryzopsis asperifolia, Wisconsin (WIS).

This variety differs from the typical in having the echinulae
of the urediniospores spaced 2-3µ on centers and narrower and
shorter teliospores.

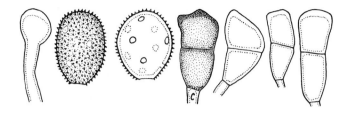

Figure 82

PUCCINIA BRACHYPODII-PHOENICOIDIS Guyot. & Malen. var.
chisosana (Cumm.) Cumm. & H. C. Greene Mycologia 58: 719.
1966. Fig. 82.

Puccinia pygmaea Eriks. var. chisosana Cumm. Southw. Nat.
8: 189. 1964.

Aecia unknown. Uredinia on adaxial leaf surface, yellowish
brown, with abundant, capitate or clavate paraphyses, to 17μ
wide in the head, wall 1μ thick below, 1.5-4(-6)μ at apex,
hyaline; spores (26-)28-33(35) x (21-)23-26(-27)μ, wall 1.5-2μ
thick, dull cinnamon-brown, densely echinulate, echinulae
spaced 0.7-1.5μ, pores obscure, 8-10(-12?), scattered. Telia
as in the species, spores 30-38(-42) x (13-)15-20(-23)μ, wall
1-1.5μ thick at sides, 3-5μ at apex.

Hosts and distribution: Bromus anomalus Rupr., B. brachyan-
thera Doell., B. sp.: southwestern U.S.A., Mexico, and Brazil.

Type: Cummins 62-415(=PUR 57364), on Bromus anomalus,
Texas, U.S.A.

Variety chisosana differs from the preceding two because of
more densely echinulate urediniospores and shorter teliospores.
Kaufmann (Mycopathol. Mycol. Appl. 32:249-261. 1967) published a photograph of teliospores of the type.

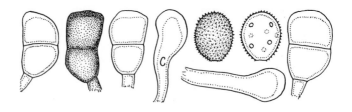

Figure 83

63. PUCCINIA BRACHYPODII Otth Mitth. Naturf. Ges. Bern 1861:
82. 1861 var. brachypodii Fig. 83.

Epitea baryi Berk. & Br. Ann. Mag. Nat. Hist. II. 13: 461.
1854. Based on uredinia.

Puccinia linearis Rob. in Desmazieres Ann. Sci. Nat. 4:
125. 1855, not Roehling 1813.

Puccinia brachypodii Fuckel Jahrb. Nassau Ver. Naturk. 23-24:
60. 1869.

Puccinia baryi Wint. Rabenh. Kryptog. Fl. Ed. 2. I. 1:178.
1882.

Aecia, localized on Berbis aristata DC., B. lycium Royle,
and B. vulgaris l.: spores (19-)22-27(-29.) x 15-21μ, wall
uniformly 1μ thick, verrucose. Uredinia mostly on adaxial
surface, conspicuously seriate, yellowish brown, paraphyses
mostly 40-70μ long, variable, cylindric-clavate or mostly
cylindric-capitate, often with a contracted "neck" below the
head, head 11-21μ diam, wall mostly 2-3.5μ thick below, 4-9μ
in the head, golden or colorless; spores 21-25 x 16-21μ, mostly
broadly ellipsoid or broadly obovoid, wall 1.5-2(-2.5)μ thick,
yellowish, closely echinulate, pores obscure, about 8, scattered.
Telia amphigenous or mostly on abaxial surface blackish, covered
by epidermis, mostly seriately arranged, with few or often no
brownish paraphyses; spores (27-)30-38(-41) x (15-)17-23(-26)μ,
variable but mostly oblong or obovoid, wall 1μ thick at sides,
2-4(-5)μ at apex, deep golden or clear chestnut-brown, smooth;
pedicels 12μ or less long, brownish.

Hosts and distribution: species of Brachypodium: Europe
and Japan.

Neotype: Otth, on Brachypodium sylvaticum (Huds.) Beauv.,
Switzerland (BERN). Neotype designation made by Cummins and
Greene (Mycologia 58:702-721. 1966.).

P. brachypodii is the oldest valid name for a world-wide
complex of rust fungi proved or presumed to produce aecia on
Berberis-Mahonia and characterized by long-covered telia,

164

abundant clavate-capitate uredinial paraphyses whose walls are
thick throughout, and closely echinulate urediospores having
numerous obscure germ pores. Many "species" have been named
within the complex but the morphological variability is nearly
continuous.

Proof that Berberis is the aecial host was provided by Mayor
in 1934 (Bull. Soc. Neuchâtel. Sci. Nat. 58:7-31 1933) and by
Payak in 1965 (Phytopathol. Z. 52:49-54).

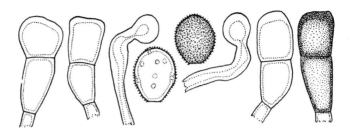

Figure 84

PUCCINIA BRACHYPODII Otth var. poae-nemoralis (Otth)
Cumm. & H. C. Greene Mycologia 58:705. 1966. Fig. 84.

Physonema minimum Bonorden Abh. Nat. Ges. Halle 5:200. 1860.
(Based on uredinia).

Uredo poae-sudeticae Westend. Bull. Roy. Acad. Belg. II.
650. 1861.

Puccinia poae-nemoralis Otth Mitth. Natur. Ges. Bern
1870:113. 1871.

Puccinia milii Eriks. Bot. Centrlbl. 64:382. 1895.

Puccinia exigua Diet. Hedwigia 36:299. 1897.

Uredo kerguelensis P. Henn. Deut. Sudpolar-Exped. 1901-
1903, VIII. Bot. p. 4. 1906.

Uredo anthoxanthina Bub. Ann. Mycol. 3:223. 1905.

Puccinia cognatella Bub. Ann. Mycol. 7:378. 1909.

Puccinia oligocarpa Syd. & Butl. Ann. Mycol. 10:262. 1912.

Puccinia narduri Gz. Frag. Trab. Mus. Nac. Cienc. Nat. Ser.
Bot. 3:13. 1914.

Uredo festucae-halleri Cruchet Bull. Soc. Vaud. Sci. Nat.
51:629. 1917.

Uredo glyceriae-distantis Eriks. Ark. Bot. 18(19):18. 1923.

Puccinia thalictri-poarum Ed. Fisch. & Mayor Mitt. Naturf.
Ges. Bern 1924:36. 1925, nom. nud.

Puccinia poae-sudeticae Joerst. Nytt. Mag. Naturv. 70:325.
1932.

Puccinia anthoxanthina Gaeum. Ber. Schweiz. Bot. Ges. 55:74.
1949.

Puccinia poae-annuae V.-Bourgin Bull. Soc. Mycol. France
84:497-498. 1968.

Aecia, localized, on Berberis jaeschkeana C. K. Schneid;

spores 20-27 x 19-23μ, wall about 1μ thick, verrucose. Uredinia
mostly on adaxial surface, yellowish or yellowish brown, with
abundant peripheral and intermixed, hyaline or yellowish,
cylindric-capitate or capitate paraphyses, mostly 50-80μ long
and 18-16μ wide, usually geniculate and with a constricted "neck",
wall 2.5-4μ thick throughout or to 7μ thick in the head; spores
(20-)22-27(-29) x (16-)18-23(-25)μ, ellipsoid, broadly ellipsoid or obovoid, wall 1.5-2(-2.5)μ thick, hyaline to pale
golden, closely echinulate, pores scattered, 8-12, obscure;
amphispores with cinnamon- to near chestnut-brown walls
sometimes formed. Telia mostly on abaxial surface, blackish,
covered by the epidermis, with brownish paraphyses scant or
numerous but sori not conspicuously loculate; spores (31-)35-50
(-64) x (14-)17-23(-25)μ, wall 1-1.5μ thick at sides, (3-)4-6
(-7)μ apically, chestnut-brown, or paler basally, smooth;
pedicels 15μ long or less, brownish.

   Hosts and distribution: On species of: Agrostis, Alopecurus,
Anthoxanthum, Arctagrostis, Calamagrostis, Catabrosa, Festuca,
Glyceria, Lolium, Melica, Milium, Nardurus, Phippsia, Phleum,
Poa, Sieglingia, Trisetum, and Vulpia; circumglobal in temper-
ate climates and at high altitudes in the tropics.

   Type: Otth, on Poa nemoralis, Switzerland (BERN).

   This variety is variable but with such intergradation that
further segregation is doubtfully desirable. In the northern
United States amphispores occur occasionally on Poa and in
Alaska in Arctagrostis latifolia (R. Br.) Griseb., often as
segments in ordinary uredia.

   The fungus is not obligately heteroecious and commonly
occurs without a Berberis associate. The Indian aecia used
in inoculations that proved the life cycle (Joshi & Payak,
Mycologia 55:247-250. 1963) were localized, but systemic
aecia occur in the Himalayan region. Aecidium montanum Butler
may belong here or with var. brachypodii.

   Variety poae-nemoralis does not have seriately arranged sori
and has longer teliospores than var. brachypodii.

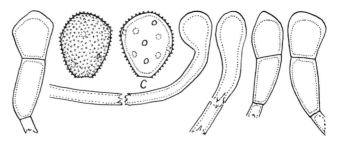

Figure 85

PUCCINIA BRACHYPODII Otth. var. arrhenatheri (Kleb.) Cumm. & H. C. Greene Mycologia 58:709. 1966. Fig. 85.

Puccinia perplexans Plowr. f. arrhenatheri Kleb. Abh. Naturw. Ver. Bremen 12:366. 1892.

Puccinia magelhaenica Peyr. ex Magnus Ber. Nat.-Med. Ver. Innsbruck 21:41. 1894.

Puccinia arrhenatheri Eriks. Beitr. Biol. Pfl. 8:14. 1898.

Puccinia koeleriae Arth. Mycologia 1:247. 1909.

Uredo paulensis P. Henn. Hedwigia 41:297. 1902.

Puccinia spicae-venti Bucholtz Ann. Mycol. 3:446. 1905.

Puccinia deschampsiae Arth. Bull. Torrey Bot. Club 37:570. 1910.

Uredo lamarckiae Kleb. Kryptogamenfl. Mark Brandenburg 5a:883. 1914.

Puccinia aerae Mayor & Cruchet in Cruchet Bull. Soc. Vaud. Sci. Nat. 51:628. 1917.

Uredo lamarckiae Cab. & Gz. Frag. Bol. R. Soc. Espan Hist. Nat. 20:309. 1920.

Puccinia distichophylli Ed. Fisch. Mitt. Naturf. Ges. Bern 1920: XLII. 1921.

Puccinia hordeicola Lindq. Rev. Fac. Agron. La Plata 33:76. 1957.

Puccinia poae-nemoralis Otth. ssp. hyparctica Savile in Savile & Parmelee Canad. J. Bot. 42:705. 1964. Based on uredinia.

Aecia localized or systemic, on Berberis; spores (20-)23-27 (-29) x (16-)19-23(-24)µ, wall 1-1.5µ thick, hyaline, verrucose. Uredinia on adaxial leaf surface, cinnamon-brown or paler, with abundant, mostly clavate or clavate-capitate, hyaline paraphyses, (7-)13-20(-28µ wide apically, to 120µ long, the "neck" constricted or not, wall uniformly (1-)2-4(-7)µ thick;

168

spores (24-)26-33(-36) x (18-)21-26(-29)μ, ellipsoid, broadly
ellipsoid, or obovoid, wall (1-)1.5-2(-2.5)μ thick, pale yellowish to cinnamon-brown, closely echinulate, pores obscure, 8-12
scattered. Telia mostly on abaxial surface, blackish, covered
by epidermis, brownish paraphyses usually scanty but sori sometimes loculate; spores (30-)36-50(-80) x (12-)15-22(-27)μ,
variable but mostly oblong or oblong-obovoid, wall 1-1.5μ thick
at sides, (2-)3-5(-7)μ apically, chestnut-brown, smooth; pedicels
15μ or less long, brownish.

Hosts and distribution: On species of Agropyron, Apera,
Arrhenatherum, Bromus, Calamagrostis, Deschampsia, Elymus,
Festuca, Helictotrichon, Hordeum, Koeleria, Lamarckia, Phalaris,
Poa, Relchella, and Trisetum: circumglobal, especially in
temperate and cooler areas.

This variety has non-seriate sori and differs from var.
brachypodii additionally because of longer urediniospores and
teliospores. It has longer, usually browner urediniospores
than var. poae-nemoralis.

The systemic aecial habit of the Arrhenatherum rust fungus
is not unique. Mains (Mycologia 25:407-417. 1933.) reported
a similar development on Berberis fendleri Gray when infected
by basidiospores from Koeleria cristata, but the systemic habit
is not typical of the North America fungus. In southern South
America there are numerous systemic and localized aecia on
Berberis. Their relationship is not known but it is suggestive
that var. arrhenatheri is common on several genera of grasses
in the area. Systemic aecia also occur in India and Pakistan.

PUCCINIA BRACHYPODII Otth var. major Cumm. & H. C. Greene
Mycologia 58:711. 1966.

Aecia unknown. Uredinia cinnamon-brown, paraphyses as in
var. poae-nemoralis; spores (30-)32-42(-49) x (24-)27-32(-34)μ,
wall 2-2.5μ thick, golden or cinnamon-brown, densely echinulate,
germ pores 9-12, scattered. Telia as in var. poae-nemoralis;
spores 46-66 x (16-)19-24μ.

Hosts and distribution: Poa horridula Pilger, P. candamoana
Pilger: Peru.

Type: Weberbauer, on Poa horridula, Peru (PUR F15673).

This variety has much larger urediniospores and somewhat
larger teliospores than the species. A fungus that is generally
similar but has nearly colorless urediospores that reach 50μ
long occurs on Hierochloë redolens Vahl in Chile.

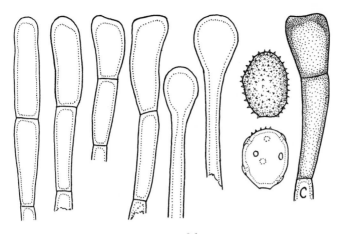

Figure 86

64. PUCCINIA MELLEA Diet. & Neger Bot. Jahrb. 24:155. 1897.
Fig. 86.

Aecia unknown. Uredinia on adaxial leaf surface, yellowish
brown, paraphyses clavate to capitate, 10-20(-25)μ wide, wall
colorless, 2-4μ thick; spores (20-)25-32(-35) x (17-)20-24(-26)μ,
mostly ellipsoid or obovoid, wall 1.5-2(-2.5)μ thick, closely
echinulate, pores 8-10, scattered. Telia on abaxial surface,
long-covered by epidermis, tending to be loculate with brown
paraphyses; spores (36-)45-85(-92) x (10-)12-18(-22)μ, mostly
cylindrical or cylindrical-clavate, wall 1-1.5(-2)μ thick at
sides, 3-5(-7)μ apically, pale golden, smooth; pedicels golden,
less than 20μ, persistent.

Hosts and distribution: Vulpia australis (Nees) Blom, V.
eriolepis (Desv.) Blom, V. megalura (Nutt.) Rydb., V. muralis
(Kunth) Nees, V. myuros (L.) Gmel.: Chile and Argentina.

Type: Neger, on Festuca muralis (=Vulpia muralis),
Concepcion, Chile (S; isotype PUR).

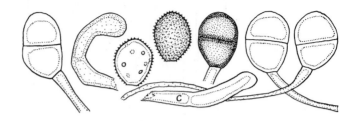

Figure 87

65. PUCCINIA ENTEROPOGONIS P. Syd. & H. Syd. Monogr. Ured.
1:751. 1904. Fig. 87.

Puccinia chloridis-incompletae Ramak. T.S., Srin. & Sund.
Proc. Indian Acad. Sci. B. 36:91. 1952.

Aecia unknown. Uredinia amphigenous or mostly in abaxial
side, cinnamon-brown, with hyaline or golden, mostly incurved,
clavate, 1-septate paraphyses, whose wall is uniformly 2-3μ
thick; spores (19-)21-24(-27) x 16-19(-20)μ, mostly ellipsoid
or oval, wall 1.5μ thick, golden to cinnamon-brown, echinulate,
pores 6 or 7, scattered. Telia blackish brown, pulvinate,
exposed; spores (23-)25-30(-33) x (17-)19-22(-24)μ, mostly
ellipsoid, wall (1.5-)2-3(-3.5)μ at sides, 3-5(-6)μ apically,
chestnut-brown, smooth; pedicels thick-walled, usually not
collapsing, hyaline to golden, persistent, to 130μ long.

Hosts and distribution: Chloris incompleta Roth, Enteropogon
monostachyus (Vahl) K. Schum.: Tanganyika, Zanzibar and India.

Type: Holst, on E. monostachyus, Usambara, Tanganyika (S).

The basally 1-septate paraphyses are unique in the genus
Puccinia.

Hennen and Cummins (Mycologia 48:126-161. 1956) published
a photograph of one teliospore of the type.

172

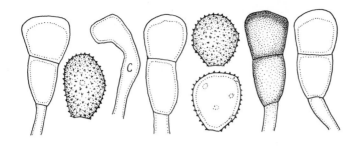

Figure 88

66. PUCCINIA DIGITARIAE-VELUTINAE V.-Bourgin Bull. Soc. Mycol.
Fr. 67:431, 1951. Fig. 88.

Puccinia digitariae-vestitae Ramachar & Cumm. Mycopathol.
Mycol. Appl. 25:18. 1965.

Aecia unknown. Uredinia amphigenous or mostly on abaxial
surface. Uredinia yellowish brown, pulverulent, with peripheral, more or less clavate and often incurved, hyaline or
yellowish, thin-walled paraphyses; spores (23-)25-32(-35) x
(18-)20-25(-28)μ, mostly oval or obovate, wall 1-1.5μ thick,
golden or pale cinnamon-brown, echinulate, pores 5-8, scattered,
Telia blackish brown, compact, early exposed; spores (36-)39-50
(-52) x 17-22(-24)μ, mostly clavate or oblong-clavate with the
apex more or less truncate, wall 1-1.5μ thick at sides, 3-5μ at
apex, smooth, deep golden to chestnut-brown; pedicels thin-
walled, golden, persistent, to 25μ long.

Hosts and distribution: Digitaria velutina P. Beauv., D.
vestita Fig. & De Not. var. scalarum (Schweinf.) Henrad: Gold
Coast, Ivory Coast, Kenya, N. Rhodesia, and Uganda.

Type: Viennot-Bourgin, on D. velutina, Ivory Coast (Herb.
Viennot-Bourgin).

The species differs from P. oahuensis in having exposed
telia and urediospores with scattered pores. Viennot-Bourgin
has republished, with extensive notes, the description in
Urediniana 4:169. 1953.

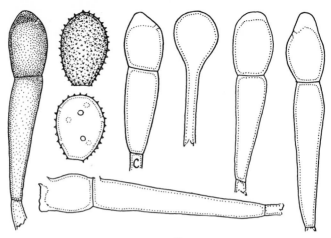

Figure 89

67. PUCCINIA AZTECA Cumm. & Hennen sp. nov. Fig. 89.

Aeciis ignotis. Urediniis plerumque epiphyllis, flavidis, paraphysibus capitatis vel clavato-capitatis, membrana plerumque uniformiter 1-1.5µ crassa, hyalina; sporae (20-)23-28(-31) x (18-)20-23(-24)µ, obovoideae vel late ellipsoideae, membrana 1-1.5µ crassa, hyalina vel flavida, echinulata, poris germinationis obscuris, 7-9, sparsis. Teliis amphigenis vel plerumque epiphyllis, erumpentibus, compactis, atro-brunneis; sporae plerumque cylindraceae, magnitudine variabili, (40-)50-95 x (11-)14-18(-20)µ, vel longiore 90-160µ ubi germinantibus, membrana ad latere 1µ crassa, ad apicem (6-)10-18(-24)µ, pallide castaneo-brunnea; pedicello plus minusve 10µ longo.

Type: Hennen 70-3, on Trisetum virletii Fourn., Desert of the Lions National Park, 10 miles west of Mexico City, Mexico, 6 June 1970 (PUR 63273).

One other collection on the same host is known from near Morelia. In Brazil, what perhaps is the same rust fungus occurs on Calamagrostis montevidensis Nees (Holway 1791, 1935, 1953). It seems probable that the lower cell elongates when teliospores germinate, thus accounting for the great variability in the lengths of the spores.

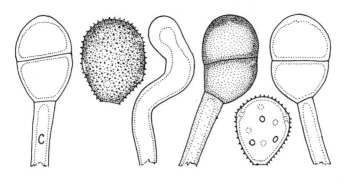

Figure 90

68. PUCCINIA ANDROPOGONIS-HIRTI Beltr. Mem. R. Soc. Espan.
Hist. Nat. 50:252. 1921. Fig. 90.

Uredo andropogonis-hirti Maire Bull. Soc. Mycol. Fr. 21:162.
1905.

Aecia unknown. Uredinia mostly on abaxial leaf surface,
cinnamon-brown, with capitate or clavate-capitate, mostly
curved and commonly geniculately curved, hyaline to brownish
paraphyses whose wall is 12-16μ thick apically becoming
progressively thinner below; spores (25-)29-35(-38) x
(20-)24-28μ, mostly broadly oval or obovoid, wall (1.5-)2-2.5
(-3.5)μ thick, cinnamon-brown or darker apically, echinulate,
pores (6-)8(-10), scattered. Telia on abaxial surface,
blackish brown, compact, becoming exposed; spores (30-)32-38(-40)
x (20-)22-28μ, mostly broadly ellipsoid, wall 1.5-2μ thick at
sides, 2-3(-4)μ at apex, smooth, chestnut-brown; pedicels
thick-walled, not collapsing, hyaline to brownish, seldom more
than 35μ long.

Hosts and distribution: Hyparrhenia hirta (L.) Stapf, the
Mediterranean region.

Lectotype: Beltrán, Castellon, Spain (MA).

Without teliospores the species is doubtfully separable
from P. eritraeensis and P. hyparrheniicola. Despite Betrán's
description and illustration, neither pore of the teliospore
is provided with a papilla and the lower pore is adjacent to
the septum.

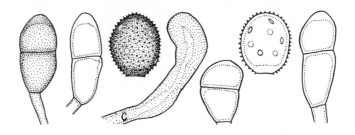

Figure 91

69. PUCCINIA HYPARRHENIICOLA Joerst. & Cumm. in Cummins Torrey Bot. Club Bull. 83:227. 1956. Fig. 91.

Aecia unknown. Uredinia on abaxial leaf surface, to 2 mm long, chestnut-brown, paraphyses 8-18μ wide apically, to 70μ long, cylindrical or cylindric-capitate, incurved, often geniculately so, thick-walled; spores (26-)28-31(-34) x (21-)24-27(-29)μ, broadly ovate or globoid, wall (1.5-)2-3μ thick, dark cinnamon- or chestnut-brown, echinulate, germ pores 6-8, scattered. Teliospores in the uredinia (28-)33-40 (-45) x 16-19(-22)μ, oblong or oblong-ellipsoid, wall 1(-1.5)μ thick at sides, (2.5-)3-3.5(-4)μ apically, golden brown with the apex paler, smooth; pedicels colorless, thin-walled, collapsing, to 15μ long.

Hosts and distribution: <u>Hyparrhenia hirta</u> (L.) Stapf; Canary Islands.

Type: I. Jørstad No. 822, Santa Cruz, Tenerife, Canary Islands, Mar. 21, 1954 (PUR; isotype O).

Cummins (loc. cit.) published a photomicrograph of 1 urediniospore and 1 teliospore.

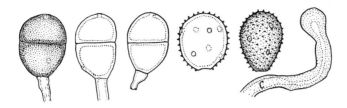

Figure 92

70. PUCCINIA KENMORENSIS Cumm. Bull. Torrey Bot. Club 72:209. 1945. Fig. 92.

Aecia unknown. Uredinia amphigenous, cinnamon-brown, with capitate or clavate-capitate, straight or curved, thick-walled, yellowish paraphyses, 50-75 x 11-16μ; spores 23-29 x 19-23μ, mostly broadly oval or obovoid, wall 1.5-2.5μ thick, dark cinnamon- or chestnut-brown, echinulate, pores 6-8, scattered. Telia not seen; teliospores 23-30 x 18-22μ, ellipsoid or broadly so, wall uniformly 2-2.5μ thick, smooth, chestnut-brown; pedicels as seen only 8-12μ but probably are longer, hyaline, thin-walled and collapsing, deciduous.

Hosts and distribution: <u>Bothriochloa decipiens</u> (Hack.) C. E. Hubb.,: Australia.

Type: Clemens, Queensland, Australia (PUR).

Cummins (loc. cit.) published a photomicrograph of 1 teliospore and 1 urediniospore.

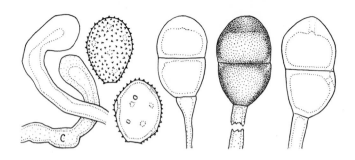

Figure 93

71. PUCCINIA ERITRAEENSIS Paz. Bot. Jahrb. 17:14. 1893.
Fig. 93.

Aecia unknown. Uredinia mostly on abaxial leaf surface,
cinnamon-brown, with clavate or capitate-clavate, incurved or
geniculate, colorless or yellowish paraphyses, the wall 6-11μ
thick apically becoming progressively thinner basally; spores
24-32(-35) x (16-)20-25(-27)μ, oval or nearly globoid, wall
1.5-2μ, cinnamon-brown or slightly darker, echinulate, germ
pores (6)7-9(-10), scattered. Telia on abaxial surface, blackish brown, exposed, compact; spores (30-)33-40(-46) x (19-)
20-27(-29)μ, mostly broadly ellipsoid, wall 2.5-3μ thick at
sides, 4-5(-6)μ apically, chestnut-brown, smooth; pedicels
rather thin-walled, usually collapsing, colorless or yellowish,
persistent, to 90μ long.

Hosts and distribution: species of Andropogon, Capillipedium,
Cymbopogon, Hyparrhenia, Trachypogon: Africa to Australia, and
in Honduras.

Type: Schweinfurth, on Andropogon sp., Haschello Kobob,
Eritraea (B).

Without teliospores it is doubtful if the species can be
distinguished from P. andropogonis-hirti and P. hyparrheniicola.

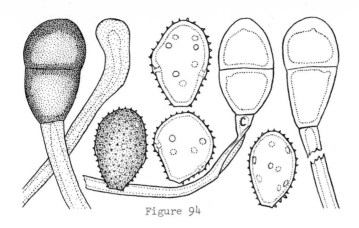

Figure 94

72. PUCCINIA PURPUREA Cke. Grevillea 5:15 1876. Fig.94.

Uredo sorghi Pass. Comm. Soc. Critt. Ital. 2:449. 1867.

Uredo sorghi Fuckel. Bot. Zeit. 29:27. 1871.

Puccinia sanguinea Diet. ex Atkinson Bull. Cornell Univ. 3:19. 1897.

Uredo sorghi-halepensis Pat. Bull. Soc. Myc. Fr. 19:253. 1903.

Puccinia prunicolor H. Syd., P. Syd. & Butl. Ann. Mycol. 4:435. 1906.

Puccinia sorghi-halepensis Speg. Anal. Mus. Nac. Buenos Aires 31:386. 1922.

Sori in leaves, mostly in abaxial side of leaves, in purple spots. Uredinia nearly chestnut-brown, pulverulent, with clavate or clavate-capitate, mostly curved, hyaline or yellowish (or purple stained from the host) paraphyses, with the wall 4-7μ apically becoming progressively paler below; spores (26-)30-40 x 23-29(-32)μ, variable, ellipsoid, obovoid, or nearly globoid, often angular, wall 2μ thick, cinnamon or slightly darker, echinulate, pores 5-8, scattered or tending to be bizonate. Telia blackish brown, compact, pulvinate, exposed; spores (37-)40-50(-55) x (22-)24-30(-33)μ, mostly ellipsoid, or oblong-ellipsoid, wall (2.5-)3-3.5μ thick at sides, 4-5(-7)μ apically, chestnut, smooth; pedicels thick-walled and mostly not collap-sing, hyaline or yellow, persistent, to 95μ long.

Hosts and distribution: Cymbopogon citratus (DC.) Stapf. ?, species of Sorghum, circumglobal in the warmer regions of the world.

Records for hosts other than sorghum need confirmation.

LeRoux and Dickson (Phytopathology 47:101-107. 1957) demon-strated that Oxalis corniculata is the, or an, aecial host but they did not publish details of morphology nor did they save specimens in WIS.

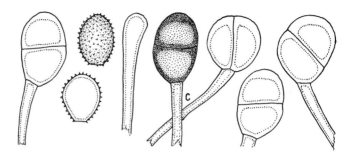

Figure 95

73. PUCCINIA ERAGROSTIDICOLA Kern, Thurst. & Whet. Mycologia
25: 469. 1933. Fig. 95.

Aecia unknown. Uredinia amphigenous or mainly in abaxial
side, yellow, with mostly clavate, hyaline paraphyses, whose
wall is uniformly 1.5-2.5µ thick or becoming thinner below;
spores (18-)20-25(-27) x (13-)15-18µ, mostly oval, wall 1-1.5µ
thick, hyaline or very pale yellowish, minutely echinulate,
pores probably scattered, obscure. Telia blackish brown,
compact, pulvinate, early exposed; spores (26-)29-33(-36) x
(20-)22-28(-30)µ, mostly broadly ellipsoid, wall (2-)3-3.5µ
thick at sides, 4-6µ apically, chestnut-brown, smooth; pedicels
moderately thick-walled, collapsing or not, hyaline or yellowish, persistent, to 75µ long.

Hosts and distribution: Eragrostis inconstans Nees: Colombia.

Type: Archer No. H69, Quebrada de la Garcia, Colombia (PAC).

This is the only species belonging in Group II that is known
on Eragrostis.

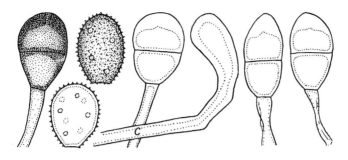

Figure 96

74. PUCCINIA NASSELLAE Arth. & Holw. in Arthur Proc. Amer. Phil.
Soc. 64:196. 1925 var. nassellae. Fig. 96.

Aecia unknown. Uredinia on abaxial leaf surface, about
cinnamon-brown, pulverulent, with mostly clavate, curved,
hyaline to golden paraphyses whose wall is 3-5µ thick apically
becoming progressively thinner below; spores (21-)26-30(-36)
x (18-)23-26(-31)µ, oval to nearly globoid, wall 1-2µ thick,
echinulate, golden, germ pores scattered, 6-8, scattered.
Telia on abaxial surface, blackish brown, compact, pulvinate,
exposed; spores (30-)36-44(-56) x (18-)21-25(-28)µ, mostly
broadly ellipsoid, wall 2-2.5µ thick at sides, 5-12µ apically,
chestnut-brown, smooth; pedicels thick-walled, hyaline to
golden, non-collapsing, persistent, to 60µ long.

Hosts and distribution: species of Nassella, Stipa brachy-
phylla Hitchc.: Argentina, Bolivia, Chile, and Peru.

Type: Holway No. 508, on Nassella caespitosa, Sorata,
Bolivia (PUR).

Greene and Cummins (Mycologia 50: 6-36. 1958) published a
photograph of paraphyses and teliospores of the type.

PUCCINIA NASSELLAE Arth. & Holw. var. platensis Lindq. Rev.
Fac. Agron. Univ. Nac. La Plata 38: 86-87. 1962.

Urediniospores 22-28 x 22-25µ; teliospores 29-40 x 18-24µ.

Type: Lindquist on Stipa neesiana Trin. & Rupr., La Plata,
Argentina (LPS 15.286; isotype PUR). Not otherwise known.

This variety has smaller spores than var. nassellae.

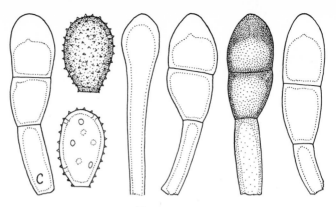

Figure 97

75. PUCCINIA MAGNUSIANA Koern. Hedwigia 15:179. 1876. Fig. 97.

Puccinia arundinacea Hedw. f. in Lam. Encycl. Meth. Bot.
8:250. 1808 (nom. confusum).

Puccinia arundinacea β epicaula Wallr. Fl. Crypt. Germ.
2:225. 1833.

Puccinia simillima Arth. Bot. Gaz. 34:17. 1902.

Puccinia alnetorum Gaeum. Hedwigia 80:139. 1941.

Aecia localized, on Anemone, Clematis and Ranunculus; spores
23-26 x 21-23μ, wall 1-1.5μ thick, verrucose. Uredinia amphigenous, yellowish brown, pulverulent, with clavate or clavate-
capitate, hyaline or yellowish paraphyses whose wall is 1.5-4μ
thick apically, becoming thinner below; spores (20-)26-35(-42)
x (13-)15-19(-21)μ, mostly ellipsoid, oblong-ellipsoid, or oval,
wall 1.5-2(-3)μ thick, hyaline to yellowish brown, echinulate,
pores obscure, 8-10, scattered or tending to be bizonate. Telia
amphigenous, blackish brown, compact, exposed, pulvinate; spores
(35-)42-56(-62) x (13-)15-24(-29)μ, variable, mostly clavate
or oblong-clavate, sometimes ellipsoid, the apex mostly rounded
or narrowly rounded, wall (1-)1.5-2(-3)μ at sides, (4-)7-10(-14)μ
apically, deep golden or chestnut-brown, smooth, pedicels thick-
walled, not or only partially collapsing, hyaline to brownish,
persistent, to 95μ long, usually about 50μ long.

Hosts and distribution: Arundo donax L., species of Phragmites:
circumglobal.

Lectotype: Koernicke, on Phragmites communis Trin., Bei
Waldau (Ostprussen) 19 Sept. 1865 (B). Lectotype designated
here.

The life cycle was demonstrated first by Cornu (Compt. Rend.
Acad. Sci. Paris 94:1731. 1882) with Ranunculus as the aecial
host.

Most American collections have urediniospores in the upper
range of measurements and tend to have dimorphic teliospores
with long, narrow, pale spores intermixed with the broader
and more pigmented spores.

182

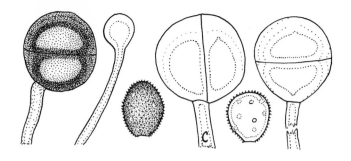

Figure 98

76. PUCCINIA EYLESII Doidge Bothalia 2(1a):201. 1927. Fig. 98.

Aecia unknown. Uredinia on adaxial side of leaves, yellowish brown, paraphyses capitate, about 50μ long, to 24μ wide, wall to 10μ thick apically, colorless or yellowish; spores 20-24 x (16-)18-20μ, mostly broadly ellipsoid, wall 1.5μ thick, pale cinnamon-brown, occasional spores (amphisporic?) with near chestnut-brown walls 2-2.5μ thick, echinulate, pores scattered, obscure, about 8. Telia mostly adaxial and on the inflorescence, blackish brown, early exposed; spores commonly or mostly dior-chidioid, (27-)30-41 x (25-)32-40μ, from broadly ellipsoid to broadly transversely ellipsoid (with reference to the hilum), wall 2-7μ thick basally, thickening progressively to 8-12μ apically, usually showing concentric lamination, golden to chestnut-brown, smooth or sometimes appearing rugose; pedicels hyaline or brownish, persistent, to 180μ long.

Hosts and distribution: Aristida aequiglumis Hack., A. junciformis Hack., A. transvaalensis Henrard: Rhodesia and South Africa.

Type: Eyles, on Aristida sp., Rhodesia, (PRE 15516).

A photograph of spores of the type was published by Cummins and Husain (Bull. Torrey Bot. Club 93:56-67. 1966).

77. PUCCINIA KWANHSIENENSIS Tai Farlowia 3:118. 1947.

Aecia unknown. Uredinia not described; paraphyses capitate or clavate, brownish, wall apparently 2.5-3.5μ thick apically becoming gradually thinner below; urediniospores 18-20μ diam, globoid, wall 2.5-3μ thick, yellowish brown, echinulate, pores 4, scattered. Telia amphigenous, blackish brown, pulvinate, exposed; spores 37-57 x 15-23μ, ellipsoid or oblong-ellipsoid, rounded or acuminate apically, wall apparently 2.5-3μ thick at sides, 3-11μ apically, pale chestnut, smooth; pedicels brownish, persistent, wall thickness not indicated, about equalling the spore in length.

Type: L. Ling on Bambusa, Kwanhsien, Szechuan, China, 11 Oct. 1936. (Pl. Pathol. Herb. No. 6852, Tsing Hua Univ., Kunming, - not seen). Not otherwise known.

With only 4 pores one would expect them to be equatorial. The pedicel length of "sporam subaequante" may not represent total length because the bamboo rusts usually have long pedicels.

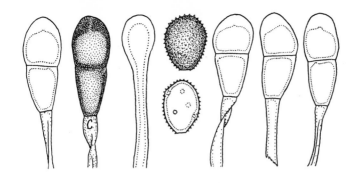

Figure 99

78. PUCCINIA SALTENSIS Cumm. Torrey Bot. Club. Bull. 83:231.
1956, var. saltensis. Fig. 99.

Aecia unknown. Uredinia in abaxial side of leaves, small,
cinnamon-brown; paraphyses capitate or clavate-capitate, 13-26μ
wide apically, to 90μ long, wall colorless or golden, more or
less evenly 2.5-4μ thick; spores 18-25(-27) x (15-)16-20(-22)μ,
mostly ellipsoid or broadly ellipsoid, wall 1.5-2(-2.5)μ thick,
dark cinnamon-brown, echinulate, germ pores 4-6, scattered.
Telia epiphyllous, pulvinate, exposed, blackish brown; spores
(30-)33-45(-50) x (14-)16-19(-22)μ, mostly clavate-ellipsoid
or oblong-ellipsoid, wall 1.5-2.5(-3)μ thick at sides, (5-)7-9
(-11)μ apically, chestnut-brown, smooth; pedicels golden or
yellowish, thin-walled and collapsing, to 55μ long.

Hosts and distribution: <u>Stipa ibarrensis</u> H.B.K.; <u>S. tucumani</u>
Parodi: Argentina and Ecuador.

Type: Hunziker No. 1844, on <u>S. tucumani</u>, Prov. Salta,
Argentina, May 2, 1942 (PUR).

Cummins (loc. cit.) published a photograph of teliospores
of the type.

PUCCINIA SALTENSIS Cumm. var. _faldensis_ H. C. Greene & Cumm. 50:11. 1958.

Similar to _saltensis_ var. _saltensis_ except the urediniospores (22-)24-29(-33) x (18-)21-25(-27)μ; the teliospores (30-)36-50 (-60) x (17-)19-24(-28)μ, wall 7-11(-15)μ thick apically; pedicels to 85μ long.

Hosts and distribution: species of _Nassella_, _Stipa_: Argentina, Bolivia, Peru, Uruguay, and perhaps Australia.

Type: Holway No. 2026, on _Stipa ichu_ (Ruiz. & Pavon) Kunth, La Falda, Argentina (PUR).

Greene and Cummins (loc. cit.) published a photograph of teliospores of the type.

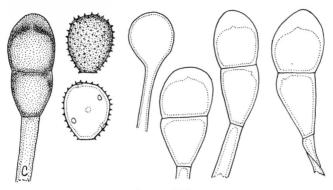

Figure 100

79. PUCCINIA CORTEZIANA Cumm. & Hennen sp. nov. Fig. 100.

Aeciis ignotis. Urediniis epiphyllis, cinnamoneo-brunneis, paraphysibus hyalinis, capitatis, 14-24μ diam, membrana uniformiter 0.5-1μ crassa; sporae 22-26(-29) x 20-22μ, plerumque obovoideae, membrana 1-1.5μ crassa, cinnamoneo-brunnea, echinulata, poris germinationis 4-6, plerumque 5, sparsis. Teliis epiphyllis, atro-brunneis, pulvinatis, compactis; sporae 33-48(-53) x (17-) 18-22(-25)μ, ellipsoideae vel obovoideae, membrana ad latere 1(1.5)μ crassa, ad apicem 5-7μ crassa, castaneo-brunnea, levi; pedicello tenui tunicati, brunneolo, usque ad 40μ longo, persistenti.

Type: Hennen 67-422 (=PUR 62783) on Brachypodium mexicanum (Roem. & Schult.) Link, road, Amecameca to Paso de Cortez, Mexico. Only known in Mexico and by this, and one other collection from Mexico State, and one collection from Michoacan State.

The germ pores are not evenly spaced and sometimes tend to be equatorial.

187

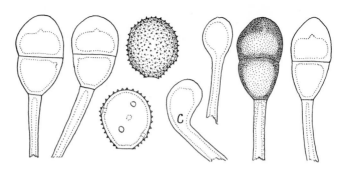

Figure 101

80. PUCCINIA DECOLORATA Arth. & Holw. in Arthur Proc. Amer.
Phil. Soc. 64:195. 1925. Fig. 101.

Aecia unknown. Uredinia on abaxial leaf surface, cinnamon-
brown, pulverulent, with hyaline or yellowish , mostly curved
or geniculate, clavate-capitate, or capitate paraphyses whose
wall is 2-5µ thick apically becoming thinner below; spores
(21-)23-27 x 19-21(-23)µ, mostly broadly oval, wall 1.5µ thick,
golden or cinnamon-brown, echinulate, pores 6-8, scattered.
Telia on abaxial surface, blackish brown, compact, early exposed,
pulvinate; spores 29-34(-38) x 18-22µ, mostly ellipsoid and only
slightly narrowed basally, wall (1.5-)2-3µ thick at sides, 5-8µ
apically, clear chestnut-brown, smooth; pedicels moderately
thick-walled, not collapsing, yellowish, persistent, to 65µ long.

Hosts and distribution: <u>Bromus coloratus</u> Steud., Bolivia.

Type: Holway No. 456, La Paz, Bolivia (PUR).

Kaufmann (Mycopathol. Mycol. Appl. 32: 249-261. 1967) published a photograph of paraphyses and teliospores of the type.

188

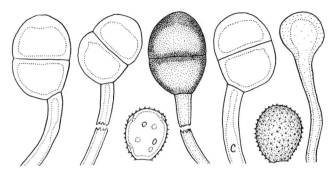

Figure 102

81. PUCCINIA PACHYPES H. Syd. & P. Syd. in Sydow & Butler Ann.
Mycol. 10:262. 1912. Fig. 102.

Aecia unknown; uredinia on abaxial leaf surface, yellowish
brown, with capitate, yellowish or golden paraphyses whose wall
is 5µ apically and thin below; spores (17-)23-26 x (16-)18-21
(-23)µ, mostly broadly ellipsoid or nearly globoid, wall 1.5µ
thick, yellowish brown, echinulate, pores (6-)8(-10), scattered.
Telia on abaxial surface, blackish brown, compact, pulvinate,
exposed; spores (27-)31-37(-40) x (21-)23-26(-28)µ, mostly
broadly ellipsoid, wall (1.5-)2-2.5µ at sides, 3-5(-7)µ apically,
chestnut-brown, smooth; pedicels thick-walled and not collapsing,
yellowish or golden, persistent, to 80µ long.

Hosts and distribution: Spodiopogon rhizophorus (Steud.)
Pilger, India.

Type: McRae (Butler No. 1609), on S. albidus (=rhizophorus),
Vayitri, Wynaad, India (S).

Cummins (Uredineana 4: Pl. VII, Fig. 40. 1953) published
a photograph of teliospores of the type.

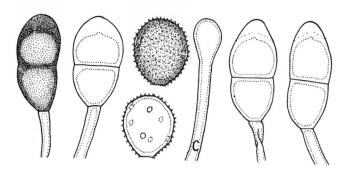

Figure 103

82. PUCCINIA DIGNA Arth. & Holw. in Arthur Proc. Amer. Phil.
Soc. 64:198. 1925. Fig. 103.

Aecia uncertain. Uredinia on adaxial leaf surface, about
cinnamon-brown, pulvinate, with capitate, hyaline paraphyses
whose wall is uniformly 1-1.5μ thick; spores (17-)26-30(-35)
x (16-)23-26(-33)μ, mostly broadly oval, wall 1-1.5(-2.5)μ
thick, yellow to golden, echinulate, pores 6-8, scattered.
Telia on adaxial surface, blackish brown, compact, pulvinate,
exposed; spores (29-)36-46(-69) x (16-)23-26(-36)μ, mostly
broadly ellipsoid or ellipsoid, wall 1.5-2.5(-5)μ thick at
sides, 6-10(-13)μ apically, chestnut-brown, smooth; pedicels
moderately thick-walled, mostly non-collapsing, hyaline to
golden, persistent, to 130μ long.

Hosts and distribution: Nassella chilensis (Trin.) Desv.,
N. pubiflora (Trin. & Rupr.) Desv., Stipa ichu (R. & P.) Kunth,
S. neesiana Trin. & Rupr.: Argentina, Bolivia, Chile, Ecuador,
and Mexico.

Type: Holway No. 451, on Stipa ichu, La Paz, Bolivia (PUR;
isotypes issued in Reliq. Holw. No. 71).

Greene and Cummins (Mycologia 50:6-36. 1958) published a
photograph of teliospores of the type.

The species was described as autoecious but the aecia
probably belong to Puccinia graminella.

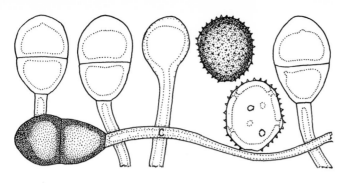

Figure 104

83. PUCCINIA UNICA Holway in Arthur and Fromme Torreya 15:263. 1915 var. unica. Fig. 104.

Aecia unknown. Uredinia mostly on adaxial leaf surface, dark cinnamon-brown; paraphyses capitate, colorless to golden brown, to 100μ long and 29μ wide, wall to 8μ thick apically; spores 26-33(-35) x (21-)23-27(-29)μ, mostly broadly ellipsoid, wall 2.5-3.5(-4)μ thick, echinulate, dark cinnamon- to nearly chestnut-brown, germ pores 8-11, scattered. Telia mostly on adaxial surface, sometimes on stems, blackish brown, exposed, compact; spores (31-)33-44(-48) x (19-)22-28(-30)μ, mostly ellipsoid or broadly ellipsoid, wall (2-)2.5-3.5(-4)μ thick at sides, 5-8(-10)μ apically, uniformly chestnut-brown, smooth; pedicels colorless, thick-walled, not collapsing, to 150μ long.

Hosts and distribution: species of _Aristida_: southwestern U.S.A. to southwestern Mexico.

Type: Holway No. 3020, on _Aristida longiramea_ Presl, Cuernavaca, Mor., Mexico (PUR).

The two following varieties differ mainly because of smaller spores.

PUCCINIA UNICA Holw. var. bottomleyae (Doidge) Cumm. & Husain Bull. Torrey Bot. Club 93:60. 1966.

Puccinia bottomleyae Doidge Bothalia 2:498. 1928.

Aecia unknown. Uredinia and paraphyses as in var. unica; spores 25-31(-33) x 21-27μ, wall dark cinnamon-brown, (1.5-)2-2.5μ thick, pores 8-11, scattered; teliospores 30-38(-40) x (21-)23-28(-30)μ, wall 2-3(-4)μ thick at sides, 4-7μ apically, uniformly chestnut-brown.

Hosts and distribution: species of _Aristida_: Spain to Ethiopia, South Africa, and India.

Lectotype: Doidge and Bottomley (PRE 29793), on _Aristida junciformis_ (as _A. welwitschiae_ Rendle), Pretoria, South Africa.

This fungus differs from var. unica in somewhat smaller spores

191

with thinner walls. In both varieties the urediniospores are
so deeply pigmented as to suggest amphispores.

Cummins and Husain (loc. cit.) published a photograph of
teliospores of the lectotype.

PUCCINIA UNICA Holw. var. chica Cumm. & Husain Bull. Torrey
Bot. Club 93:60. 1966.

Aecia unknown. Uredinia and paraphyses about as in var.
unica; spores 20-24(-26) x 18-21µ, wall 1-1.5µ thick, cinnamon-
brown, echinulate, pores 6-8, scattered, teliospores (25-)28-32
(-34) x (22-)24-26µ, wall 2-3(-4)µ thick at sides, 4-6µ at apex,
chestnut-brown.

Hosts and distribution: Aristida longiramea Presl, A.
ternipes Cav.: eastern Mexico.

Type: Cummins No. 63-158 (PUR 59375), on Aristida ternipes,
Tamaulipas State.

Cummins and Husain (loc. cit.) published a photograph of
teliospores of the type.

Variety chica not only has smaller spores than the other
varieties but the urediniospores have thin walls and, apparently,
no tendency toward amphispores.

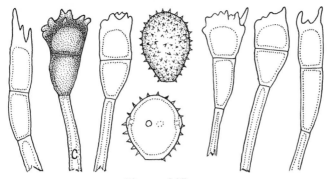

Figure 105

84. PUCCINIA DIARRHENAE Miyabe & Ito in Ito J. Coll. Agr. Tohoku
Imp. Univ. 3:190. 1909. Fig. 105.

Aecia unknown. Uredinia not seen; spores in the telia
germinated and mostly collapsed, 24-28(-31) x (17-)19-22(-24)μ,
mostly obovoid or broadly ellipsoid, wall 1.5μ thick, pale
yellowish, echinulate, germ pores 4(5?), equatorial. Telia
on abaxial leaf surface, early exposed, compact, blackish;
spores (30-)34-45(-58) x (10-)12-18(-20)μ (digitations excluded),
mostly elongately obovoid, sometimes cylindrical and then
usually paler, wall 1-1.5(-2.5)μ thick at sides, 3-5μ apically
excluding digitations, the apex with 2 to several digitations
2-10μ long; pedicels yellowish, rather thick-walled, collapsing
or not, to 50μ long.

Hosts and distribution: Diarrhena manshurica Maxim., D.
japonica Franch. & Sav.: China, Japan, Korea and the U.S.S.R.

Type: Yamada, on Diarrhena japonica, Morioka, Prov. Rikuchu,
Japan, 21 Oct. 1906 (SAPA; isotype PUR). This specimen was
received from Ito marked "Type collection", although neither
of the 2 specimens originally listed was so designated.

Because of the long pedicels of the teliospores and the
equatorial germ pores, it is obvious that this coronate species
is not related to P. coronata.

85. PUCCINIA HORDEINA Lavrov Bestimmungschluessel Pflanzenparas.
Kult. Wildwachs. Nutzpfl. Sibir. 1:126. 1932. Not seen.

Aecia unknown. Uredinia not described; spores 16-27 x
16-22μ, nearly globoid, germ pores 3 or 4. Telia mostly in
linear series on the sheaths, covered by the epidermis, with
brown paraphyses; spores 37-89 x 11-27μ, the apical wall 5-8μ
thick; pedicels very short; 1-celled spores few.

Type: On Hordeum vulgare L., western Siberia. Not seen.

The description is adapted from Tranzschel (Conspectus
Uredinalium URSS. p. 112. 1939). It is doubtful that the
germ pores are few and, not improbably, the fungus is P.
striiformis or P. hordei.

86. PUCCINIA TRISETICOLA Tranz. Trudy Bot. Inst. Akad. Nauk SSSR 4:328. 1940.

Aecia unknown. Uredinia mostly epiphyllous, yellowish; spores 19-22 x 16-17.5, subglobose or ellipsoid, wall colorless or pale yellowish, loosely echinulate, germ pores 3 or 4, subequatorial. Telia mostly hypophyllous, blackish, covered by the epidermis, weakly loculate with few paraphyses; spores 35-48.5 x 13.5-18μ, clavate or subcylindrical, mostly truncate and 2.5-6.5μ thick apically, pale brown; pedicels short, persistent.

Type: Tranzschel, on Trisetum sibiricum Pupr., Primorskaja and Ussurijskaja, Far Eastern USSR (LE; not seen).

The description is adapted from the original.

Tranzschel (loc. cit.) obtained spermogonia on Actaea alba and Cimicifuga daurica by inoculation. Necrotic spots terminated the infections on Actaea and the plants of Cimicifuga died before aecia developed.

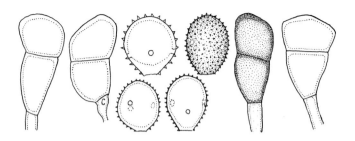

Figure 106

87. PUCCINIA CHAETII Kern & Thurst. Mycologia 36:511. 1944.
Fig. 106.

Aecia unknown. Uredinia amphigenous, cinnamon-brown or when
amphisporic chestnut-brown; spores 23-29 x 17-23µ, mostly
obovoid or broadly ellipsoid, wall 1-1.5µ thick, pale cinnamon-
brown, echinulate, germ pores 3, equatorial or slightly subequatorial; amphispores 27-35 x 26-29µ, mostly obovoid, wall
2.5-3µ thick, dark cinnamon-brown or nearly chestnut-brown,
echinulate, germ pores 3, subequatorial and often near the
hilum. Telia blackish-brown, covered by the epidermis; spores
(33-)38-44(-47) x (18-)20-26(-29)µ, wall 1-1.5µ thick at sides,
2-3.5µ apically, golden or pale chestnut-brown, smooth; pedicels
brownish, about 15µ long; 1-celled teliospores common.

Hosts and distribution: Chaetium festucoides Nees:
Venezuela.

Type: Chardon No. 3885, El Sombrero, Est. Guarico,
Venezuela (PAC; isotype PUR).

196

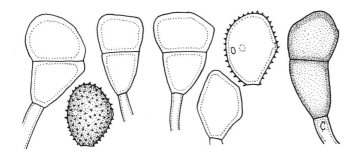

Figure 107

88. PUCCINIA PASPALINA Cumm. Bull. Torrey Bot. Club 72:211.
1945. Fig. 107.

Uredo paspali-scrobiculati H. Syd. & P. Syd. in Sydow &
Butler Ann. Mycol. 4:444. 1906.

Uredo paspalina H. Syd. & P. Syd. Ann. Mycol. 15:177. 1917.

Aecia unknown. Uredinia amphigenous or mainly on abaxial
surface, cinnamon-brown or paler; spores 24-31(-34) x (17-)20-24
(-27)μ, mostly broadly ellipsoid or obovoid, frequently angular,
wall 1.5μ thick, golden or cinnamon-brown, echinulate, germ
pores 3, equatorial. Telia amphigenous or mostly on the sheaths,
covered by the epidermis, greyish black; spores (33-)38-46 x
(17-)24-26(-30)μ, variable but mostly clavate, wall 1.5-3μ thick
at sides, 3-3.5μ apically, yellowish or golden, smooth; pedicels
colorless, to 10μ long; 1-celled spores numerous.

Hosts and distribution: species of Paspalum: Nyasaland
and Uganda to Ceylon, Australia, and Japan.

Type: Clemens, on Paspalum orbiculare Frost, Brisbane,
Australia (PUR F10873).

A photograph of teliospores of the type was published with
the diagnosis.

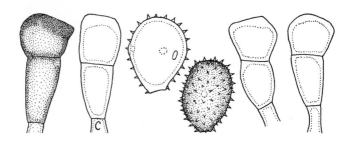

Figure 108

89. PUCCINIA CENCHRI Diet. & Holw. in Holway Bot. Gaz. 24:28. 1897 var. cenchri. Fig. 108.

Uredo cenchriphila Speg. Mus. Nac. Buenos Aires Anal. 19:316. 1909.

Aecia unknown. Uredinia amphigenous or mainly on adaxial leaf surface, cinnamon-brown; spores (27-)31-34(-37) x (20-)24-27(-31)μ, mostly broadly ellipsoid or ellipsoid, wall 2-3μ thick, prominently echinulate, cinnamon-brown, germ pores 2 or sometimes 3, equatorial. Telia on abaxial leaf surface, covered by the epidermis, blackish brown, inconspicuous; spores 37-44(-51) x (17-)20-24μ, mostly oblong or clavate, wall 1.5μ thick at sides, 3-7μ apically, golden or chestnut-brown, smooth; pedicels colored, thin-walled, to 15μ long.

Hosts and distribution: species of Cenchrus: southern United States and the West Indies southward to Argentina, and in the Islands of the Pacific.

Type: E. W. D. Holway, on C. multiflorus, Guadalajara, Mexico, 12 Oct. 1896 (S; isotype PUR).

PUCCINIA CENCHRI Diet. & Holw. var. africana Cumm. Torrey Bot. Club. Bull. 79:217. 1952.

Uredo cenchricola P. Henn. Mus. Congo Anal. 2(3): 223. 1908.

Generally similar to P. cenchri var. cenchri. Urediniospores (29-)31-37(-41) x (20-)23-28μ, mostly broadly ellipsoid, germ pores 4 or 5 equatorial; teliospores (34-)37-45 x (17-)20-25μ, oblong or clavate, wall 1.5μ thick at sides, to 7μ apically; pedicels to 30μ long but usually shorter.

Hosts and distribution: Cenchrus ciliaris Fig. & De Not.: Central Africa.

Type: C. G. Hansford No. 3517, Kawanda, Uganda, July 1941 (PUR; isotype IMI).

The variety differs from the typical mainly in the greater number of germ pores of the urediniospores.

198

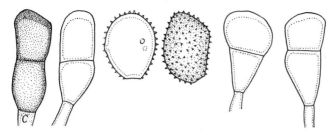

Figure 109

90. PUCCINIA DOLOSOIDES Cumm. Torrey Bot. Club Bull. 72: 212.
1945. Fig. 109.

Aecia unknown. Uredinia on abaxial leaf surface, yellowish
brown; spores (27-)33-36 x 21-27μ, ovate or broadly ellipsoid,
wall 1.5μ thick, golden or light cinnamon-brown, echinulate,
germ pores 3 or 4, equatorial. Telia in abaxial leaf surface,
covered by the epidermis, blackish brown; 34-43(-50) x
(17-)19-23(-25)μ, mostly clavate or oblong, wall 1-1.5μ thick
at sides, 2-4μ apically, chestnut-brown, smooth; pedicels brown,
to 10μ long.

Hosts and distribution: Paspalum commersonii Lam.; central
Africa and Ceylon.

Type: F. C. Deighton No. 32, Njala, Sierra Leone, 24 Sept.
1926 (PUR; isotype IMI).

Cummins (loc. cit.) published a photograph of teliospores
of the type.

91. PUCCINIA SETARIAE-FORBESIANAE Tai in Wang Acta Phytotax. Sinica 10:295. 1965.

Aecia unknown. Uredinia amphigenous, dark brown; spores 27-31 x 20-24µ, ovoid or irregularly globoid, wall 1-1.5µ thick, finely echinulate, brownish yellow, germ pores 3 or 4 equatorial. Telia covered by the epidermis, then exposed by a slit, blackish brown; spores 25-31 x 16-24µ, irregularly ellipsoid, often angular, apex truncate or narrowed, wall 1-1.5µ thick or rarely thicker apically, (presumably more or less chestnut-brown), smooth; pedicels colored, short, often inserted laterally; 1-celled spores few.

Type: Tai, on _Setaria_ _forbesiana_ (Nees) Hook. f., Tapugi, Kunming, Yunnan, China (Plant Pathol. Herb. No. 7631, Tsing Hua Univ. =Inst. Microbiol., Peking 3631; not seen). One other collection was recorded.

Tai did not describe paraphyses in the uredinia but otherwise the species appears similar to P. _dolosa_ and its variants.

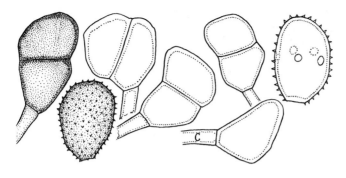

Figure 110

92. PUCCINIA POLYSORA Underw. Torrey Bot. Club Bull. 24:86.
1897. Fig. 110.

Aecia unknown. Uredinia amphigenous, cinnamon-brown; spores
29-36(-40) x (20-)23-29µ, mostly ellipsoid or obovoid, often
angular, wall 1-1.5µ thick, echinulate, golden or yellowish,
germ pores 4 or 5, aequatorial. Telia amphigenous, covered by
the epidermis, indehiscent, small, blackish brown; spores 29-41 x
(18-)20-27µ, usually angularly ellipsoid or oblong but highly
variable, wall evenly 1.5µ thick or very slightly thicker
apically, chestnut-brown, smooth, very brittle; pedicels yellow
or brownish, thin-walled, to 30µ long; 1-celled teliospores
often abundant.

Hosts and distribution: Erianthus alopecuroides (L.) Ell.,
Euchlaena mexicana Schrad., Tripsacum dactyloides L., T.
lanceolatum Rupr., T. laxum Nash, T. pilosum Scribn. & Merr.,
Zea mays L.: United States southward to Peru and eastward
across central Africa to Thailand and the Philippine Islands.

Type: B. M. Duggar, on T. dactyloides, Auburn, Alabama, U.S.A.,
Oct. 1891 (Isotype PUR).

201

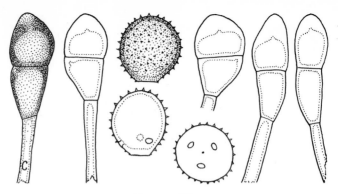

Figure 111

93. PUCCINIA SPOROBOLI Arth. Iowa Agr. Coll. Dept. Bot. Bull. 1884: 159. 1884. var. sporoboli. Fig. 111.

Aecia occur on species of <u>Allium</u> and <u>Lilium</u>; spores (19-)21-25(-27) x (16-)18-21(-23)μ, globoid, broadly ellipsoid, or oblong, wall (0.5-)1μ thick, finely verrucose, hyaline. Uredinia amphigenous, cinnamon-brown; spores (24-)26-30(-32) x (20-)24-28(-32)μ, mostly broadly obovoid, wall 2-2.5μ thick or thinner basally, cinnamon-brown, finely echinulate, germ pores 3 or 4, near the hilum. Telia amphigenous, early exposed, blackish, compact; spores (25-)30-44(-50) x (14-)17-21(-23)μ, mostly oblong-ellipsoid or narrowly obovoid, wall 1-3(-4)μ thick at sides, (4-)6-10(-12)μ apically, chestnut-brown or often paler below, smooth; pedicels yellowish, thick-walled but often collapsing, to 50μ long.

Hosts and distribution: <u>Sporobolus</u> <u>asper</u> (Michx.) Kunth, <u>S</u>. <u>heterolepis</u> A. Gray: Wisconsin west to North Dakota and Nebraska, U.S.A.

Type: Holway, on <u>Sporobolus</u> <u>heterolepis</u>, Decorah, Iowa (PUR).

Cummins and Greene (Brittonia 13:271-285. 1961) published a photograph of teliospores of the type.

Arthur (Mycologia 9: 294-312. 1917) proved the life cycle by inoculation.

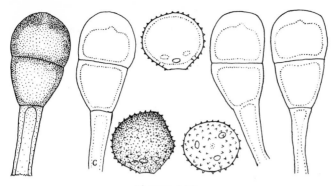

Figure 112

PUCCINIA SPOROBOLI Arth. var. robusta Cumm. & H. C. Greene
Brittonia 13:272. 1961. Fig. 112.

Aecia, Aecidium yuccae Arth., occur on species of Leucocrinum,
Smilacina, and Yucca; spores (22-)24-29(-33) x (18-)20-24(-26)μ,
mostly broadly ellipsoid, wall 1-1.5μ thick, verrucose, hyaline;
urediniospores (23-)25-29(-32) x 23-27(-30)μ, mostly broadly
obovoid, wall 1.5-2.5μ thick, cinnamon-brown, paler basally,
echinulate, germ pores (3-)5 or 6, around the hilum; teliospores
(38-)42-54(-62) x (19-)22-29(-35)μ, oblong or oblong-obovoid,
wall 1.5-2(-3)μ thick at sides (5-)7-10(-13)μ apically, chestnut-
brown, smooth; pedicels yellowish, thick-walled, mostly not
collapsing, to 50μ long.

Hosts and distribution: Calamovilfa gigantea (Nutt.) Scribn.
& Merr., C. longifolia (Hook.) Scribn., Sporobolus asper (Michx.)
Kunth, S. heterolepis Gray: Ontario and Alberta to Colorado
and Oklahoma, U.S.A.

Type: Baxter, on Calamovilfa longifolia, Burns, Wyoming (PUR.)

Bethel's inoculation of Leucocrinum montanum Nutt. (reported
by Arthur, Manual of the Rust in United States and Canada, under
Puccinia amphigena) first proved the life cycle. Subsequent and
successful inoculations are summarized by Cummins and Greene
(Brittonia 13: 271-285. 1961) who also published a photograph
of teliospores of the type.

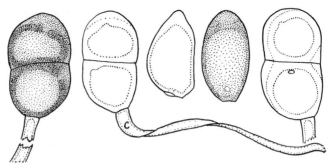

Figure 113

94. PUCCINIA TRIPSACICOLA Cumm. Torrey Bot. Club Bull. 79:225.
1952. Fig. 113.

Uromyces tripsaci Kern & Thurst. Mycologia 35:445. 1943
(based on uredinia).

Aecia unknown. Uredinia in adaxial side of leaves, oblong
or linear; spores 42-66 x 19-26μ, mostly oblong, wall 1.5-2μ
thick at sides, 7-10μ apically, golden, smooth, germ pore 1,
basal. Telia in abaxial side of leaf, early exposed, compact,
to 3 mm long, blackish brown; spores (34-)39-50 x (19-)21-28μ,
mostly ellipsoid, sometimes tending diorchidioid, wall 3-4μ
thick at sides, 5-7(-9)μ apically, smooth, chestnut-brown;
pedicels colorless, mostly thin-walled and collapsing, to 100μ
long.

Hosts and distribution: Tripsacum dactyloides L.: Ecuador
and Venezuela.

Type: A. S. Hitchcock, Chimborazo, Ecuador, July 17, 1923
(BPI).

A photograph of teliospores of the type was published by
Cummins (loc. cit.).

The spores that Kern and Thurston described as teliospores
of Uromyces are interpreted here to be urediniospores.

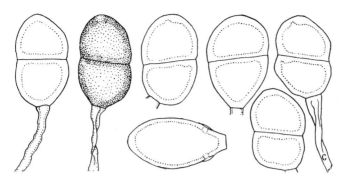

Figure 114

95. PUCCINIA ADVENA H. Syd. Ann. Mycol. 22:419. 1924. Fig. 114.

Aecia unknown. Uredinia on adaxial leaf surface, yellowish brown, rather long covered by the epidermis; spores (35-)38-50 (-57) x (17-)19-24(-27)μ, ellipsoid, oblong, or fusoid, wall 1-1.5μ thick at sides, similar apically or often 2-7μ, yellowish brown, smooth, germ pores 2, next the hilum. Telia on abaxial surface, early exposed, pulvinate or subpulverulent, blackish-brown; spores (33-)35-40(-43) x 22-26(-28)μ, ellipsoid or ovate-ellipsoid, wall 2μ thick at sides, 2-3μ apically, chestnut-brown, smooth; pedicels colorless, thin-walled and collapsing, to 100μ long.

Hosts and distribution: Oplismenus africanus Beauv.: Union of South Africa.

Type: Van der Bijl No. 1537, Woodbush, Transvaal (STE-VB).

Only the one collection is known.

Ramachar and Cummins (Mycopath. Mycol. Appl. 25:7-60. 1965) published a photograph of teliospores of the type.

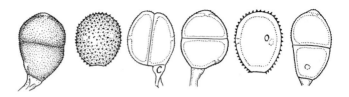

Figure 115

96. PUCCINIA BRACHYCARPA H. Syd. in Sydow & Petrak Ann. Mycol. 29:148. 1931. Fig. 115.

Aecia unknown. Uredinia not seen but spores in the telia 23-27 x 18-26µ, broadly ellipsoid or globoid, wall 2.5-3.5µ thick, dark cinnamon-brown or chestnut-brown, finely echinulate, germ pores 3 or 4, equatorial; (Sydow described cylindrical or clavate, thin-walled, brownish paraphyses which I have not seen). Telia on abaxial leaf surface, chestnut-brown, early exposed, pulverulent; spores 24-30(-33) x (17)20-24µ, wall uniformly 1.5µ thick, chestnut-brown, closely and minutely punctate-verrucose, germ pore apical in upper cell, near the hilum in lower cell; pedicels colorless, very fragile, broken near the hilum.

Hosts and distribution: Pseudoraphis aspera (Koenig) Pilger: the Philippines.

Neotype: Clemens No. 1599, on Chamaeraphis aspera, (=Pseudoraphis aspera), Manila, Del Norte, Luzon (PUR). Neotype designated by Ramachar and Cummins Mycopath. Mycol. Appl. 25: 51. 1965.

This is one of the few grass rust fungi that have verrucose teliospores and the germ pore of the lower cell depressed. When the teliospores dry, they characteristically collapse from the poles toward the septum.

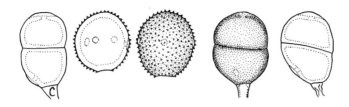

Figure 116

97. PUCCINIA SUBCENTRIPORA Arth. & Cumm. Philip. J. Sci. 59:439. 1936. Fig. 116.

Puccinia praecellens H. Syd. in Cummins Ann. Mycol. 35:99. 1937.

Aecia unknown. Uredinia on abaxial leaf surface, cinnamon-brown; spores 27-34(-39) x (20-)25-30μ, mostly broadly ellipsoid, wall 3-4μ thick, cinnamon-brown or pale chestnut-brown, closely echinulate, germ pores 3 or 4, equatorial. Telia mainly on abaxial surface, chestnut-brown, rather pulverulent; spores (20-)25-34(37) x (20-)24-27μ, mostly oblong or irregularly ellipsoid, wall uniformly 1.5-2μ thick, golden or clear chestnut-brown, smooth, the germ pore of the lower cell located midway to the pedicel; pedicels colorless, thin-walled, delicate and collapsing, seen to 20μ, perhaps longer but always broken short.

Hosts and distribution: Panicum punctatum Burm., Pennisetum clandestinum Hochst. (?): Philippine Islands.

Type: M. S. Clemens No. 5898, on P. punctatum, Gapan, Nueva Ecija Prov., Luzon, Philippine Islands (PUR).

This is one of the few rust fungi of grasses in which the lower pore is depressed in the teliospores.

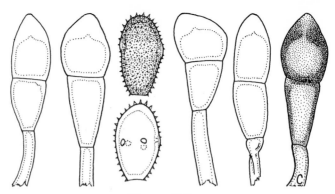

Figure 117

98. PUCCINIA GRAMINIS Pers. Syn. Meth. Fung. p. 228. 1801 ssp. graminis. Fig. 117.

Puccinia linearis Roehl. Deutschl. Fl. Ed. 2.III. 3:132. 1813.

Puccinia cerealis H. Mart. Prodr. Fl. Mosq. Ed. altera. p. 227. 1817.

Puccinia anthistiriae Barcl. J. Asiatic Soc. Bengal 58:246. 1889.

Puccinia jubata Ellis & Barth. Erythea 4:2. 1896.

Puccinia megalopotamica Speg. An. Mus. Nac. Buenos Aires 6:224. 1898.

Puccinia vilis Arth. Bull Torrey Bot. Club 28:663. 1901.

Puccinia elymina Miura Fl. Manchuria & East. Mongolia 3:280. 1928.

Puccinia brizae-maximi Ramakr., T. S. & Sund. Indian Phytopathol. 6:30. 1953.

Puccinia favargeri Mayor Rev. Mycol. 22:273. 1957.

Puccinia albigensis Mayor Rev. Mycol. 22:279. 1957.

Aecia (Aecidium berberidis Pers.) occur on species of Berberis, cupulate or cylindrical, in groups; spores 16-23 x 15-19µ, globoid or more or less oblong, wall 1-1.5µ thick at sides, 5-9µ apically, verrucose. Uredinia amphigenous or most commonly on sheaths and stems, about cinnamon-brown; spores (22-)26-40(-45) x (13-)16-22(-24)µ, mostly oblong-ellipsoid, wall mostly 1.5-2µ, rarely to 3µ or even 4µ, the apex usually thicker, yellowish to golden brown, echinulate, strongly so toward the ends and usually less so equatorially, germ pores (3)4 or 5, equatorial. Telia most commonly on sheaths and stems, early exposed, blackish brown, compact; spores (33-)40-60(-66;-76) x (13-)16-23(-25)µ, ellipsoid, oblong-ellipsoid, or narrowly obovoid, wall (1-)1.5-2(-2.5)µ thick at sides, (5-)7-10(-12)µ apically, chestnut-brown, smooth;

208

pedicels usually brownish, usually collapsing, to 80µ long, usually about 50µ long.

Hosts and distribution: On species of Aegilops, Agropyron (incl. Elytrigia and Roegneria), Alopecurus, Avena, Bothriochloa, Briza, Bromus, Cinna, Cynodon, Echinochloa, Elymus, (incl. Hordelymus), Glyceria, Heteranthelium, Hierochloe, Hordeum, Koeleria, Lamarckia, Limnodea, Leersia, Melica, Milium, Oryza, Secale, Setaria, Sitanion, Triticum, Vulpia: circumglobal.

Lectotype: Persoon, "praesertim in culmis graminum varii generis"; unquestionably= Triticum (L 910.263-499); designated by Jørstad (Blumea 9:1-20. 1958).

I am following Urban (Ceska Mykol. 21:12-16. 1967) in recognizing two subspecies, based primarily on the length of the urediniospores. The subspecies are reasonably distinct, but there is some intergradation. The species itself is remarkably distinctive, despite variability in spore sizes.

Urban, again based on the sizes of urediniospores, recognizes var. graminis, with spores (20-)26-36(-45) x (13-)16-21(-22)µ and var. stakmanii Guyot, Massen. & Saccas, with spores (20-)33-36(-39) x (13-)14-21(-23)µ. The rust of Triticum, Aegilops, and Elymus, is ssp. graminis var. graminis, that of Avena, Hordeum, Secale, and various other genera, is ssp. graminis var. stakmanii.

PUCCINIA GRAMINIS Pers. ssp. graminicola Urban Ceska Mykol. 21:14. 1967.

Puccinia anthoxanthi Fuckel Jahrb. Nass. Ver. Nat. 27:15. 1873.

Puccinia phlei-pratensis Eriks. & Henn. Z. Pflanzenkr. 4:140. 1894.

Puccinia subandina Speg. An. Mus. Nac. Buenos Aires III. 1:65. 1902.

Puccinia sesleriae-coeruleae Ed. Fisch. Beitr. Kryptog. Schweiz 2:259. 1904.

Puccinia culmicola Diet. Bot. Jahrb. 37:100. 1905.

Puccinia avenae-pubescentis Bub. Ann. Mycol. 4:107. 1906.

Puccinia heimerliana Bub. in Bubák & Kabat Ann. Mycol. 5:40. 1907.

Puccinia ikaoensis Hara Trans. Agr. Soc. Shizuoka Pref. 286: 47. 1921.

Puccinia dactylidis Gaeum. Ber. Schweiz. Bot. Ges. 55:79. 1945.

Uredo deschampsiae-caespitosae Wang Acta Phytotax. Sinica 10:298. 1965.

Sori as in ssp. graminis. Urediniospores (18-)20-30(-34) x (12-)14-20(-22)μ, wall 1.5-2.5 (rarely -3.5)μ, thicker apically, yellowish to golden brown, echinulate, germ pores 3 or 4(5); teliospores (27-)34-60(-64;-75) x (11-)16-23(-25)μ, variable in shape as in ssp. graminis, varying in length from the general range to as short as 27-34μ in some collection on Anthoxanthum and 30-43μ on Dichelachne.

Hosts and distribution: On species of Agropyron, Agrostis, Aira, Alopecurus, Ammophila, Amphibromus, Anthoxanthum, Apera, Arrhenatherum, Avenochloa, Beckmannia, Brachypodium, Briza, Calamagrostis, Catabrosa, Cynosurus, Dactylis, Deschampsia,

210

Deyeuxia, Diarrhena, Dichelachne, Echinopogon, Festuca, Glyceria, Hierochloe, Koeleria, Lamarckia, Lolium, Melica, Milium, Muhlenbergia, Neostapfia, Orcuttia, Phalaris, Phleum, Poa, Polypogon, Scleropoa, Sesleria, Sphenopholis, Trisetum, Vulpia: circumglobal.

Type: Urban, on Dactylis glomerata, Bohemia: Vysenske kopce near Cesky Krumlov 13 July 1960 (PRC).

Urban (loc. cit.) recognizes no varieties of ssp. graminicola and assigns ssp. minor and media and vars. eriksonii, calamagrosteos, lolii, vulpiae (all nomina nuda) of Guyot, Massenot & Saccas and ssp. lolii nom. nud. of Waterhouse to ssp. graminicola.

Some hosts cannot be placed because of lack of adequate data. Included are Aristida, Chrysopogon, Coleanthus, Corynephorus, Danthonia, Gastridium, Haynaldia, Holcus, Hystrix, Lagurus, Molinia, Panicum, Psilurus, Puccinellia, Sporobolus, Stipa, Tridens, and Ventenata. Most of these are apparently only occasionally rusted and may not regularly support a population of P. graminis. It is obvious, for example, that grasses of the tribes Andropogoneae and Paniceae rarely support P. graminis.

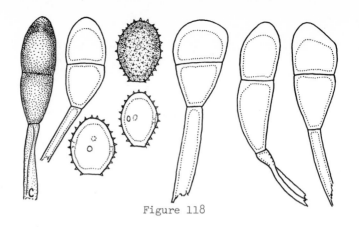

Figure 118

99. PUCCINIA SESLERIAE Reichardt Verh.- Bot. Ges. Wien 1877:
842. 1877. Fig. 118.

Puccinia avenastri Guyot Uredineana 3:67. 1951.

Aecia occur on Rhamnus saxatilis Jacq.; spores 18-26 x
16-21μ, globoid or polyhedral, wall thin, colorless, verrucose
(from Gäumann and Terrier, 1952). Uredinia amphigenous,
yellowish brown, spores (23-)26-30(-34) x (16-)18-22(-24)μ,
mostly obovoid or ellipsoid, wall (1.5)2-3μ thick at sides, 3-4μ
apically, yellowish to golden brown, echinulate, germ pores 3 or
4(5) usually equatorial but sometimes scattered in shorter
spores. Telia amphigenous, early exposed, blackish brown,
compact; spores (30-)38-50(-58) x (15-)18-23(-25)μ, mostly
ellipsoid or elongately obovoid, wall 1.5-2(-3)μ thick at sides,
6-10(-12)μ apically, clear chestnut-brown, smooth; pedicels
persistent, brownish, rather thick-walled but usually collapsing,
to 80μ long but usually 40-60μ.

Lectotype: Reichardt, on Sesleria coerulea, Weixeltal u.
Baden, Rakousko, Austria, Sept. 1876 (BRNU; isolectotype W)
designated here following a selection by Z. Urban but not yet
published.

Reichardt (loc. cit.) first demonstrated the life cycle and
this was verified by Gäumann and Terrier (Ber. Schweiz. Bot. Ges.
62: 297-306. 1952), who also reviewed the negative results.

It is doubtful if the species is separable from P. graminis,
and Treboux (Ann. Mycol. 12:480-483. 1914) and Fischer (Mitt.
Naturf. Ges. Bern 1916: 125-163. 1917) successfully inoculated
Berberis with a fungus that seems to be indistinguishable from
Reichardt's species.

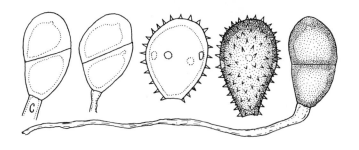

Figure 119

100. PUCCINIA BELIZENSIS Mains Contrib. Univ. Michigan Herb. 1.8. 1939. Fig. 119.

Aecia unknown. Uredinia amphigenous, near chestnut-brown; spores (32-)34-46(-58) x (25-)28-32(-34)μ, mostly obovoid, wall 2-3μ thick at sides, 3-5(-6)μ apically, dark cinnamon-brown or nearly chestnut-brown, coarsely echinulate, germ pores 3-5, equatorial. Telia amphigenous and on stems and inflorescence, often extensively confluent on stems, early exposed, pulvinate, chocolate-brown; spores (30-)36-45(-48) x 20-24(-28)μ, mostly ellipsoid or obovoid, the septum often oblique, wall 2-3μ thick, 3-5(-6)μ apically, golden or clear chestnut-brown, smooth; pedicels thin-walled and collapsing, hyaline or yellowish, tapering, to 200μ long but usually broken much shorter.

Hosts and distribution: <u>Olyra</u> <u>latifolia</u> L., <u>O</u>. <u>yucatana</u> Chase: British Honduras and Southeastern Mexico.

Type: Mains No. 3781, on <u>Olyra</u> <u>latifolia</u>, Cohune Ridge, El Cayo Distr., British Honduras (MICH).

Mains (<u>loc. cit.</u>) noted closely associated aecia on <u>Sebastiana</u> <u>standleyana</u>.

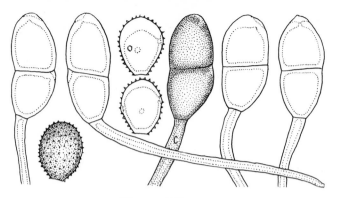

Figure 120

101. PUCCINIA ERYTHROPUS Diet. Bot. Jahrb. 37:101. 1905.
Fig. 120.

Aecia unknown. Uredinia on abaxial leaf surface, cinnamon-brown; spores 25-33 x 18-23μ, broadly ellipsoid or obovoid, wall 2-3μ thick at sides, usually 4-8μ apically, cinnamon- or chestnut-brown, echinulate, germ pores 3 or 4, equatorial. Telia on abaxial surface, often confluent, early exposed, pulverulent, blackish brown; spores (30-)33-45(-50) x (14-)16-20 (-22)μ, mostly ellipsoid, wall 1.5-2.5μ thick at sides, 3-5μ apically, the area over the germ pore pale and almost papilla-like, chestnut-brown, smooth; pedicels colorless to brownish (or purple from the host), thick-walled, not collapsing, to 130μ long.

Hosts and distribution: Erianthus maximus Brogn., Miscanthus sacchariflorus (Maxim.) Hack., M. sinensis Anderss.: U.S.S.R. southward to China, Japan, and the Philippine Islands.

Type: Yoshinaga, on M. sinensis, Umaji-mura, Tosa, Japan (S).

A photograph of teliospores of the type was published by Cummins (Urediniana 4:Pl. VI, Fig. 33. 1953).

214

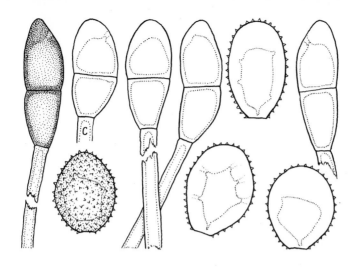

Figure 121

102. PUCCINIA SEYMOURIANA Arth. Bot. Gaz. 34:11. 1902. Fig. 121.

Puccinia cephalanthi Barth. N. Amer. Ured. No. 261 emend. 1922.

Aecia, Aecidium cephalanthi Seym., occur on species of Amsonia, Apocynum, Asclepias, and Cephalanthus; spores 32-42 x 28-35μ, wall irregularly 5-12μ thick, the lumen stellate, finely verrucose. Uredinia on adaxial leaf surface, yellow; spores (27-)30-40(-45) x (19-)21-27(-32)μ, obovoid or broadly ellipsoid, wall 2-3μ thick laterally, 9-15μ apically, colorless or yellowish, echinulate, germ pores obscure, probably equatorial. Telia on adaxial surface, exposed, dark brown, pulvinate; spores (35-)38-53(-58) x (15-) 18-23(-26)μ, cylindrical, oblong-ellipsoid, or ellipsoid, wall 1.5μ thick laterally, 5-9μ apically, chestnut-brown, smooth; pedicels colorless, thick-walled and mostly not collapsing, to 100μ long.

Hosts and distribution: species of Spartina: southern Canada and the United States east of the Rocky Mountains.

Type: Davis, on Spartina pectinata Link, Racine, Wisconsin, U.S.A. (PUR; isotypes Arth. & Holw. Ured. exsic. icon. No. 53a).

Arthur (J. Mycol. 12:24. 1906) first proved the life cycle by inoculation. Hennen and Cummins published a photograph of teliospores of the type (Mycologia 48:126-162. 1956).

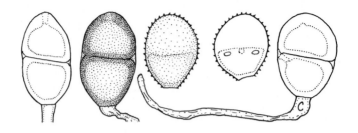

Figure 122

103. PUCCINIA HYPARRHENIAE Cumm. Bull. Torrey Bot. Club 83:226.
1956. Fig. 122.

Aecia unknown. Uredinia mostly on abaxial leaf surface,
yellow; spores (25-)27-33(-35) x (22-)24-27(-29)μ, ovoid or
obovoid, wall 1-1.5μ thick at sides, 10-19μ apically, finely
echinulate, hyaline, germ pores 3 or 4, equatorial, just below
the apical thickening. Telia mostly on abaxial surface, exposed,
pulvinate, blackish brown; spores (36-)38-40(-46) x (23-)25-28
(-30)μ, wall 3-4μ thick at sides, to 5.5μ apically, golden or
clear chestnut-brown, smooth; pedicels colorless, thin-walled
and collapsing, to 90μ long.

Hosts and distribution: _Hyparrhenia_ _rufa_ (Nees) Stapf:
Nyasaland.

Type: P.O. Wiehe No. 222, Zomba, Nyasaland (PUR; isotype IMI).

A photograph of teliospores of the type was published by
Cummins (loc. cit.)

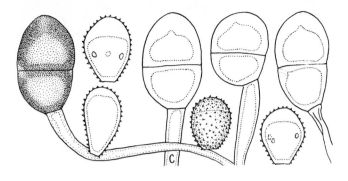

Figure 123

104. PUCCINIA EUCOMI Doidge Bothalia 3:497. 1939. Fig. 123.

Aecia unknown. Uredinia amphigenous but mostly on abaxial leaf surface, often confluent; yellow; spores (24-)26-32(-35) x 18-22(-24)μ, mostly obovoid, wall 2-2.5μ thick at sides, usually 5-8μ apically, colorless or pale yellowish, echinulate, germ pores 3, equatorial, obscure. Telia like the uredinia but pulvinate and blackish brown, early exposed; spores (30-)35-44 (-47) x (22-)24-30(-33)μ, mostly broadly ellipsoid, wall 2.5-4μ thick at sides, 5-9μ apically, chestnut-brown but not densely so, smooth; pedicels yellowish or colorless, thick-walled and not collapsing, to 100μ long.

Hosts and distribution: Andropogon eucomus Nees, A. huillensis Rendl.: South Africa.

Type: Doidge & Bottomley, on A. eucomus, Donkerpoort, Pretoria District, Union of South Africa (PRE 30129; isotype PUR).

A photograph of teliospores of the type was published by Cummins (Urediniana 4: Pl. V, Fig. 31. 1953).

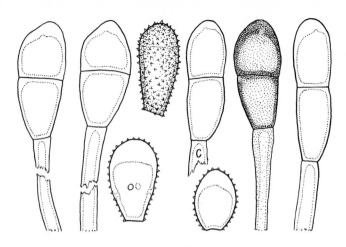

Figure 124

105. PUCCINIA SPARGANIOIDES Ell. & Barth. Erythea 4:2. 1896.
Fig. 124.

Uredo peridermiospora Ell. & Tracy J. Mycol. 6:77. 1890.

Puccinia peridermiospora Arth. Science II. 10:569. 1899.

Puccinia fraxinata Arth. Bot. Gaz. 34:6. 1902.

Aecia, Aecidium fraxini Schw., occur on species of Forestiera
and Fraxinus; spores 26-35 x 21-27μ, globoid or ellipsoid, wall
2-3μ thick at sides, 7-13μ apically, finely verrucose, colorless.
Uredinia mostly on abaxial leaf surface, yellow; spores (27-)30-
43(-47) x (16-)20-27(-30)μ, mostly ellipsoid or oblong, wall
1.5-3μ thick laterally, 8-10μ apically, colorless, echinulate,
pores 4, equatorial, obscure. Telia mostly on abaxial surface,
exposed, pulvinate, blackish; spores (37-)40-58(-64) x (14-)17-
23(-25)μ, ellipsoid or oblong-ellipsoid, wall 1.5μ thick at
sides, 5-7μ apically, chestnut-brown, smooth; pedicels colorless
or yellowish, rather thick-walled but usually partially collapsing,
to 100μ long.

Hosts and distribution: species of Spartina: southern
Canada, the United States east of the Rocky Mountains, and in
Brazil.

Type: Bartholomew, on Spartina pectinata Link (mistaken for
Carex sparganioides, hence the specific epithet), Rooks County,
Kansas (FH; isotype PUR).

Arthur (Bot. Gaz. 29:275. 1900) first proved the life cycle
by inoculation. Hennen and Cummins (Mycologia 48:126-162. 1956)
published a photograph of the teliospores of the type.

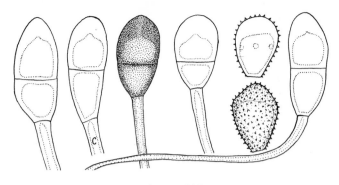

Figure 125

106. PUCCINIA WIEHEI Cumm. Bull. Torrey Bot. Club 79:226. 1952.
Fig. 125.

Aecia unknown. Uredinia amphigenous, elliptical, small,
yellow; spores (22-)25-28(-31) x 19-23μ, mostly obovoid, wall
1.5-2μ thick at sides, 5-10μ apically, finely echinulate,
colorless, germ pores 3 or 4, equatorial, obscure. Telia
amphigenous, early exposed, pulvinate, blackish brown; spores
(34-)40-48(-51) x (20-)22-24(-26)μ, mostly ellipsoid or oblong-
ellipsoid, wall 2-3.5μ thick at sides, 8-12μ apically, chestnut-
brown, smooth; pedicels colorless, thick-walled and not collapsing,
to 120μ long.

Hosts and distribution: Setaria splendida Stapf: Nyasaland.

Type: P. O. Wiehe No. 369, Vipya, Nyasaland (PUR; isotype
IMI).

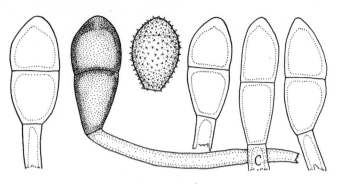

Figure 126

107. PUCCINIA VILFAE Arth. & Holw. Univ. Iowa Lab. Nat. Hist. Bull. 4:388. 1898 var. vilfae Fig. 126.

Puccinia sydowiana Diet. Hedwigia 36:299. 1897, not Zopf 1879.

Puccinia verbenicola Arth. Bot. Gaz. 35:16. 1903.

Aecia, Aecidium verbenicolum Ell. & Kell., occur on species of Verbena; spores mostly angularly globoid or ellipsoid, (20-)24-28(-35) x (16-)19-24(-26)μ, wall colorless, 0.5-1.5 (-2.5)μ thick at sides, 3-12μ apically, finely verrucose. Uredinia amphigenous, yellowish; spores (22-)26-33(-40) x (18-)22-26(-28)μ, mostly ellipsoid or obovoid, wall colorless, 1-1.5(-2.5)μ thick at sides, (3-)6-10(-15)μ apically, echinulate, pores very obscure, probably 3 or 4, equatorial. Telia amphigenous, blackish, pulvinate, compact; spores (35-)40-53 (-63) x (16-)21-28(-32)μ, mostly ellipsoid or oblong-ellipsoid, wall chestnut-brown, 1.5-2.5(-3)μ thick at sides, 3-7(-10)μ apically, smooth; pedicels usually yellowish, thick-walled but often collapsing, to 140μ long.

Hosts and distribution: species of Sporobolus: the United States east of the Rocky Mountains and in South Africa.

Type: Bartholomew, on Sporobolus asper (Michx.) Kunth, Rockport, Kansas, U.S.A. (S).

Arthur (Bot. Gaz. 29:274. 1900) first proved the life cycle by inoculation. Cummins and Greene (Brittonia 13:271-285.) published a photograph of teliospores of the type.

PUCCINIA VILFAE Arth. & Holw. var. mexicana Cumm. Southw. Nat. 12:83. 1967.

Urediniospores (24-)26-30(-34) x (18-)20-24(-25)μ, wall 1-1.5μ thick at sides (4-)7-10(-12)μ at apex, hyaline, pores obscure. Teliospores (28-)31-40(-42) x (20-)22-26(-30)μ, wall 1.5-2.5(-3.5)μ thick at sides, (3.5-)5-7(-8)μ at apex.

Hosts and distribution: Sporobolus buckleyi Vasey: Mexico and U.S.A. (Texas).

Type: Cummins 62-210(=PUR 60274), Ciudad Mante, Tamps., Mexico.

The variety differs from the typical because of shorter teliospores. A photograph of teliospores of the type was published by Cummins (loc. cit.).

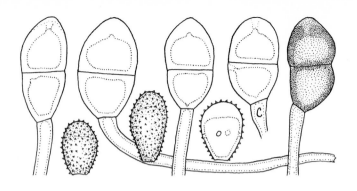

Figure 127

108. PUCCINIA IMPERATAE Poirault Assoc. Nat. Nice Bull. 1:105. 1913. Fig. 127.

Uredo imperatae Magn. Zool.-Bot. Ges. Wein Verhandl. 50:439. 1900.

Puccinia imperatae Beltr. Roy. Soc. Espan. Hist. Nat. Mem. 50:251. 1921.

Puccinia imperatae Doidge Bothalia 2:474. 1928.

Aecia unknown. Uredinia amphigenous, often confluent, yellow; spores (20-)23-30(-34) x 18-22(-25)μ, mostly globoid or obovoid, wall 1-2μ thick at sides, 3-8(-10)μ apically, colorless or yellowish, echinulate, germ pores 4, just below the apical thickening, obscure. Telia amphigenous, often confluent, early exposed, pulvinate, blackish brown; spores (30-)34-50(-60) x 19-26(-29)μ, mostly ellipsoid or oblong-ellipsoid, wall (2.5-)3-3.5(-4)μ thick at sides, 5-8(-12)μ apically, golden or clear chestnut-brown, smooth; pedicels colorless or nearly so, thick-walled, mostly not collapsing, to 160μ long.

Hosts and distribution: Imperata cylindrica (L.) Beauv. and varieties: Mediterranean region and South Africa.

Type: Poirault, Juan-les-Pins, near Nice, France (not seen).

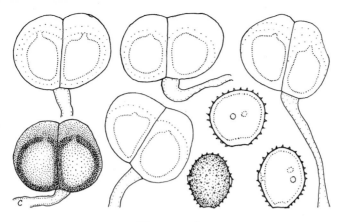

Figure 128

109. PUCCINIA LEVIS (Sacc. & Bizz.) Magn. Deuts. Bot. Ges. Ber. 9:190. 1891 var. levis. Fig. 128.

Diorchidium leve Sacc. & Bizz. Michelia 2:648. 1882.

Puccinia paspali Tracy & Earle Bull. Torrey Bot. Club 22:174. 1895.

Aecia unknown. Uredinia amphigenous, dark cinnamon- or chestnut-brown; spores (23-)25-31 x (20-)23-27μ, globoid or broadly ellipsoid with pores face-view, wall 1.5-2μ thick, echinulate, cinnamon- or near chestnut-brown, germ pores 2, in flattened sides, equatorial. Telia amphigenous, or mostly on abaxial surface, pulvinate, early exposed, blackish brown; spores 31-43(-46) x (22-)26-33(-36)μ, varying from ellipsoid to broadly ellipsoid, usually tending to be diorchidioid and often strongly so, wall (1.5-)2.5-4(-5)μ thick at sides, (5-)7-12(-14)μ over pores, dark chestnut or darker except over pores, smooth; pedicels colorless, thick-walled and not collapsing, to 175μ long.

Hosts and distribution: Axonopus chrysoblepharis (Lag.) Chase, A. scoparius (Fluegge) Kuhlm., Entolasia marginata (R. Br.) Hughes, Hackelochloa granularis (L.) Kuntze, species of Paspalum, Thraysia campylostachya (Hack.) Chase, T. paspaloides H. B. K.: Southern United States to Argentina and possibly in Australia.

Type: Bizzozero, on an herbarium specimen of Manisuris granularis (=Hackelochloa granularis) from Brazil (PAD).

Ramachar and Cummins (Mycopath. Mycol. Appl. 25:7-60. 1965) published a photograph of teliospores of the type.

PUCCINIA LEVIS Sacc. & Bizz. var. goyazensis (P. Henn) Ramachar & Cumm. Mycopath. Mycol. Appl. 25:43. 1965.

Puccinia goyazensis P. Henn. Hedwigia 34:94. 1895.

Urediniospores 26-31 x 22-24μ with pores face-view, wall 1.5-2μ thick, cinnamon- or dark cinnamon-brown, echinulate,

germ pores 2, in flattened sides, equatorial. Teliospores
(26-)29-35(-42) x (22-)26-30(-32)μ, broadly obovoid, broadly
ellipsoid, cuboidal, or rarely ellipsoid, mostly diorchidioid,
wall (2-)2.5-3(-4)μ thick at sides, 5-7(-9)μ over the pores,
dark chestnut-brown except over the pores; pedicels thin- or
thick-walled, collapsing or not, to at least 150μ long.

Hosts and distribution: Panicum millegrana Poir., P.
missionum Mez,. P. schiffneri Hack.: Brazil and Mexico.

Type: Ule No. 1928, on Panicum sp., Goyaz, Brazil (B; isotype
PUR).

Ramachar and Cummins (loc. cit.) published a photograph of
teliospores of the type.

PUCCINIA LEVIS Sacc. & Bizz. var. tricholaenae (H. Syd. &
P. Syd.) Ramachar & Cumm. Mycopath. Mycol. Appl. 25:44. 1965.

Diorchidium tricholaenae H. Syd. & P. Syd. Ann. Mycol. 10:33.
1912.

Uromyces tricholaenae Gz. Frag. & Cif. Bol. Roy. Soc. Esp.
Hist. Nat. 25:357. 1925.

Puccinia tricholaenae (H. Syd. & P. Syd.) Ramak. T. & K.
Ramak. Proc. Indian Acad. Sci. B. 28:63. 1948.

Urediniospores (24-)26-33 x (21-)23-27(-29)μ with pores face
view, wall 2μ thick, dark cinnamon-brown, echinulate, germ pores
2, in flattened sides, equatorial. Teliospores 37-47(-55) x
29-33μ, wall (2.5-)3-4μ thick at sides, (4-)5-7(-8)μ over the
pores, chestnut-brown, not much paler over the pores; pedicels
thick-walled, mostly not collapsing, to 175μ long.

Hosts and distribution: Rhynchelytrum repens (Willd.) C. E.
Hubb.: circumglobal in the warmer areas.

Type: Burtt Davy (Pole- Evans No. 286), on Tricholaena rosea
(=Rhynchelytrum repens), Barberton, Transvaal, So. Africa (S).

PUCCINIA LEVIS Sacc. & Bizz. var. panici-sanguinalis (Rangel)
Ramachar & Cumm. Mycopath. Mycol. Appl. 25:44. 1965.

Puccinia rottboelliae P. Syd. & H. Syd. Monogr. Ured. 1:800.
1904.

Uromyces panici-sanguinalis Rangel Arch. Mus. Rio de Janeiro
18:159. 1916.

Uredo paspali-perrottetii Petch Ann. Roy. Bot. Gard. Peradeniya
6:216. 1917.

Puccinia setariae-viridis Diet. Ann. Mycol. 15:493. 1917.

Puccinia kimurai Hirat. f. & Yosh. Mem. Tottori Agr. Coll.
3:314. 1935.

Puccinia jaagii Boed. Bull. Jard. Bot. Buitenzorg Ser. II.
16:264. 1940.

Diorchidium brachiariae Wakef. & Hansf. Proc. Linn. Soc.
London 161:167. 1949.

<u>Diorchidium</u> <u>digitariae</u> Ahmad Biologia 2:31. 1956.

Urediniospores (23-)25-28(-30) x (18-)20-25μ, wall 1.5-2(-3)μ, dark cinnamon-brown, echinulate, germ pores 3 (rarely 4), equatorial. Teliospores (25-)29-37(-40) x (22-)23-30(-32)μ, mostly broadly ellipsoid or broadly obovoid, mostly diorchidioid, wall 2-3μ thick at sides, (4-)5-7(-9)μ over the pores, dark chestnut-brown except usually paler over the pores; pedicels mostly thick-walled and not collapsing, to at least 140μ long.

Hosts and distribution: <u>Brachiaria</u> sp., species of <u>Digitaria</u>, <u>Eriochloa</u> <u>procera</u> C. E. Hubb., <u>Hemarthria</u> <u>compressa</u> (L. f.) R. Br., <u>Hyparrhenia</u> <u>newtonii</u> Stapf, <u>Ichnanthus</u> <u>minarum</u> (Nees) Doell., species of <u>Panicum</u> and <u>Paspalum</u>, <u>Pennisetum</u> <u>mutilatum</u> Hack. ex Kuntze, <u>Reimarochloa</u> <u>brasiliensis</u> (Spreng.) Hitchc., <u>Rottboellia</u> <u>exaltata</u> L. f. species of <u>Setaria</u>, <u>Sorghum</u> <u>plumosum</u> (R. Br.) Beauv.: Ceylon and Pakistan eastward to Central and South America, Florida and the West Indies.

Type: Rangel No. 1103, on <u>Panicum</u> <u>sanguinale</u> (= <u>Digitaria</u> <u>sanguinalis</u> (L.) Scop., Cubango near Niteroy, Brazil (R; isotype (PUR).

Ramachar and Cummins (loc. cit.) published a photograph of teliospores of the type.

<u>Puccinia</u> <u>levis</u> comprises a complex of somewhat variable forms but having similar principal features, e.g. dark brown urediniospores, dark brown, often nearly opaque, strongly diorchidioid, long-pedicelled teliospores. Most collections lack teliospores and this, together with ignorance of the aecial stages, renders the present treatment tentative.

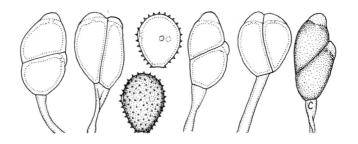

Figure 129

110. PUCCINIA FLACCIDA Berk. & Br. in Berkeley J. Linn. Soc.
14:91. 1873. Fig. 129.

Diorchidium flaccidum (Berk. & Br.) Kuntze Rev. Gen. 3:468.
1898.

Diorchidium levigatum H. Syd., P. Syd. & Butl. Ann. Mycol.
5:500. 1907.

Puccinia levigata (H. Syd., P. Syd. & Butl.) Hirat. f. Tottori
Agr. Coll. Mem. 3:315. 1935.

Aecia unknown. Uredinia amphigenous, dark cinnamon-brown;
spores 23-30 x (17-)23-27µ, obovoid or ellipsoid, wall 1.2-2.5µ
thick, dark cinnamon- or chestnut-brown, echinulate, germ pores
3, equatorial. Telia amphigenous, exposed, pulvinate, blackish
brown; spores 25-44 x 15-23µ, ellipsoid or oblong, tending to
be strongly diorchidioid, wall 1-1.5µ thick at sides, 2-4µ
apically, golden or cinnamon-brown, smooth; pedicels colorless,
thin-walled, collapsing, to 60µ long; germination occurs without dormancy.

Hosts and distribution: Oplismenus burmanii (Retz.) Beauv.,
O. compositus (L.) Beauv., O. undulatifolius (Ard.) Beauv.,
Panicum chionachne Mez: Ceylon, India, and Japan.

Type: Thwaites No. 1136, on Panicum sp. (error for Oplismenus,
possibly compositus, according to C. E. Hubbard in litt.),
Peradeniya, Ceylon (K).

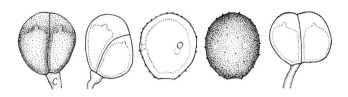

Figure 130

111. PUCCINIA NYASAENSIS Cumm. Bull. Torrey Bot. Club 83:228.
1956. Fig. 130.

Aecia unknown. Urediniospores in the telia (26-)28-32 x
(22-)24-26(-28)μ, broadly ovate or globoid, wall (2-)2.5-3μ
thick, very dark chestnut-brown, finely echinulate, perhaps
sometimes smooth, germ pores 3, equatorial. Telia amphigenous,
exposed, pulvinate, blackish brown; spores 24-33μ wide, 24-28(-32)μ
high, strictly diorchidioid, obovoid or nearly globoid, wall
1.5-2μ thick at sides, 5-7μ apically, chestnut-brown, smooth;
pedicels colorless, fragile, to 45μ long but mostly deciduous.

Hosts and distribution: Panicum pectinatum Rendle: Nyasaland.

Type: P. O. Wiehe No. 467, Mlanje, Chambe plateau, Nyasaland
(PUR; isotype IMI).

A photograph of spores of the type was published by Cummins
(loc. cit.).

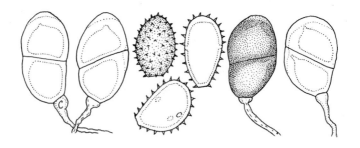

Figure 131

112. PUCCINIA DEFORMATA Berk. & Curt. J. Linn. Soc. 10:357. 1869. Fig. 131.

Puccinia amianthina H. Syd. & P. Syd. Bot. Jahrb. 45:260. 1910.

Puccinia olyrae-latifoliae V.-Bourgin Bull. Soc. Mycol. France 70:417. 1954.

Aecia unknown. Uredinia amphigenous and in inflorescence, pale yellowish, probably bright yellow when fresh; spores (24-)27-32(-36) x (19-)21-27(-30)µ, mostly obovoid, wall 1.5-2µ thick, occasionally slightly thicker at apex, yellowish, echinulate, germ pores, 2 or 3 (4?), obscure. Telia amphigenous and in inflorescence, early exposed, chocolate-brown, moderately compact; spores (26-)30-40(-44) x (19-)21-28(-30)µ, variable but mostly ellipsoid or obovoid, varying from puccinioid to diorchidioid, mostly with only a somewhat oblique septum, wall (1.5)2.5-3.5(-4)µ at sides, (2.5-)3-6(-8)µ apically, golden brown or chestnut-brown, smooth; pedicels yellowish or color-less, thin-walled and collapsing, to 150µ but often less than 100.

Hosts and distribution: Olyra cordifolia H.B.K., O. latifolia L.: Central America to Venezuela, Brazil, Trinidad, West Central Africa, and Uganda.

Type: Wright, on Olyra latifolia, Cuba (FH; isotype PUR).

Teliospores from leaves have slightly thinner and paler walls than those from inflorescences but urediniospores do not differ.

228

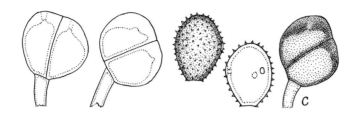

Figure 132

113. PUCCINIA LOPHATHERI (H. Syd. & P. Syd.) Hirat. f. J. Jap. Bot. 14:36. 1938. Fig. 132.

Diorchidium lophatheri H. Syd. & P. Syd. Ann. Mycol. 12:107. 1914.

Uredo lophatheri Petch Ann. Roy. Bot. Gard. Peradeniya 7:296. 1922.

Aecia unknown. Uredinia amphigenous, brownish; spores (22-)24-28(-31) x (18)20-23(-25)μ, mostly obovoid, wall 1.5-2(-2.5)μ thick, yellowish to cinnamon-brown, echinulate, germ pores 3, equatorial. Telia amphigenous, blackish brown, early exposed, rather pulverulent; spores (24-)26-31(-33) x (20-)23-26(-28)μ, mostly strongly diorchidioid, mostly broadly ellipsoid, wall (1-)1.5-2(-2.5)μ thick at sides, (5-)6-8(-9)μ thick over each germ pore, golden brown or clear chestnut-brown, smooth; pedicels thin-walled, collapsing, colorless, to 75μ but usually broken short.

Hosts and distribution: Centotheca lappacea (L.) Desv., Lophatherum gracile Brong.: China, Japan, and Taiwan.

Type: Fujikuro No. 110, on Lophatherum gracile var. elatum, Taihoku, Taiwan (S).

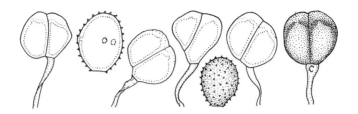

Figure 133

114. PUCCINIA NEGRENSIS P. Henn. Hedwigia 43:159. 1904. Fig. 133.

Triphragmium graminicola Beeli Bull. Jard. Bot. Bruxelles 8:5. 1923.

Aecia unknown. Uredinia amphigenous, cinnamon-brown; spores (19-)24-27 x (15-)17-21µ, mostly broadly ellipsoid, wall 1.5-2.5µ thick, golden or cinnamon-brown, echinulate, germ pores 3, equatorial. Telia on abaxial leaf surface, exposed, compact, blackish brown; spores (20-)24-26 x (17-)19-22µ, diorchidioid, mostly broadly ellipsoid, or globoid, wall 1-1.5µ thick at sides, 2-4µ apically, golden or chestnut-brown, smooth; pedicels colorless, thin-walled and collapsing, to 50µ long but fragile and broken short.

Hosts and distribution: Panicum millegrana Poir., P. aff. (Brachiaria) ramosum L.: Brazil and Congo.

Type: E. Ule, on Panicum sp., Moura, Rio Negro, Est. Amazonas, Brazil (B; isotype PUR).

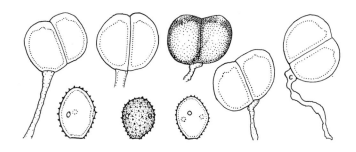

Figure 134

115. PUCCINIA TAIWANIANA Hirat. f. & Hashioka in Hiratsuka Tottori Soc. Agr. Sci. Trans. 5:240. 1935. Fig. 134.

Puccinia pangasinensis H. Syd. in Cummins Ann. Mycol. 35:99. 1937.

Aecia unknown. Uredinia amphigenous or mostly on the abaxial leaf surface, golden; spores (17-)20-24 x (14-)17-19μ, mostly obovoid or ellipsoid, wall 1-1.5μ thick, yellow to pale brownish, echinulate, germ pores 3(4), equatorial. Telia on abaxial surface, exposed, pulvinate, blackish brown; spores (20-)23-27 (-31) x (18-)20-24μ, ellipsoid or obovoid, tending to be diorchidioid, wall 1.5-2.5μ thick at sides, 3-5μ apically, chestnut-brown, smooth; pedicels colorless, thin-walled and collapsing, to 70μ long.

Hosts and distribution: Cyrtococcum patens (L.) A. Camus: China, Japan, and the Philippines.

Type: Hiratsuka, on Panicum patens (=C. patens), Loochoo Island, Okinawa, Japan (herb. Hiratsuka).

Ramachar and Cummins (Mycopathol. Mycol. Appl. 25:7-60. 1965) published a photograph of teliospores of the type.

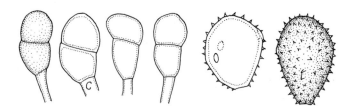

Figure 135

116. PUCCINIA PANICI-MONTANI Ramachar & Cumm. Mycopath. Mycol.
Appl. 25:49. 1965. Fig. 135.

Puccinia panici-montani Fujik. in Sawada Descr. Cat.
Formosan Fungi. 4:64. 1928, nomen nudum.

Uredo panici-plicati Saw. J. Taihoku Soc. Agr. Forst. 7:42.
1943, nomen nudum.

Aecia unknown. Uredinia amphigenous, yellowish brown; spores
(31-)34-37(-41) x (22-)27-31μ, mostly obovoid, usually angularly
so, wall (1-)1.5-2μ thick, cinnamon-brown or near it, rather
sparsely echinulate, germ pores 3 or 4, equatorial. Telia
unknown; teliospores in the uredinia 26-31 x 15-19μ, mostly
oblong or clavate, wall uniformly 1.5μ thick, wall pale golden
or almost colorless, smooth; pedicels thin-walled, fragile
and collapsing, to 18μ long; the spores probably germinate without a dormant period.

Hosts and distribution: Setaria palmifolia (Koenig) Stapf.,
S. plicata (Lam.) Cooke: Taiwan.

Type: Fujikuro, on Panicum plicatum (=S. plicata), Taipei,
Taiwan, 22 Feb. 1914 (TAI).

The species is poorly known and only a few teliospores have
been seen.

232

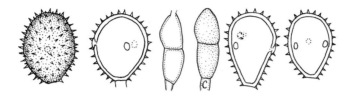

Figure 136

117. PUCCINIA ICHNANTHI Mains Bull. Torrey Bot. Club 66:619.
1939. Fig. 136.

Aecia unknown. Uredinia mostly on abaxial leaf surface,
cinnamon-brown; spores (29-)32-38(-42) x 23-27(-29)μ, broadly
ellipsoid or obovoid, wall 1.5-2μ thick, golden or cinnamon-
brown, echinulate, germ pores 2(3), equatorial. Telia on
abaxial leaf surface, yellowish, probably bright orange when
fresh, early exposed; spores 28-34 x 12-14μ, very delicate,
narrowly ellipsoid or fusoid, wall uniformly 0.5-1μ thick,
colorless, smooth; pedicels colorless, thin-walled and collapsing, to 30μ long but usually broken short; the spores germinate without dormancy and collapse.

Hosts and distribution: Ichnanthus candicans (Nees) Doell:
Brazil (only the type known).

Type: Chase No. 12143A, on Ichnanthus candicans, Tijuca,
Brazil (MICH).

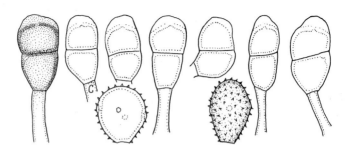

Figure 137

118. PUCCINIA PUTTEMANSII P. Henn. Hedwigia 41:105. 1902.
Fig. 137.

Aecia unknown. Uredinia mainly on abaxial leaf surface,
cinnamon-brown; spores (20-)22-24 x (17-)20-24µ, mostly broadly
ellipsoid or obovoid, wall 1.5µ thick, golden or pale cinnamon-
brown, echinulate, germ pores 4, rarely 3, equatorial. Telia
mainly on abaxial surface, exposed, blackish brown, pulvinate;
spores (27-)34-37 x (17-)20-24µ, mostly clavate or oblong-
ellipsoid, wall 1.5µ thick at sides, 4-7µ apically, deep golden
or clear chestnut-brown, smooth; pedicels yellowish, thin-
walled, mostly collapsing, to 30µ long.

Hosts and distribution: Panicum millegrana Poir., P.
sciurotis Trin., P. sellowii Nees: Brazil and Trinidad.

Type: A. Puttemans No. 140, on Panicum sp., Brazil (B;
isotype PUR).

Cummins (Mycologia 34:669-695. 1942) published a photograph of teliospores of the type.

234

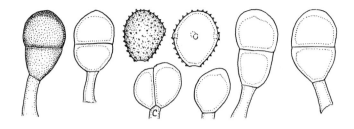

Figure 138

119. PUCCINIA HUBERI P. Henn. Hedwigia Beibl. 39:76. 1900.
Fig. 138.

Aecial stage unknown. Uredinia amphigenous, pale cinnamon-brown; spores (20-)24-27 x (17-)20-24μ, mostly obovoid or broadly ellipsoid, wall 1.5μ thick, pale cinnamon-brown or golden, echinulate, germ pores 3 or 4, equatorial. Telia amphigenous, exposed, blackish brown, compact; spores (27-)31-39 x (17-)20-26μ, mostly ellipsoid or ellipsoid-clavate, wall 2μ thick at sides, 3-5μ apically, chestnut-brown, smooth; pedicels golden, thin-walled but mostly not collapsing, frequently inserted somewhat laterally, to 15μ long; 1-celled spores numerous.

Hosts and distribution: Panicum ovalifolium Poir., P. trichoides Sw.: Brazil, Costa Rica, and Puerto Rico.

Type: Huber No. 3, on P. ovalifolium, Para, Botan. Garten, Brazil, 1896 (B; isotype PUR).

The sori of this species always are located in brown necrotic spots.

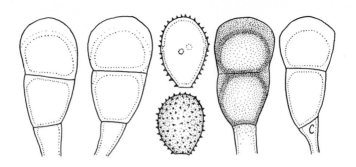

Figure 139

120. PUCCINIA ARAGUATA Kern Mycologia 30:544. 1938. Fig. 139.

Puccinia paspalicola Kern, Thurst. & Whetz. Univ. P. Rico Monogr. B. 2:284. 1934 (Oct.), not Arthur 1934 (June).

Aecia unknown. Uredinia amphigenous, pale cinnamon-brown; spores (24-)27-31(-34) x (17-)19-21(-24)μ, mostly obovoid or ellipsoid, wall 1-1.5μ thick, golden-brown, echinulate, germ pores 4 where seen with certainty, obscure, equatorial. Telia on adaxial leaf surface, early exposed, pulvinate, blackish brown; spores (40-)44-51(-62) x 24-27μ, broadly clavate or oblong-clavate, wall 2-2.5μ thick at sides, 5-9μ apically, golden or clear chestnut, smooth; pedicels colorless, thin-walled, short and always broken near the hilum.

Hosts and distribution: Paspalum microstachyum Presl: Venezuela.

Type: Chardon & Toro No. 600, Aragua, Venezuela (PAC; isotype PUR).

Cummins (Mycologia 34:669-695. 1942) published a photograph of teliospores of the type.

236

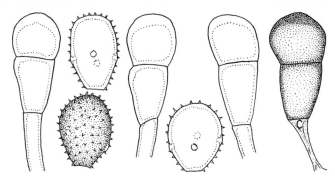

Figure 140

121. PUCCINIA SUBSTRIATA Ell. & Barth. Erythea 5:47. 1897 var.
substriata Fig. 140.

Puccinia pilgeriana P. Henn. Bot. Jahrb. 40:226. 1908.

Uredo cubangoensis Rangel Mus. Rio de Janeiro Arg. 18:160.
1916.

Puccinia tubulosa Arth. Amer. J. Bot. 5:464. 1918, in part.

Puccinia paspalicola Arth. Manual Rusts U.S. & Can. p. 127.
1934, in part.

Uredo setariae-onuri Diet. Rev. Sudamer. Bot. 4:81. 1937.

Aecia, Aecidium tubulosum Pat. & Gaill., occur on species of
Solanum; cupulate, spores (23-)26-31(-37) x 18-23μ, wall 1-1.5μ
thick, verrucose. Uredinia amphigenous or mainly on abaxial
surface, cinnamon-brown; spores 24-31(-37) x (20-)24-27(-31)μ,
mostly broadly ellipsoid or obovoid, wall 1.5-2μ thick, cinnamon-
brown, echinulate, germ pores (3 or)4(or 5), equatorial. Telia
mostly on abaxial surface, exposed, compact, dark brown; spores
(29-)34-50 x 20-26(-29)μ, mostly oblong-ellipsoid, or clavate,
wall 1.2-2μ thick at sides, 3-7μ apically, clear chestnut-brown
or golden, smooth; pedicels colorless or yellowish, thin-walled
and mostly collapsing, to 30μ long.

Hosts and distribution: species of Digitaria, Paspalum, and
Setaria: southern U.S.A. southward to Panama, Trinidad, Brazil
and Bolivia, and in Hawaii and Uganda (?).

Type: Bartholomew, on Paspalum setaceum Michx., Kansas (FH;
isotypes Ellis. & Ev. N. Amer. Fungi No. 3577; Barth. Fungi
Columb. No. 1186).

A photograph of teliospores of the type was published by
Cummins (Mycologia 34:669-695. 1942). The first inoculation
that proved the life cycle was by Thomas (Phytopathology 8:163-
164. 1918).

PUCCINIA SUBSTRIATA Ell. & Barth. var. imposita (Arth.)
Ramachar & Cumm. Mycopathol. Mycol. Appl. 25:26. 1965.

<u>Puccinia</u> <u>imposita</u> Arth. Bull. Torrey Bot. Club 46:112. 1919.

Aecia occur on <u>Solanum</u> ssp.. Urediniospores (26-)29-36(-39) x (20-)22-25(-27)μ, mostly ellipsoid or broadly ellipsoid, wall 2μ thick, echinulate, cinnamon-brown, germ pores 3 or 4, equatorial. Telia exposed; spores (34-)38-50(-56) x (18-)23-28(-30)μ, wall 1.5-2μ thick at sides, 4-7(-8)μ apically, chestnut-brown, smooth; pedicels colorless or brownish, mostly less than 15μ long.

Hosts and distribution: species of <u>Digitaria</u>: southern U.S.A. to Cuba, Puerto Rico, Guatemala, and in Argentina and Bolivia.

Type: Atkinson No. 1586, on <u>Leptoloma</u> <u>cognatum</u> (=<u>D</u>. <u>cognata</u> (Benth.) Henrard, Auburn, Alabama, U.S.A. (PUR 18556).

Ramachar and Cummins (loc. cit.) reported successful inoculation of <u>Solanum</u> <u>carolinense</u> L. and <u>S</u>. <u>melongena</u> L. Field evidence in <u>Texas</u> indicated that S. <u>elaeagnifolium</u> Cav. is the common aecial host in the southwestern <u>U.S.A.</u>

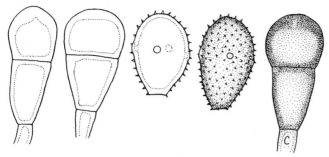

Figure 141

PUCCINIA SUBSTRIATA Ell. & Barth. var. penicillariae (Speg.)
Ramachar & Cumm. Mycopathol. Mycol. Appl. 25:26. 27. 1965.
Fig. 141.

Puccinia penniseti Zimm. Ber. Land.-u. Forstwirt. Deut.
Ostafr. 2:11-37. 1904-1906, not Barclay 1891.

Puccinia penicillariae Speg. Anal. Mus. Nac. B. Aires 26:119.
1914.

Puccinia penniseti-spicati Petrak Sydowia 13:223. 1959.

Aecia unknown. Uredinia mainly on abaxial leaf surface,
cinnamon-brown; spores 34-41(46) x (19-)22-26µ, mostly broadly
ellipsoid or obovoid, wall (1.5-)2-2.5(-3)µ thick, cinnamon-
brown, echinulate, germ pores 4 or 5, equatorial. Telia mainly
on abaxial surface, exposed, blackish brown; spores (34-)44-58
(-65) x (20-)24-27µ, mostly oblong-ellipsoid or clavate, wall
(1.5)2-3(-4)µ thick at sides, 4-8µ apically, chestnut-brown,
smooth; pedicels yellowish, thin-walled, collapsing or not, to
20µ long.

Hosts and distribution: Beckeropsis uniseta (Nees) Stapf;
species of Pennisetum: Africa, including Madagascar.

Type: Spegazzini, on Penicillaria typhoideum (Pennisetum
typhoides), Dakar, Senegal (LPS 8513).

Ramachar and Cummins (loc. cit.) published photographs
of teliospores of the types of P. penicillariae (Fig. 18) and
P. penniseti (Fig. 17). Unfortunately the legends are the
reverse of this.

This variety differs from the typical only in having larger
spores. Aecia on Solanum are not uncommon in Africa but it
has not been demonstrated that they belong in the life cycle
of this variety.

PUCCINIA SUBSTRIATA Ell. & Barth. var. indica Ramachar &
Cumm. Mycopathol. Mycol. Appl. 25:30. 1965.

Uredinia amphigenous, cinnamon-brown; spores (25-)27-34 x

239

(20-)22-24(-28)μ, mostly broadly ellipsoid or obovoid, wall
1.5-2μ thick, golden or pale cinnamon-brown, echinulate, germ
pores (3 or)4(or 5), equatorial. Telia mainly on abaxial leaf
surface, rather tardily exposed but becoming pulvinate, blackish
brown; spores (41-)51-71 x (14-)17-20(-24)μ, mostly oblong or
clavate, wall 1.5-2μ thick at sides, 4-8μ apically, golden or
clear chestnut-brown, smooth; pedicels yellowish, thin-walled
and collapsing or not, to 20μ long.

Hosts and distribution: Pennisetum typhoides (Burm.) Stapf:
India.

Type: M. J. Thirumalachar, on Pennisetum typhoides,
Goribidnur, Mysore, India (PUR).

Ramakrishnan and Soumini (Indian Phytopathol. 1:97-103.
1948) demonstrated that the aecial stage occurs on Solanum
melongena L.

PUCCINIA SUBSTRIATA Ell. & Barth. var. insolita (P. Syd.
& H. Syd.) Ramachar & Cumm. Mycopath. Mycol. Appl. 25:31. 1965.

Puccinia insolita Syd. Flora Bas- et Moy. Congo 3(1):11.
1909.

Puccinia elgonensis Wakef. Linn. Soc. Lond. Proc. 161:178.
1949.

Puccinia kigeziensis Wakef. & Hansf. Linn. Soc. Lond. Proc.
161:182. 1949.

Aecia unknown. Uredinia on abaxial leaf surface, cinnamon-
brown; spores (26-)32-40(-42) x (20-)25-27(-29)μ, mostly oval
or ellipsoid, often angular, wall 1.5-2μ thick, golden or pale
cinnamon-brown, echinulate, germ pores 3 or 4 (or 5), equatorial. Telia mostly hypophyllous, pulvinate, blackish brown;
spores (27-)30-37(-48) x 17-20(-24)μ, mostly clavate, wall
1.5-2μ thick at sides, 3-6μ apically, clear chestnut-brown,
smooth; pedicels yellowish, thin-walled, mostly collapsing, to
15μ long.

Hosts and distribution: Panicum antidotale Retz., P.
maximum Jacq., Setaria barbata (Lam.) Kunth, S. orthosticha
Schum., S. sphacelata (Schum.) Stapf & Hubb.: Equatorial
Africa.

Type: Vanderyst, on Panicum maximum, Kisantu, Yindu,
Congo (S).

This variety has the smallest teliospores and is the least
known.

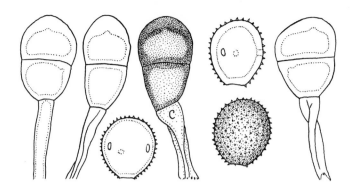

Figure 142

122. PUCCINIA TRIPSACI Diet. & Holw. in Holway Bot. Gaz. 24:27-28.
1897. Fig. 142.

Puccinia ceanothi Arth. Mycologia 2:233. 1910.

The aecia (Aecidium ceanothi Ell. & Kell.) occur on species of
Ceanothus: spores 19-24 x 18-21μ, globoid or broadly ellipsoid,
wall 2-2.5μ thick, finely verrucose, hyaline. Uredinia mostly on
abaxial leaf surface, cinnamon-brown; spores 26-30(33) x 26-30
(-31)μ, globoid, wall 1.5-2μ thick in ordinary urediniospores or
3-5μ thick in amphispores, golden-brown to cinnamon-brown,
echinulate, germ pores (3)4, equatorial. Telia mostly on abaxial
surface, blackish brown, early exposed, pulvinate; spores
(28-)30-40(-45) x (19-)22-27(-31)μ, mostly obovoid or broadly
ellipsoid, wall 2-3(-4)μ thick at sides, (5-)6-8(-9)μ apically,
chestnut-brown, smooth; pedicels yellowish, thick- or thin-
walled, mostly collapsing, to 90μ long.

Hosts and distribution: Andropogon gerardi Vitman, A. hallii
Hack., Trisacum lanceolatum Rupr., T. pilosum Scribn. & Merr.:
South Dakota (U.S.A.) south to Jalisco and Mexico state, Mexico.

Type: Holway, on Tripsacum dactyloides (=error for T.
lanceolatum), near Mexico City, Mexico (S; isotypes Arth. &
Holw. Ured. Exsic. Icones No. 35a).

The life cycle was demonstrated by Arthur (Mycologia 2:233.
1910) using basidiospores from Andropogon hallii. spores from
Tripsacum have not been tried but aecia have been collected near
Mexico City. The species, as defined here, has not been collec-
ted on Andropogon in Mexico nor on Tripsacum in the United
States.

A photograph of teliospores of the type was published by
Cummins in 1953 (Urediniana 4: Pl. V, Fig. 29).

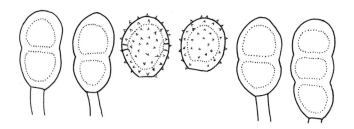

Figure 143

123. PUCCINIA ENNEAPOGONIS Korbon. Akad. Nauk Tadzhik SSR. 22:29. 1957. Fig. 143.

Aecia unknown. Urediniospores in the telia 25-33 x 18-22μ, broadly ellipsoid, ellipsoid, or ovate, wall 3-4μ (?) thick, sparcely echinulate, germ pores 2, equatorial, color not stated. Telia on leaves or rarely on stems, blackish brown, exposed, pulvinate, velvet-like; spores 40-50 x 28μ, broadly ellipsoid, sometimes clavate, sometimes 3- or 4-celled, wall 3-4μ (?) thick at sides, slightly thicker apically, pale brown, smooth; pedicels thick, probably not collapsing, to 100μ long.

Type: Linczevskij, on Enneapogon persicus Boiss., Pjandzh river valley near the village of Bogarac, Southern Tadzhik SSR (TAD?; not seen). Not otherwise reported.

The description is adapted from the original diagnosis and figure.

This species is remarkably similar to P. isiacae.

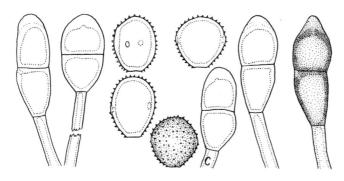

Figure 144

124. PUCCINIA ERIANTHICOLA Cumm. Uredineana 4:42. 1953.
Fig. 144.

Uredo rubida Arth. & Holw. in Arthur Amer. Philos. Soc.
Proc. 64:216. 1925.

Aecia unknown. Uredinia mostly on abaxial leaf surface,
cinnamon-brown; spores (20-)23-28 x (17-)20-24μ, broadly oval
or globoid, flattened on the pore-bearing sides, wall 2-2.5μ
thick, usually 2-3.5μ on the pore-bearing sides, cinnamon-
brown, echinulate, germ pores 2, equatorial. Telia like the
uredinia but pulvinate and blackish brown; spores (29-)32-42
(-47) x (14-)16-20μ, mostly ellipsoid or oblong-ellipsoid,
wall 2-2.5μ thick at sides, 5-8μ apically, clear chestnut-
brown, smooth; pedicels yellowish or brownish, thin-walled
and usually collapsing, to 40μ long.

Hosts and distribution: Andropogon condensatus (Nees)
Kunth (?), Erianthus angustifolius Nees, E. asper Nees: Brazil.

Type: E. W. D. & Mary M. Holway No. 1954, on E. angustifolius,
Garlagua near Taipas, Brazil, (PUR).

A photograph of teliospores of the type was published by
Cummins (loc. cit.).

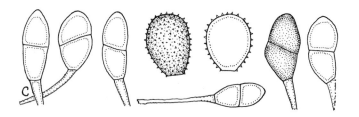

Figure 145

125. PUCCINIA BAMBUSARUM Arth. Bot. Gaz. 65:467. 1918. Fig. 145.

Uredo olyrae P. Henn. Hedwigia 43:164. 1904 (telia present but not described).

Aecia unknown. Uredinia mostly on abaxial leaf surface, yellowish, probably brightly so when fresh; spores (22-)24-32 (-34) x (16-)18-22μ, ellipsoid or obovoid, wall 1.5μ thick, colorless or yellowish, echinulate, germ pores obscure. Telia on abaxial surface, cinnamon-brown, early exposed; spores 20-28(-30) x (10-)12-15(-17)μ, mostly ellipsoid or narrowly obovoid, septum often oblique but diorchidioid spores rare, wall (1-)1.5-2μ at sides, (2-)2.5-4μ apically, yellowish or pale golden brown, smooth; pedicels thin-walled, delicate, colorless, to 80μ long but usually broken short.

Hosts and distribution: Arundinaria (?) sp.: Peru.

Lectotype: Ule No. 3161 (=PUR F4977), on Olyra sp. =error for Arundinaria sp., Rio Amazonas, Iquitos, Peru.

No other specimen has been available. Arthur (loc. cit.: Proc. Amer. Phil. Soc. 64:168-169. 1925) discussed the identity of the hosts. The host involved here is doubtless a member of the Bambusoideae, as determined by Mrs. Chase (see Arthur, 1918).

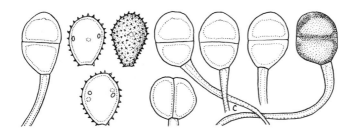

Figure 146

126. PUCCINIA LASIACIDIS Kern Mycologia 30:456. 1938. Fig. 146.

Aecia unknown. Uredinia on abaxial surface, cinnamon-brown or paler; spores (18-)20-24 x (14-)17-20μ, mostly globoid, wall 1-1.5μ thick, yellow or golden, echinulate, germ pores 3 or 4 equatorial, obscure. Telia mainly on abaxial surface, tardily dehiscent but becoming pulvinate, blackish brown; spores (22-)27-29 x 18-20μ, mostly oblong or oblong-ellipsoid, wall 1.5-2μ thick at sides, 2-3.5(-4)μ apically, golden brown, smooth; pedicels colorless, thin-walled and collapsing, to 95μ long.

Hosts and distribution: Lasiacis divaricata (L.) Hitchc.: Venezuela.

Type: F. D. Kern & R. Toro No. 1718, Reservoir, Chaco, Dist. Federal, Venezuela (PAC).

Ramachar and Cummins (Mycopath. Mycol. Appl. 25:7-60. 1965) published a photograph of teliospores of the type.

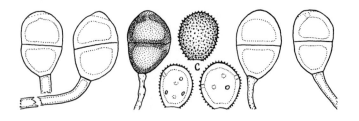

Figure 147

127. PUCCINIA GUARANITICA Speg. Anal. Soc. Cient. Argent.
26:12. 1888. Fig. 147.

Puccinia chichenensis Mains Carnegie Inst. Wash. Publ.
461:100. 1935.

Aecia unknown. Uredinia amphigenous, cinnamon-brown;
spores (few seen) 20-25(-27) x (15-)17-21(-23)μ, broadly
ellipsoid or obovoid, wall (2-)2.5-3.5(-4)μ, pale yellow to
cinnamon-brown, echinulate, pores 4 or 5 equatorial or 4-6,
scattered. Telia amphigenous, early exposed, blackish, compact;
spores (26-)28-31(-33) x 20-23(-25)μ, broadly ellipsoid,
(tending to be diorchidioid on G. virgata), wall 2-3(-4)μ
laterally, 4-7μ apically, chestnut-brown, smooth; pedicels thin-
walled and collapsing, golden, attaining a length of 100μ.

Hosts and distribution: Gouinia guatamalensis (Hack.)
Swallen, G. latifolia (Griseb.) Vasey, G. ramosa Swallen, G.
virgata (Presl) Scribn.: Mexico to Bolivia and Paraguay.

Type: Balansa, on Tricuspis latifolia (=Gouinia latifolia),
Guarapi, Paraguay (LPS; isotype PUR).

A photograph of teliospores of the type was published by
Hennen and Cummins (Mycologia 48:126-162. 1956).

The urediniospore in late season collections usually are
larger, have thicker walls, and are darker in color than
those in early season. Possibly the thick-walled, darker
spores are amphisporic or perhaps the material is heterogenous.

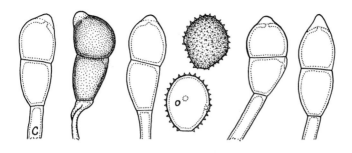

Figure 148

128. PUCCINIA PIPTOCHAETII Diet. & Neger Bot. Jahrb. 27:3. 1899.
Fig. 148.

Aecia unknown. Uredinia mostly on abaxial leaf surface
and on sheaths, cinnamon-brown; spores (17-)19-25(-27) x
(14-)17-21(-23)μ, mostly obovoid, wall (1-)1.5-2(-2.5)μ thick
but thicker when immature, cinnamon-brown, echinulate, germ
pores (2)3(4), equatorial. Telia mostly on abaxial surface
and sheaths, blackish brown, early exposed, pulvinate; spores
(25-)30-43(-45) x (14-)16-21(-23)μ, variable but mostly ellipsoid, or narrowly obovoid, wall (1-)1.5-2μ thick at sides,
(3-)4-7μ apically, the apex usually conical, deep golden or
clear chestnut-brown except the conical apex paler, smooth;
pedicels thin-walled and collapsing, hyaline, to 50μ long.

Hosts and distribution: Piptochaetium montevidensis (Spreng.)
Parodi, P. stipoides (Trin. & Rupr.) Hack. ex Arech.: Argentina,
Bolivia, Chile, and Uruguay.

Type: Neger, on Piptochaetium sp., near Concepcion, Chile
(S).

247

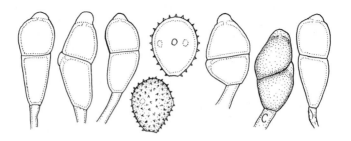

Figure 149

129. PUCCINIA MILLEGRANAE Cumm. Bull. Torrey Bot. Club 83:228. 1956. Fig. 149.

Aecia unknown. Uredinia on abaxial leaf surface, pale brownish; spores (24-)26-30(-34) x (17-)21-24μ, ovate or ellipsoid, wall 1μ thick, colorless or yellowish, finely echinulate, germ pores 3, equatorial. Telia on abaxial surface, exposed, brown; spores (30-)35-43(-46) x 15-19(-21)μ, variable but mostly oblong-ellipsoid or ellipsoid, wall 1μ thick at sides, 4-7μ apically, yellow or golden with the apical umbo paler; pedicels colorless, thin-walled and collapsing, to 45μ long; the spores germinate without dormancy.

Hosts and distribution: Panicum millegrana Poir.: Brazil.

Type: Holway No. 1834, Reserva Florestal, Itatiaya, Rio de Janeiro (PUR; isotypes Reliq. Holw. No. 144 as P. flaccida).

A photograph of teliospores of the type was published with the diagnosis.

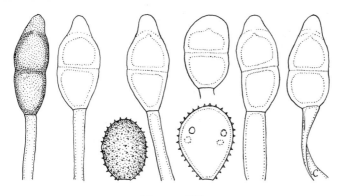

Figure 150

130. PUCCINIA GYMNOTHRICHIS P. Henn. Hedwigia 35:242. 1896.
Fig. 150.

Puccinia burmeisteri Speg. Anal. Mus. Nac. B. Aires 6:222.
1899.

Aecia unknown. Uredinia amphigenous, cinnamon-brown, or
paler; spores (24-)26-32(-34) x (20-)22-26µ, ellipsoid, broadly
ellipsoid, or obovoid, wall 1.5-2µ thick, golden or cinnamon-
brown, echinulate, germ pores 3 or 4, equatorial. Telia
amphigenous, early exposed, compact, blackish brown; spores
(26-)32-45(-52) x (14-)16-21(-26)µ, mostly ellipsoid or
narrowly ellipsoid with a differentiated pale umbo apically,
wall 2µ thick at sides, (4-)5-9(-13)µ apically, golden or
chestnut-brown, smooth; pedicels colorless, thin-walled and
collapsing, to 80µ long.

Hosts and distribution: species of Pennisetum: Ecuador and
Brazil to Argentina.

Type: Lorentz, on Gymnothrix latifolia (=Pennisetum latifolium Spreng.), Siambon, Sierra de Tucuman, Argentina (B;
isotype PUR).

Ramachar and Cummins (Mycopathol. Mycol. Appl. 25:7-60.
1965) published a photograph of teliospores of the type.

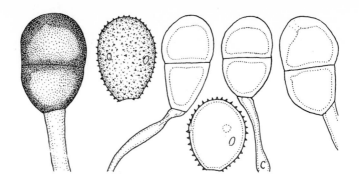

Figure 151

131. PUCCINIA OPIPARA Cumm. Bull. Torrey Bot. Club 68:468. 1941. Fig. 151.

Aecia unknown. Uredinia on abaxial leaf surface, yellowish; spores 31-40(-43) x (22-)25-31μ, mostly broadly ellipsoid, wall 1-1.5μ thick, colorless or yellowish, echinulate, germ pores 3 or 4, equatorial. Telia on abaxial surface, exposed, pulvinate, blackish brown; spores (27-)33-44 x 23-28(-31)μ, mostly ellipsoid, wall 2-3μ thick at sides, 4-7μ apically, opaque chestnut-brown, smooth; pedicels pale yellow, thin-walled and mostly collapsing, to 70μ long.

Hosts and distribution: Oplismenus minarum Nees: Bolivia.

Type: E. W. D. & Mary M. Holway No. 541, Sorata, Coroico, Prov. de Nor Yungas, Bolivia (PUR; isotypes: Reliq. Holw. No. 82 as Puccinia levis).

The large urediniospores and the teliospores, which are almost opaque and have a smoky tint, characterize the species.

A photograph of teliospores of the type was published by Cummins (loc. cit.).

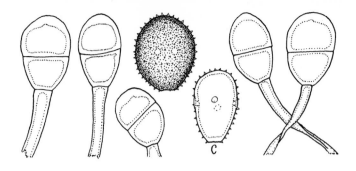

Figure 152

132. PUCCINIA PAPPOPHORI Cumm. Torrey Bot. Club Bull. 83:229.
1956. Fig. 152.

Aecia unknown. Uredinia amphigenous or mostly on adaxial
leaf surface, cinnamon-brown; spores 31-38 x (22-)24-29µ ovate,
ellipsoid, or broadly ellipsoid, wall 1.5µ thick, cinnamon-
brown, echinulate, germ pores 4, equatorial. Telia like the
uredinia but pulvinate and blackish brown; spores (26-)29-36(-39)
x (16-)18-23(-25)µ, mostly ellipsoid or oblong-ellipsoid, wall
2-3µ thick at sides, 3-6µ apically, chestnut-brown, smooth;
pedicels yellowish; moderately thick-walled, collapsing or
not, to 85µ long.

Hosts and distribution: Pappophorum mucronulatum Nees:
Bolivia.

Type: E. W. D. & Mary M. Holway No. 367 Cochabamba, Bolivia
(PUR; isotypes Reliq. Holw. No. 59 as P. gymnotrichis P. Henn.).

Cummins (loc. cit.) published a photograph of teliospores
of the type.

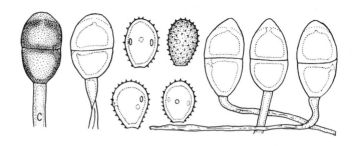

Figure 153

133. PUCCINIA POLLINIICOLA H. Syd. in Sydow & Petrak Ann. Mycol. 29:156. 1931. Fig. 153.

Aecia unknown. Uredinia amphigenous or mostly on abaxial leaf surface, yellow; spores (17-)19-25 x 14-18μ, wall 1.5-2.5μ thick, occasionally to 5μ at the apex, yellowish or golden, echinulate, germ pores obscure, probably 3, equatorial. Telia amphigenous and on the stems, exposed, pulvinate, chocolate-brown; spores 27-36 x 17-22μ, mostly ellipsoid, wall 3-3.5μ thick at sides, 4.5-6μ apically, deep golden or clear chestnut-brown, smooth; pedicels colorless, mostly thin-walled and collapsing, to 90μ long.

Hosts and distribution: Microstegium glabratum (Brogn.) Hosok., M. vimineum (Trin.) A. Camus: Formosa, Japan, and the Philippines.

Isotype: M. S. Clemens No. 7226, on Pollinia viminea (=M. vimineum), Baguio, Luzon, Philippine Islands (BPI).

A photograph of teliospores of the isotype was published by Cummins (Uredineana 4: Plate V, Fig. 27. 1953).

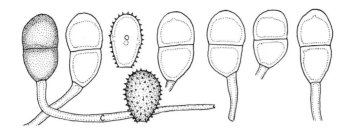

Figure 154

134. PUCCINIA FACETA H. Syd. Ann. Mycol. 32:289. 1934. Fig. 154.

Aecia unknown. Uredinia mostly on abaxial leaf surface, about cinnamon-brown; spores (22-)23-26(-28) x (16-)18-20(-21)μ, mostly ellipsoid or obovoid, wall 1.5-2μ thick, cinnamon-brown, echinulate, germ pores 4, equatorial. Telia on abaxial surface, blackish brown, early exposed, pulvinate; spores (26-)29-35 x (16-)18-20μ, ellipsoid or oblong-ellipsoid, wall 2-2.5μ thick at sides (2.5-)3-4μ apically, clear chestnut-brown, smooth; pedicels thin-walled and collapsing, yellowish, to 100μ long.

Hosts and distribution: Olyra heliconia Lindm.: Brazil.

Type: Chase, No. 12047, on Olyra heliconia, Santa Rita do Araguaya on Rio Araguaya, Goyaz, Brazil (holotype lost; isotypes BPI, PUR).

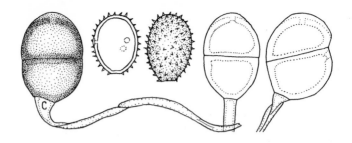

Figure 155

135. PUCCINIA INCLITA Arth. Bull. Torrey Bot. Club 46:115.
1919. Fig. 155.

Aecia unknown. Uredinia mostly on abaxial leaf surface,
yellowish, probably brightly so when fresh; spores (25-)27-34
(-40) x (20-)22-26(-28)μ, ellipsoid or broadly obovoid, wall
1-1.5μ thick, colorless, echinulate, germ pores 3, equatorial,
obscure. Telia mostly on abaxial surface, blackish brown,
exposed; spores 35-42(-50) x (23-)26-29μ, mostly broadly
ellipsoid or broadly obovoid, wall 2-3(-3.5)μ thick at sides,
3-5(-6)μ apically, chestnut-brown, smooth; pedicels brownish,
thin-walled and collapsing, to 60μ long but usually broken
short; 1-celled and incompletely septate spores are common in
the type.

Hosts and distribution: species of Ichnanthus, Oplismenus:
British Honduras and Puerto Rico to Brazil and Ecuador.

Type: Whetzel and Olive No. 397, on Ichnanthus pallens
(Swartz) Munro, El Junque, Puerto Rico (PUR).

Ramachar and Cummins (Mycopathol. Mycol. Appl. 25:7-60.
1965) published a photograph of teliospores of the type.

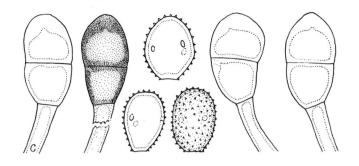

Figure 156

136. PUCCINIA MISCANTHIDII Doidge Bothalia 3:500. 1939. Fig. 156.

Aecia unknown, uredinia amphigenous, often confluent, yellow; spores 24-30 x (17-)19-23(-25)μ, mostly obovoid or broadly oval, wall 1.5-2.5μ thick, colorless, echinulate, germ pores 3 (or 4) equatorial, obscure. Telia like the uredinia but pulvinate and blackish brown; spores (30-)33-46(-50) x 20-27(-30)μ, mostly ellipsoid or oblong-ellipsoid, wall 2-3μ thick at sides, 5-9μ apically, chestnut, smooth; pedicels brownish or nearly colorless, thin-walled, collapsing, to 90μ long.

Hosts and distribution: Miscanthus capensis (Nees) Anderss., M. junceus (Stapf) Pilger, M. sorghum (Nees) Pilger: South Africa.

Type: E. M. Doidge No. 30104, on M. sorghum, Lundie's Hill, Umkomaas Valley, Natal, Union of South Africa (PRE).

Cummins (Urediniana 4: Plate V, Fig. 32. 1953) published a photograph of teliospores of the type.

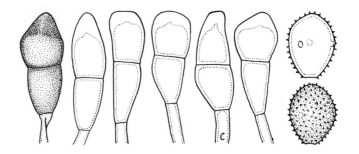

Figure 157

137. PUCCINIA CHISOSENSIS Cumm. Southw. Nat. 8:184. 1964. var. chisosensis. Fig. 157.

Aecia unknown. Uredinia mostly adaxial, cinnamon-brown; spores (23-)26-30(-38) x (18-)20-24(-26)μ, mostly obovoid or broadly ellipsoid, wall (1-)2(-2.5)μ thick, echinulate, cinnamon-brown, pores 3(4), equatorial. Telia mostly abaxial, blackish brown, compact, early erumpent; spores (30-)36-50(-56) x (14-)16-21(-24)μ, mostly oblong-ellipsoid or narrowly obovoid, wall (1.5-)2(-3)μ thick at sides, (4-)6-10(-12)μ at apex, chestnut-brown, smooth; pedicels yellowish, persistent, to 45μ long.

Hosts and distribution: Piptochaetium fimbriatum (H.B.K.) Hitchc: southern Texas, U.S.A. south to San Luis Potosi and Zacatecas, Mexico.

Type: Cummins No. 62-388 (PUR 57365) on Piptochaetium fimbriatum, Chisos Mts., Texas.

This species differs from Puccinia piptochaetii Diet. & Neger in having larger urediniospores and longer teliospores whose apical wall is thicker and lacks a differentiated umbo.

The following variety has longer spores than var. chisosensis.

PUCCINIA CHISOSENSIS Cumm. var. longa Cumm. Southw. Nat. 12:75. 1967.

Aecia unknown. Uredinia abaxial; spores (23-)25-30 x (18-)20-23(-25)μ, mostly broadly ellipsoid; wall 2-3μ thick or to 4μ at apex, dark cinnamon-brown, echinulate, pores 3(4), equatorial. Telia adaxial, early exposed, pulvinate, blackish brown; spores (35-)42-60(-68) x (15-)17-23(-25)μ, mostly ellipsoid or elongate obovoid, wall 1.5-2(-3)μ thick at sides, (4-)6-10(-14)μ at apex, chestnut-brown, smooth; pedicels yellow to brownish, to 75μ, usually shorter.

Hosts and distribution: _Piptochaetium fimbriatum_ (H.B.K.) Hitchc.: Mexico.

Type: Cummins 62-131 (=PUR 60054), Saltillo, Coahuila.

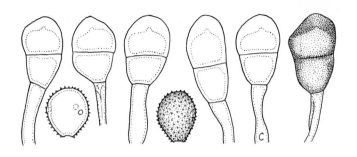

Figure 158

138. PUCCINIA EMACULATA Schw. Amer. Philos. Soc. Proc. II. 4:295.
1832. Fig. 158.

Puccinia graminis Pers. var. brevicarpa Pk. N.Y. State Mus.
Ann. Rept. 25:122. 1873.

Uredo sphaerospora Berk. & Curt. in Cooke Grevillea 20:110.
1892.

Puccinia panici Diet. Erithea 3:80. 1895.

Uredo panici-urvilleani Diet. & Neg. Bot. Jahrb. 27:15. 1899.

Puccinia pammellii Arth. J. Mycol. 11:56. 1905.

The aecial stage, Aecidium pammellii Trel., occurs on species
of Euphorbia; spores 20-32 x 16-23μ; globoid or ellipsoid, wall
1.5-2μ thick, finely verrucose, hyaline. Uredinia mostly on
adaxial leaf surface, cinnamon-brown; spores (19-)21-27(-30) x
(17-)20-24μ, mostly broadly ellipsoid or globoid, wall 1.5-2μ
thick, echinulate, cinnamon-brown, germ pores 3 or sometimes 4,
equatorial. Telia on adaxial surface, early exposed, pulvinate,
blackish brown; spores (27-)33-44(-49) x (15-)17-21(-24)μ mostly
ellipsoid or narrowly obovoid, wall 2.5-3.5μ thick at sides, 3-9μ
apically, chestnut-brown, smooth; pedicels colorless, thin-walled
and mostly collapsing, to 80μ long.

Hosts and distribution: species of Panicum, Paspalum stramineum
Nash, Sacciolepis striatus Nash: the United States east of the
Continental Divide, Northern Mexico, and (?) Chile.

Type: von Schweinitz, on Panicum pubescens (=Panicum capillare
L.), Philadelphia, Pennsylvania, U.S.A. (PH).

Stuart (Indiana Acad. Sci. Proc. 1901:284. 1902) first
demonstrated the life cycle by inoculation, using Puccinia panici.
Similar attempts with P. emaculata, strict sense, have been
negative.

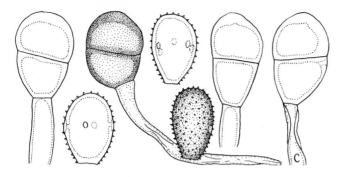

Figure 159

139. PUCCINIA KAWANDENSIS Cumm. Uredineana 4:44. 1953. Fig. 159.

Aecia unknown. Uredinia amphigenous, dark cinnamon-brown; spores (26-)28-33(-35) x 20-25(-28)μ, mostly broadly ellipsoid or obovoid, wall 2.5-3μ thick, chestnut- or dark cinnamon-brown, echinulate, germ pores 4 or 5, equatorial. Telia like the uredinia but pulvinate and blackish brown; spores 33-43(-49) x (23-)25-29(-31)μ, mostly ellipsoid or broadly ellipsoid, occasionally diorchidioid, wall 2-3μ thick at sides, 4-7μ apically, chestnut-brown, smooth; pedicels colorless or yellowish, thinwalled and mostly collapsing, to 90μ long.

Hosts and distribution: <u>Chrysopogon aucheri</u> (Boiss.) Stapf: Uganda.

Type: C.G. Hansford No. 3513, Kawanda, Uganda (PUR; isotype IMI).

Cummins (loc. cit.) published a photograph of teliospores of the type.

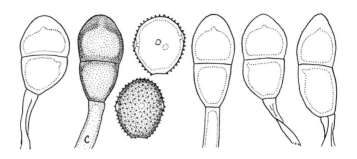

Figure 160

140. PUCCINIA SORGHI Schw. Trans. Amer. Phil. Soc. II. 4:295.
1832. Fig. 160.

Puccinia maydis Berenger Atti Soc. Ital. 6:475. 1845.

Puccinia zeae Berenger in Klotzsch Herb. Viv. Mycol. Suppl.
No. 18. 1851.

Aecia, Aecidium oxalidis Thuem., occur on species of Oxalis;
spores 18-26 x 13-19μ, mostly globoid or ellipsoid, wall 1-1.5μ
thick, pale yellowish, verrucose. Uredinia amphigenous, cinnamon-
brown; spores (24-)26-31(-33) x (21-)24-28(-30)μ, mostly broadly
ellipsoid or broadly obovoid, wall 1.5-2μ thick, golden or
cinnamon-brown, echinulate, germ pores 3 or 4 equatorial or
approximately so. Telia amphigenous, early exposed, blackish
brown, compact; spores (28-)30-42(-46) x (14-)18-23(-25)μ, oblong,
ellipsoid or obovoid, wall (1-)1.5-2(-3)μ thick at sides (4-)5-7(9)
apically, chestnut-brown or the longer narrower spores usually
golden brown, smooth; pedicels mostly thin-walled and collapsing,
pale yellowish to brownish, to 80μ long.

Hosts and distribution: Euchlaena mexicana Schrad., Zea mays
L.: worldwide where maize is grown.

Lectotype: Schweinitz, on Zea mays, Bethlehem, Pennsylvania
(PH; isotype PUR). Because Schweinitz listed "in foliis Sorghi
et Zeae cultae" designation of the lectotype is required.

Arthur (Bot. Gaz. 38:64-67. 1904) first demonstrated the life
history by inoculation. Cummins (Phytopathology 31:856-857.
1941) published a photograph of teliospores of the type.

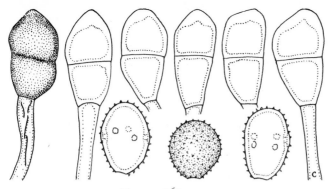

Figure 161

141. PUCCINIA ARTHURI P. Syd. & H. Syd. Monogr. Ured. 1:775.
1904. Fig. 161.

Aecia unknown. Uredinia amphigenous, yellowish brown; spores
(24-)27-33 x (18)21-24(-25)μ, mostly broadly ellipsoid or oval,
wall golden to cinnamon-brown, 1.5-2μ thick, echinulate, germ
pores 4-6, equatorial. Telia amphigenous, pulvinate, blackish
brown; spores (29-)33-43(-48) x (17-)20-24(-28)μ, mostly ovate-
oblong, or elongate obovoid, wall 2-3μ thick at sides, 4-7μ
apically, chestnut, smooth; pedicels thick-walled, collapsing
or not, colorless or yellowish, to about 100μ long.

Hosts and distribution: Pennisetum crinitum (H.B.K.) Spreng.:
Mexico.

Type: E. W. D. Holway No. 3629, Patzcuaro, Mexico (S; isotype PUR).

Ramachar and Cummins (Mycopath. Mycol. Appl. 25:7-60. 1965)
published a photograph of teliospores of the type.

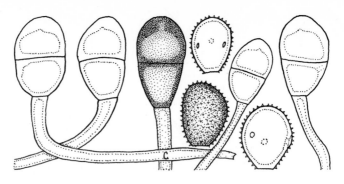

Figure 162

142. PUCCINIA CACABATA Arth. & Holw. in Arth. Amer. Phil. Soc. Proc. 64:179. 1925. Fig. 162.

Uredo chloridis-berroi Speg. Rev. Argent. Bot. 1:135. 1925.

Puccinia stakmanii Presley in Presley & King Phytopathology 33:385. 1943.

Uredo chloridis-polydactylidis Viégas Bragantia 5:82. 1945.

The aecia, Aecidium gossypii Ell. & Ev., occur on species of Gossypium; spores 16-21 x 15-16μ, wall 1-1.5μ thick, verrucose. Uredinia amphigenous, cinnamon-brown; spores (22-)24-30(-32) x (17-)19-23(-25)μ, obovoid or broadly ellipsoid, wall 1.5-2μ, cinnamon-brown, often darker apically, echinulate, pores 3, rarely 4, equatorial. Telia amphigenous and on stems, early exposed, blackish, pulvinate; spores (27-)34-40(-44) x (17-)20-24(-26)μ, ellipsoid, oblong, or broadly ellipsoid, wall 2-3(-4)μ thick laterally, 4-9 apically, mostly chestnut-brown, smooth; pedicels thick-walled, not collapsing, colorless to golden, usually to about 90μ long, much longer in occasional collections.

Hosts and distribution: species of Bouteloua, Cathesticum, Chloris: southwestern U.S.A. to the Bahamas, Bolivia, and Argentina.

Type: Holway No. 721, on Chloris ciliata, Sur Yungas, Bolivia (PUR; isotypes Reliq. Holw. No. 88).

Presley and King (Phytopathology 33:382-389. 1943) first proved the life cycle by inoculation. Hennen and Cummins (Mycologia 48:126-162. 1956) published a photograph of teliospores of the type.

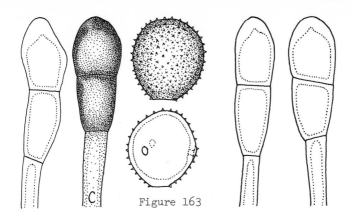

Figure 163

143. PUCCINIA PATTERSONIAE P. Syd. & H. Syd. Monogr. Ured.
1:820. 1904. Fig. 163.

Aecia unknown. Uredinia on abaxial surface of leaves,
confluent, cinnamon-brown; spores (28-)30-33(-36) x (25-)
27-32μ, mostly globoid, wall 2-3(-3.5)μ thick, echinulate,
cinnamon-brown, germ pores 3, equatorial. Telia mostly on
abaxial surface, extensively confluent, blackish brown, early
exposed, pulvinate; spores (38-)40-54(-62) x (15-)18-22(-25)μ,
mostly oblong-ellipsoid or narrowly obovoid, wall 1.5-2(-3)μ
thick at sides, (4-)6-8(-10)μ apically, chestnut-brown, smooth;
pedicels rather thick-walled, collapsing or not, yellowish to
golden, to 70μ long.

Hosts and distribution: Tripsacum dactyloides L.: Maryland
and North Carolina west to Indiana and Texas.

Type: Varney, on Tripsacum dactyloides, Manhattan, Kansas
(S; isotype Sydow Uredineen No. 1729).

This species has been submerged since Arthur (N. Amer.
Flora 7:279. 1920) treated it as a synonym of Puccinia
tripsaci, but it has longer and narrower teliospores.

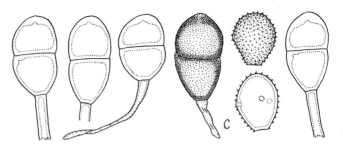

Figure 164

144. PUCCINIA KAKAMARIENSIS Wakef. & Hansf. Linn. Soc. Lond.
Proc. 161:181. 1949. Fig. 164.

Aecia unknown. Uredinia amphigenous or mostly on adaxial
surface, yellowish; spores (18-)20-24(-27) x (16-)17-20(-22)μ,
mostly pyriform or ellipsoid, wall 1-1.5μ thick, colorless,
finely echinulate, pores 3 or 4, equatorial. Telia amphigenous,
blackish, early exposed, pulvinate, compact; spores (25-)32-42
(-47) x (18-)22-26(-28)μ, mostly ellipsoid or oblong-ellipsoid,
wall chestnut-brown, (2-)2.5-3(-4)μ thick at sides, (3-)4-6
(-8)μ apically, smooth; pedicels nearly colorless, thin-walled
and collapsing, sometimes inserted obliquely, to 150μ long;
occasional spores strongly diorchidioid.

Hosts and distribution: Sporobolus filipes Stapf, S.
fimbriatus Nees, S. fimbriatus var. latifolius Stent, S.
panicoides A. Rich.: Uganda, Nyasaland, Kenya, and Union of
South Africa.

Type: Liebenberg No. 1774, on Sporobolus sp. (=S. filipes),
Kakamari Uganda (K).

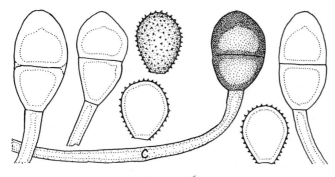

Figure 165

145. PUCCINIA ARUNDINELLAE Barcl. J. Asiatic Soc. Bengal
58:245. 1889. Fig. 165.

Uredo pretoriensis H. Syd. & P. Syd. Ann. Mycol. 10:34.
1912.

Aecia unknown. Uredinia mostly adaxial, pale yellow when
dry, doubtless bright yellow when fresh; spores 23-30(-32) x
(18)20-24μ, mostly broadly ellipsoid or obovoid, wall colorless, 1.5-2(-2.5)μ thick, echinulate, pores very obscure but
possibly equatorial. Telia amphigenous, blackish brown,
compact, early erumpent; spores (35-)40-56(-62) x (18-)22-30
(-32)μ, wall (2-)2.5-4(-5)μ thick at sides, (5-)7-9(-11)μ
apically, clear chestnut-brown, often progressively paler
externally at apex, smooth; pedicels colorless or yellowish,
thick-walled, persistent, to 160μ long.

Hosts and distribution: Arundinella bengalensis (Spreng.)
Druce, A. ecklonii Nees, A. nepalensis Trin., A. sp.: South
Africa, India and Burma.

Neotype: Barclay, on Arundinella bengalensis (as A.
wallichii Nees), Simla, India (S); designated by Cummins and
Greene (Trans. Jap. Mycol. Soc. 7:52-57. 1966) who published
a photograph of the teliospores.

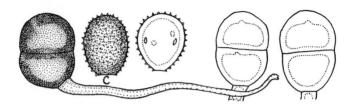

Figure 166

146. PUCCINIA BURNETTII Griff. Bull. Torrey Bot. Club 29:298.
1902. Fig. 166.

Aecia, Aecidium eurotiae Ell. & Ev., occur on species of
Eurotia; spores 19-24 x 16-19µ, globoid or ellipsoid, wall 1-1.5µ
thick, finely verrucose, hyaline. Uredinia unknown; spores
(24-)28-30(-34) x (20-)21-25(-30)µ, mostly broadly ellipsoid,
wall 3-3.5µ thick, yellow or golden, echinulate, germ pores 4
or 5, equatorial. Telia on adaxial leaf surface, exposed,
deeply pulvinate, chocolate-brown, to 2 cm long; spores (32-)
26-41(-47) x (22-)25-28(-31)µ, mostly oblong-ellipsoid, wall
2-3µ thick at sides, 4-6µ apically, clear chestnut-brown,
smooth; pedicels colorless, thick-walled but often collapsing
laterally, to 200µ long but usually broken shorter.

Hosts and distribution: Oryzopsis hymenoides (Roem. & Schult.)
Ricker, species of Stipa: western U.S.A. and in the U.S.S.R. and
Iran.

Type: Griffiths, on Stipa comata Trin. & Rupr., Buffalo,
Wyoming, U.S.A. (WIS; isotypes Griff. West Amer. Fungi No. 387).

A photograph of teliospores of the type was published by
Greene and Cummins (Mycologia 50:6-36. 1958).

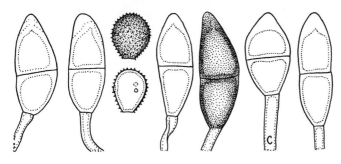

Figure 167

147. PUCCINIA ENTRERRIANA Lindq. Bol. Soc. Argent. Bot. 5:157.
1954. Fig. 167.

Aecia unknown. Uredinia not seen; spores ellipsoid or
obovoid, (16-)20-24 x (16-)18-21μ, wall 1.5-2μ, cinnamon-brown,
echinulate, pores 3, equatorial. Telia on adaxial leaf surface,
early exposed, deeply pulvinate, chocolate, attaining a length
of 1 cm, spores ellipsoid or fusiform-ellipsoid, (33-)38-52(-63)
x (14-)16-22(-24)μ, wall 1.5-2(-2.5)μ at sides, 4-10(-12)μ
apically, golden or clear chestnut-brown, smooth; pedicel colorless or pale yellowish, thin-walled and collapsing laterally,
attaining a length of 150μ but usually broken shorter.

Hosts and distribution: Stipa sp.: Argentina.

Type: Hirschhorn (LaPlata Museum No. 6100), Parera, Prov.
Entre Rios (LPS).

A photograph of teliospores of the type was published by
Greene and Cummins (Mycologia 50:6-36. 1958).

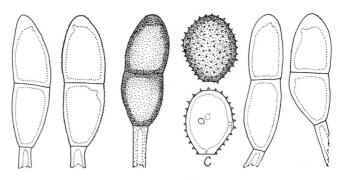

Figure 168

148. PUCCINIA ARUNDINARIAE Schw. Schr. Nat. Ges. Leipzig 1:72. 1822. Fig. 168.

The aecia (Aecidium smilacis Schw.) occur on species of Smilax; spores (23-)25-29(-32) x (19-)23-28μ, globoid or broadly ellipsoid, wall 1-1.5μ thick, colorless, verrucose. Uredinia on abaxial leaf surface, cinnamon-brown; spores (26-)30-36 x (22-) 26-30(-34)μ, mostly broadly obovoid, wall 2-2.5(-3)μ thick, golden becoming dark cinnamon-brown, rather sparsely echinulate, pores 3 or 4, equatorial. Telia on abaxial surface, early exposed, blackish brown; spores tending dimorphic with the shorter spores usually broader and darker brown, 38-65(-75) x (16-)20-26μ, mostly ellipsoid or oblong-ellipsoid, wall 2-2.5 (-3)μ thick at sides, (4-)5-7(-10)μ apically, minutely punctate-verrucose, especially in short dark brown spores, less so or smooth in elongate pale spores, from golden to clear chestnut-brown; pedicels colorless, not collapsing, to 160μ long.

Hosts and distribution: Arundinaria tecta (Walt.) Muhl.: the United States from North Carolina west to Texas.

Type: von Schweinitz, on Arundinaria (= A. tecta, Salem, North Carolina (PH).

Cummins (unpublished; data in PUR) demonstrated the life cycle using telia from South Carolina to inoculate Smilax in an out-of-doors experiment in Indiana.

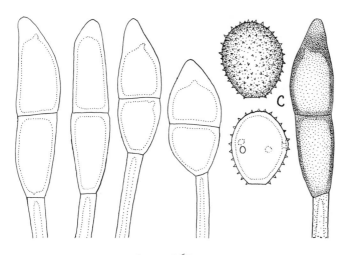

Figure 169

149. PUCCINIA KUSANOI Diet. Bot. Jahrb. 27:568. 1899. Fig. 169.

The aecia (<u>Aecidium</u> <u>deutziae</u> Diet.) occur on species of <u>Deutzia</u>, spores 22-28 x (17-)19-24μ, wall 1-1.5μ thick, nearly colorless, verrucose. Uredinia on abaxial leaf surface, cinnamon-brown; spores (27-)29-34(-36) x (22-)24-28(-30)μ, broadly ellipsoid or obovoid, wall 2-3(-4)μ thick, golden brown, echinulate, pores (3)4(5), equatorial. Telia on abaxial surface, early exposed, blackish brown; spores tending dimorphic, (44-)50-78(-86) x (14-)17-21(-24)μ, wall (1.5-)2-2.5(-3.5)μ at sides, 6-12(-15)μ apically, clear chestnut-brown or the shorter broader spores darker, minutely punctate-verrucose (expecially the robust spores) or smooth; pedicels colorless or yellowish, thick-walled, not collapsing, to 200μ long.

Hosts and distribution: species of <u>Nipponobambusa</u>, <u>Phyllosta-chys</u>, <u>Pleioblastus</u>, <u>Pseudosasa</u>, <u>Sasa</u>, <u>Sasaella</u>, <u>Semiarundinaria</u>, <u>Sinobambusa</u>: Japan and China to Taiwan.

Type: S. Kusano No. 10, on <u>Arundinaria</u> <u>simoni</u> (=<u>Pleioblastus</u> <u>simoni</u>), Bot. Gard., Tokyo, 13 Dec. 1897 (S.). This is the only specimen in the Dietel Herbarium marked "n. sp." in Dietel's script, hence is considered to be the holotype.

S. Uchida (Mem. Mejiro Gakuen Woman's Junior Coll. 2:21-28. 1965) has listed the many species of host plants.

The species is similar to <u>P. arundinariae</u>. Asuyama (Ann. Phytopathol. Soc. Jap. 6:27-29. 1936.) proved the life history.

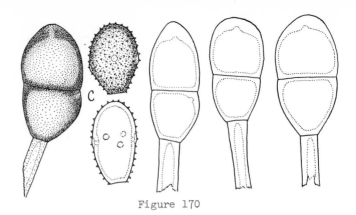

Figure 170

150. PUCCINIA CRYPTANDRI Ell. & Barth. Erythea 5:47. 1897 var. cryptandri. Fig. 170.

Uromyces simulans Pk. Bot. Gaz. 4:127. 1897 (based on uredia).

Puccinia simulans (Pk.) Barth. N. Am. Ured. No. 32. 1922.

Spermogonia and aecia unknown. Uredinia on adaxial surface and caulicolous, oblong, cinnamon-brown; spores (24-)28-36(-45) x (17-)21-26(-30)µ, mostly ellipsoid or oblong, wall yellowish to cinnamon-brown, (1-)1.5-3(-4)µ thick, rather coarsely echinulate germ pores 4-6(-8), mainly equatorial but scattered in occasional spores. Telia mostly on adaxial surface, oblong, blackish, early exposed, pulvinate, compact; spores (32-)38-46(-56) x (22-)25-30 (-36)µ, mostly ellipsoid or oblong-ellipsoid, wall chestnut-brown, 1.5-2.5(-3.5)µ thick at sides, 4-8(-10)µ apically, smooth; pedicels colorless or tinted, thick-walled, not collapsing, to at least 125µ long; 1-celled teliospores often present.

Hosts and distribution: Sporobolus contractus Hitchc., S. cryptandrus (Torr.) Gray: U.S.A. from Wisconsin to Montana and south to Texas, Arizona, and northern Mexico.

Type: E. Bartholomew No. 2264, on S. cryptandrus, Rockport, Kansas, 16 Sept., 1896 (FH).

Variety luxurians (see p. 383) is similar except that it has scattered pores.

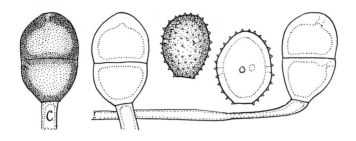

Figure 171

151. PUCCINIA MOLINIAE L. Tul. Ann. Sci. Nat. Bot. IV. 2:141. 1854. Fig. 171.

Puccinia nemoralis Juel Oefvers. Kongl. Vetensk.-Akad. Foerh. 51:506-507. 1894.

Puccinia brunellarum-moliniae Cruchet Centrlb. Bakt. II. 13:96. 1904.

Aecia (Aecidium melampyri Kunze & Schm.) occur on species of Melampyrum, Origanum, and Prunella; spores (15-)16-20 x (12-)14-18μ, mostly nearly globoid, wall 1-1.5μ thick, colorless, verrucose. Uredinia mostly on adaxial leaf surface, yellowish brown; spores (23-)25-30(-33) x (20-)22-26(-29)μ, mostly broadly obovoid or globoid, wall 3.5-4.5(-5)μ thick, golden or pale yellowish brown, echinulate, germ pores 3 or 4, equatorial. Telia mostly on adaxial surface, early exposed, confluent, chocolate-brown, rather loose, almost pulverulent; spores (32-)36-48(-56) x (20-)24-28(-32)μ, mostly broadly ellipsoid, wall (2.5-)3-4(-5)μ thick at sides, (4-)5-8(-10)μ apically, chestnut-brown, smooth; pedicels yellowish or colorless, thin-walled and collapsing, to 200μ long but usually shorter.

Hosts and distribution: Molinia coerulea (L.) Moench.: Europe; also recorded for China.

Neotype: Specimen in PC with original script label "Puccinia graminis Pers. in foliis variorum graminum. Autumno 1831--. ubiqui" and at bottom of the label: "Puccinia Moliniae Tul. in Ann. Sc. Nat. Ser. 4, t. 2, 1854, cum descript. Leveille." Neotype designated now, there being no assurance that a more authentic specimen exists.

Juel (loc. cit.) first demonstrated the life cycle, using Melampyrum pratense as the aecial host.

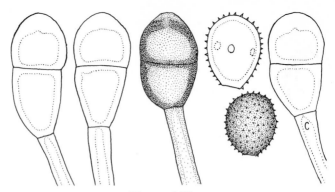

Figure 172

152. PUCCINIA SETARIAE-LONGISETAE Wakef. & Hansf. Linn. Soc.
Lond. Proc. 161:186. 1949. Fig. 172.

Aecia unknown. Uredinia amphigenous or mainly on abaxial surface, cinnamon-brown; spores (26-)27-31 x (20-)22-24µ, mostly
broadly ellipsoid, wall 2-2.5µ thick, cinnamon-brown, echinulate,
germ pores 3 or 4, equatorial. Telia mostly on abaxial surface,
exposed, pulvinate, blackish brown; spores (37-)40-50(-57) x
(20-)24-27(31)µ, ellipsoid or oblong-ellipsoid, wall 2-3µ thick
at sides, 5-10µ apically, chestnut-brown, smooth; pedicels golden,
thick-walled, mostly not collapsing, to 120µ long.

Hosts and distribution: Setaria kagerensis Mez, S.
longiseta Beauv.: Uganda.

Type: C. G. Hansford No. 960, on S. longiseta, Kabale,
Kigezi, Uganda (K).

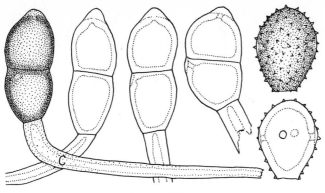

Figure 173

153. PUCCINIA PHRAGMITIS (Schum.) Koern. Hedwigia 15:179. 1876 var. phragmitis. Fig. 173.

Uredo phragmitis Schum. Enum. Pl. Saell. 2:231 1803; telia described.

Puccinia arundinacea Hedw. in Lam. Encycl. Meth. Bot. 8:250. 1808 in part (nomen confusum).

Puccinia trailii Plowr. Monogr. Brit. Ured. Ustil. p. 176. 1889.

Puccinia desmazieresi Constan. Ann. Mycol. 14:251. 1916.

Aecia, Aecidium rubellum Pers., occur on species of Fagopyrum, Polygonum, Reynoutria, Rumex, and Rheum; spores 18-23 x 15-19μ, ellipsoid or broadly so, wall (1-)1.5(-2)μ thick, colorless, prominently verrucose, commonly in a band. Uredinia amphigenous, cinnamon-brown; spores (23-)26-33(-36) x (18-)20-24(-26)μ, ellipsoid or obovoid, wall 2.5-4μ thick, yellow to golden brown, echinulate, germ pores (3)4 or 5(6), equatorial. Telia amphigenous, exposed, large, deeply pulvinate, chocolate-brown; spores (36-)40-60(-66;-74) x (16-)19-24(-28)μ, ellipsoid, wall (2-)2.5-3.5(-4)μ thick at sides, 5-8(10)μ apically, the apex usually a paler umbo, deep golden brown to clear chestnut-brown, long narrow spores usually are paler than the robust ones, smooth; pedicels persistent, colorless or tinted, thick-walled, not collapsing, to 200μ long.

Hosts and distribution: Species of Phragmites: circumglobal.

Neotype: Koernicke, on Phragmites communis, Waldau (Ostprussen) (B). Neotype designated here because original specimen is not in the Schumacher herbarium (in C), according to Prof. Skovsted (in litt.).

Winter (Hedwigia 14:113-115. 1875) first demonstrated the life cycle, using Rumex hydrolapathum as the aecial host.

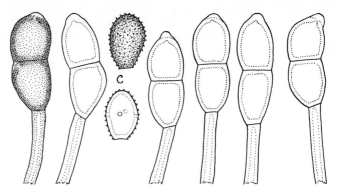

Figure 174

PUCCINIA PHRAGMITIS (Schum.) Koern. var. longinqua (Cumm.) Cumm. stat. nov. Fig. 174.

Puccinia longinqua Cumm. Mycologia 43:91. 1941.

Urediniospores (19-)21-26(-28) x (13-)16-19μ, mostly ellipsoid, wall 2.5-3μ thick, rarely to 3μ apically, cinnamon-brown, echinulate, germ pores 3 or 4, equatorial. Telia amphigenous and on sheaths, spores (33-)40-54 x (16-)18-21μ, ellipsoid, wall 2.5-3μ thick at sides, 4-6μ apically, chestnut-brown, smooth; pedicels brownish or nearly hyaline, thick-walled, not collapsing, to 200μ long.

Hosts and distribution: Phragmites sp.: China.

Type: S. Y. Cheo No. 2431, Ta Tseh Shan, Yung Hsien, Kwangsi Prov., China (PUR).

The small urediniospores and narrow teliospores distinguish this variety from the typical.

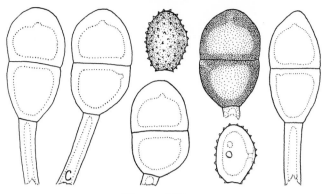

Figure 175

154. PUCCINIA ISIACAE Wint. in O. Kuntze Plantae orient.-ross.
p. 127. 1887. Fig. 175.

Puccinia arundinacea var. obtusata Otth in Trog Mitth. Naturf.
Ges. Bern 1857:46. 1857.

Uredo isiacae Thuem. Grevillea 8:50. 1879.

Puccinia sinkiangensis Wang Acta Phytotax. Sinica 10:295.
1965.

Puccinia obtusata (Otth) Ed. Fisch. Beitr. Kryptog. Schweiz
1(1):57. 1898.

Puccinia inulae-phragmiticola Tranz. Trav. Mus. Bot. Acad.
Imp. Sci. St. Petersb. 3:53. 1906.

The aecia (Aecidium ligustri Strauss) occur on Oleaceae,
Cruciferae and some 20 other families; spores 16-24 x 12-17μ,
ellipsoid or broadly ellipsoid, wall 1.5μ thick, finely verrucose,
hyaline. Uredinia amphigenous, confluent in large groups; spores
(22-)24-29(-31) x (17-)20-23μ, ellipsoid or broadly ellipsoid,
wall (3.5-)4-5μ except 5-7μ around the pores and at the apex,
golden or pale cinnamon-brown, echinulate with short broad-
based spines spaced 3-4μ on centers, pores 3, equatorial. Telia
amphigenous and on the sheaths, early exposed, about chocolate-
brown, confluent in large areas up to some 10 cm long; spores
(33-)37-48(-53) x (21-)24-29(-32)μ, mostly ellipsoid, wall
(3-)3.5-4(-5)μ thick at sides, (4.5-)5-7(-8)μ at apex, uniformly
chestnut-brown or the apex slightly paler, smooth; pedicels
thick-walled, not collapsing, hyaline, to 200μ long.

Hosts and distribution: Phragmites communis Trin., P. maximus
(Forsk.) Chiov.: Spain and Morocco to Germany and southern
U.S.S.R.

Type: Kaernbach, on Arundo phragmites (= Phragmites communis),
Kasandschick, Turkmenia, U.S.S.R. (S).

A comparison of the type specimens of P. isiacae and P.
obtusata indicates that the two species are indistinguishable on
the basis of urediniospores and teliospores. The two species

275

usually have been maintained because basidiospores from European
telia (see Bock, Centralbl. Bakt. II. 20:564-592. 1908) infect
Ligustrum but not the aecial hosts (of 8 families) that Tranzschel (Trav. Mus. Bot. Acad. Imp. Sci. St. Petersbourg 3:37-55.
1906) successfully infected using telia from southern Russia.
Guyot and Malençon (Trav. Inst. Sci. Chérifien Série Bot. No. 11.
181 pp. 1957) also produced aecia on a number of hosts, other
than _Ligustrum_, using telia from Morocco. For a summary see
Gäumann (Die Rostpilze Mittleuropas, pp. 751-755. 1959) who
used the name _Puccinia trabutii_. His figure 615 does not apply
to either _P. isiacae_ or _P. trabutii_. More recently, Mayor (Bull.
Soc. Bot. Suisse 77:128-155. 1967), using Swiss telia, successfully infected several species of _Forsythia_, _Fraxinus_, _Ligustrum_,
and _Syringa_.

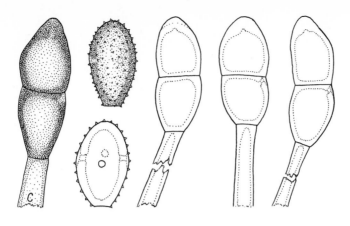

Figure 176

155. PUCCINIA TOROSA Thuem. Mycoth. Univ. No. 1725. 1880.
Fig. 176.

Aecial stage unknown. Uredinia (few seen) mostly on adaxial
surface, about cinnamon-brown; spores (27-)29-35(-37) x (16-)19-
23µ, wall 3-4(-5)µ thick, golden brown, echinulate with short,
broad-based spines spaced about 3-3.5µ on centers, pores 4,
equatorial. Telia amphigenous and on sheaths, early exposed in
large blackish or chocolate-brown confluent groups up to 3 cm
long; spores (44-)50-64(-70) x (19-)22-28(-30)µ, ellipsoid,
wall 2.5-3µ thick at sides, (4-)6-8(-9)µ apically, golden or
clear chestnut-brown, smooth; pedicels persistent, thick-walled
and not collapsing, brownish, to 250µ long.

Hosts and distribution: Arundo donax L.: South Africa.

Type: MacOwan, on A. donax (as Donax arundinacea), Somerset
East, South Africa (Thuem. Mycoth. Univ. No. 1725).

The species is generally similar to Puccinia phragmitis but
the teliospores are broader.

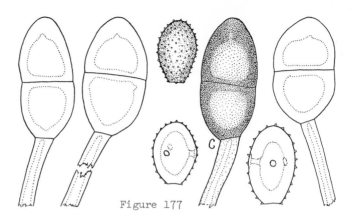

Figure 177

156. PUCCINIA TRABUTII Roum. & Sacc. in Saccardo Michelia 7:307.
1880 var. trabutii. Fig. 177.

Aecia unknown. Uredinia not seen; spores in the telia
(23-)26-32(-36) x (18-)20-24(-27)μ, mostly ellipsoid or broadly
ellipsoid, wall (3-)3.5-4.5(-5)μ thick, yellowish or golden,
echinulate, germ pores 3(4), equatorial. Telia amphigenous and
on sheaths, early exposed, confluent in areas to 8 cm, felt-like,
chocolate-brown; spores (40-)48-60(-68) x (20-)24-30(-33)μ,
mostly ellipsoid, wall (4-)5-7(-8)μ thick at sides, (9-)10-12(-14)μ
at apex, clear chestnut-brown or golden brown, smooth; pedicels
hyaline, thick-walled, not collapsing, to at least 250μ long.

Hosts and distribution: Arundo donax L., Phragmites communis
Trin., P. gigantea J. Gay, P. karka (Retz.) Trin., P. maximus
(Forsk.) Chiov.: Morocco to southern U.S.S.R. and West Pakistan.

Type: Trabut, on Phragmites gigantea, Algeria (PAD; isotype
S).

There is no acceptable evidence that the aecial stage is known
but the species has been confused with P. isiacae and, hence,
aecial hosts have sometimes been assigned to it.

Tranzschel (Trav. Mus. Bot. Acad. Imp. Sci. St. Petersbourg
3:37-55. 1906) compared "Original-Exemplaren" of P. obtusata,
P. isiacae, and P. trabutii and concluded that, although the
teliospores varied in size and shape, the three were not distinguishable. There is considerable similarity but the teliospores
of P. trabutii average about 10μ longer and the wall nearly
twice as thick as in P. isiacae.

The following variety, for which uredinia and urediniospores
are not known, has equally conspicuous telia, and similar but
narrower teliospores.

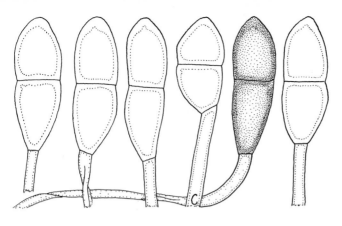

Figure 178

PUCCINIA TRABUTII Roum. & Sacc. var. abei (Hirat. f.) comb.
nov. Puccinia abei Hiratsuka f. J. Jap. Bot. 13:249. 1937.
Fig. 178.

PUCCINIA TRABUTII Roum. & Sacc. var. abei (Hirat. f.) comb.
nov. Fig. 178.

Puccinia abei Hiratsuka f. J. Jap. Bot. 13:249. 1937.

Aecia, uredinia, and urediniospores unknown. Telia on the
culms, early exposed, compact, chocolate-brown, confluent in a
group to 7 cm long; spores (44-)48-66(-70) x 20-27μ; wall uniformly (2.5-)3-4μ thick, golden or chestnut-brown; pedicels
hyaline, collapsing laterally, slender, seen to 325μ long.

Hosts and distribution: Phragmites longivalvis Steud.: Japan.

Type: G. Yamada, Nonodake-mura, prov. Rikuzen, Japan (Herb.
Hiratsuka; isotype PUR).

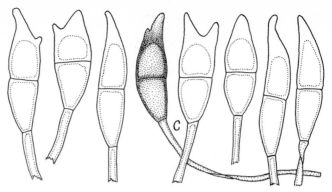

Figure 179

157. PUCCINIA ASPERELLAE-JAPONICAE Hara Trans. Agr. Soc. Shizuoka Pref. No. 286:47. 1921. Fig. 179.

Aecia unknown. Uredinia not seen; spores in the telia old and collapsed, approximately 18-20 x 16-18μ, broadly ellipsoid or globoid, wall 1.5μ thick, yellowish, echinulate, germ pores obscure, probably scattered. Telia on adaxial leaf surface, early exposed, compact, from cinnamon-brown to chocolate-brown; spores (32-)40-65(-72) x (12-)14-22(-24)μ, variable, the long spores mostly ellipsoid or fusiform-ellipsoid, short spores wedge-shaped or obovoid, wall (1-)1.5-2(-3)μ thick at sides, usually thicker in the short than in the long spores, mostly 4-6μ apically (excluding digitations) and with 2-5 digitations up to 12μ long, the long spores typically have an elongate solid apex up to 20μ long but may have digitations in addition, golden brown or clear chestnut-brown, the pigmentation apparently developing slowly, smooth; pedicels persistent, yellowish, narrow, collapsing or not, to 150μ long but mostly less than 100μ.

Hosts and distribution: <u>Hystrix japonica</u> (Hack.) Ohwi: Japan.

Type: Hara, on <u>Asperella japonica</u> (=<u>Hystrix japonica</u>) Kawakami-mura, Prov. Mino, Japan, 1913 (SAPA; isotype PUR).

Although many spores have digitate processes, the species obviously has no relationship with <u>P</u>. <u>coronata</u>.

Ito and Murayama (Trans. Sapporo Nat. Hist. Soc. 17:167. 1943) provided a Latin diagnosis but Hara's publication in Japanese is valid.

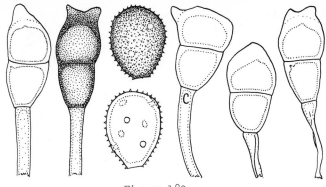

Figure 180

158. PUCCINIA NEOCORONATA H. C. Greene & Cumm. Mycologia 50:25. 1958. Fig. 180.

Aecia unknown. Uredinia amphigenous; spores (25-)28-33(-36) x (21-)23-26(-29)μ, wall 1.5-2(-2.5)μ thick, cinnamon-brown, echinulate, pores 5-7, scattered. Telia amphigenous, early exposed, pulvinate, blackish; spores (30-)37-46(-60) x (14-) 17-23(-26)μ, mostly clavate or oblong-clavate, wall 1.5-2.5(-3.5)μ thick at sides, (4-)6-9(-13)μ apically, chestnut-brown, smooth, the apex typically with (0-)2-3(-5) pale projections; pedicels yellowish, mostly 20-40μ in length.

Hosts and distribution: Piptochaetium fimbriatum (H. B. K.) Hitchc., Stipa pringlei Scribn.: southern Arizona and northern Mexico.

Type: W. G. and Ragnild Solheim No. 2453, Santa Catalina Mts., Pima County, Arizona (PUR52460; isotypes Solheim Mycofl. Saximont. Exs. No. 589, as P. stipae Arth.)

Greene and Cummins (loc. cit.) published a photograph of teliospores of the type.

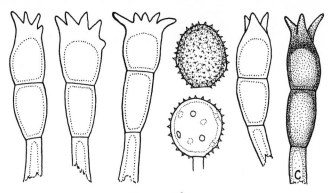

Figure 181

159. PUCCINIA FESTUCAE Plowr. Grevillea 21:109. 1893. Fig. 181.

Puccinia festucae Plowr. Gard. Chron. III. 8:139. 1890, nom. nudum.

Uredo festucae DC. Fl. Fr. 6:82. 1815.

Uredo festucae-ovinae Eriks. Ark. Bot. 18:13. 1923.

Aecia (Aecidium periclymeni Schum.) occurs on species of Lonicera; spores 17-21(-23) x (18-)20-27(-29)μ, wall 1.5μ thick, verrucose. Uredinia on the adaxial leaf surface, yellow; spores 24-29(-32) x (18-)22-25(-28)μ, broadly ellipsoid or globoid, wall 1.5-2μ thick, yellowish to golden, echinulate, germ pores 6-8, scattered; pedicels often tend to be persistent. Telia on adaxial surface, early exposed, blackish brown; spores (38-)42-58(-62) x 14-18μ excluding digitations, wall 1-1.5(-2)μ thick at sides, 2-5μ apically excluding digitations, digitations usually 3-5, mostly 8-20μ long, occasionally the apex merely elongate, chestnut-brown, smooth; pedicels brown, rather thick-walled,· not collapsing, to about 20μ long.

Hosts and distribution: Festuca altaica Trin., F. ovina L., F. rubra L.: Europe to India, Korea, and Alaska.

Type: Plowright, on Festuca ovina (=error for F. rubra according to C.E. Hubbard), Ashwicken Fen, Norfolk, England (K).

Plowright (Gard. Chron. III. 8:139. 1890) first demonstrated the life history of this species and the type is the telial specimen used.

The North American records consist of a few collections of aecia and one specimen (uredinia only) on Festuca altaica from Alaska in which the spores are like European material.

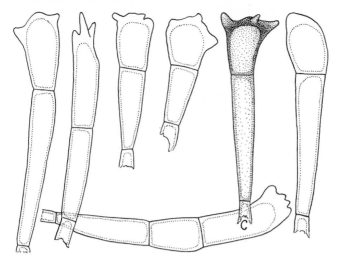

Figure 182

160. PUCCINIA LEPTOSPORA Ricker J. Mycol. 11:114. 1905. Fig. 182.

Aecia unknown. Uredinia unknown; spores occasionally in the telia 24-29 x 17-20μ, wall 1.5-2μ thick, colorless, finely echinulate, pores about 8, scattered, obscure. Telia mostly on abaxial leaf surface, blackish brown, early exposed, compact; teliospores (62-)85-140(-165) x (12-)16-20(-23)μ, cylindrical or cylindrical-clavate, wall (1-)1.5(-3)μ thick at sides, 3-5(-7)μ apically, provided apically with digitate processes 3-10(-13)μ long; chestnut-brown or golden, paler in lower cell, smooth; pedicels yellowish, 20μ or less long.

Hosts and distribution: Trisetum virletii Fourn.: Mexico.

Type: C. A. Purpus, on Trisetum virletii, Ixtaccihuatl, Federal Distr., Mexico, 1903 (WIS; isotype PUR).

The teliospores are coronate but their length greatly exceeds those of Puccinia coronata Cda. sensu lat.

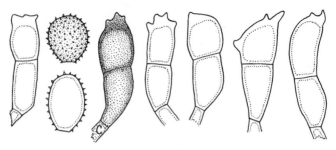

Figure 183

161. PUCCINIA PRAEGRACILIS Arth. Bull. Torrey Bot. Club 34:585.
1907 var. praegracilis.   Fig. 183.

Aecia (Aecidium graebnerianum P. Henn.) occur on species of
Habenaria and Orchis, cupulate; spores 16-21(-23) x (13-)15-18μ,
mostly globoid or broadly ellipsoid, wall about 1μ thick, colorless, verrucose. Uredinia mostly on adaxial leaf surface, yellow;
spores (17-)19-22(-23) x (12-)16-19(-21)μ, mostly broadly
ellipsoid or broadly obovoid, wall 1.5μ thick, pale yellowish
or nearly colorless, echinulate, germ pores scattered, about 6
or 7, very obscure. Telia amphigenous or mostly on abaxial
surface, covered by the epidermis, with brownish stromatic
paraphyses tending to divide the sorus into locules; spores
(28-)35-48(-56) x (11-)13-17μ, mostly nearly cylindrical or
elongate-obovoid, wall 0.5-1μ thick at sides, 2.5-4(-6)μ
apically excluding the digitations, golden brown or clear
chestnut-brown apically, smooth except for a few digitations
2-5μ long at the summit of the spore; pedicels mostly less
than 15μ long.

Hosts and distribution: Agrostis thurberiana Hitchc.:
Western Canada. Aecia are known from adjacent U.S.A.

Type: Holway, Glacier, B. C., Canada (PUR 21988).

The following varieties have been established but it is
probable that, with additional collections, the slight distinc-
tions may disappear.

PUCCINIA PRAEGRACILIS Arth. var. cabotiana Savile Can. J.
Bot. 35:199. 1957.

Aecia as in var. praegracilis. Urediniospores as in var.
praegracilis. Telia as in var. praegracilis; spores slightly
longer, (33-)37-56(-61) x 11.5-15μ.

Hosts and distribution: Hierochlöe odorata (L.) Beauv.:
Eastern Canada.

Type: Savile 3291, Sugar Loaf, Victoria County, Nova Scotia
(DAOM).

PUCCINIA PRAEGRACILIS Arth. var. connersii (Savile) Savile
Mycologia 43:458. 1951.

Puccinia connersii Savile Mycologia 42:665. 1950.

Aecia as in var. praegracilis. Uredinia as in var. praegracilis.
Telia as in var. praegracilis; spores somewhat shorter, (23-)25-
38(-42) x (12-)13-17(-20)µ.

Hosts and distribution: Deschampsia atropurpurea (Wahl.)
Steele: Eastern Canada.

Type: Savile, Great Whale River, Quebec (DAOM 23446; isotype
PUR).

The relationship of the aecial and telial stages was suggested
by Holway when he collected the type of P. praegracilis.
Subsequent field observations by Savile have substantiated
Holway's suggestion.

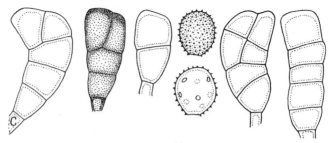

Figure 184

162. PUCCINIA TOMIPARA Trel. Trans. Wisconsin Acad. Sci. Arts, Letters 6:127. 1885. Fig. 184.

Rostrupia tomipara (Trel.) Lagerh. J. Bot. Fr. 3:189. 1889.

Aecia on species of Thalictrum; spores (16-)20-24(-27) x (14-)16-19(-21)μ, mostly broadly ellipsoid or globoid, wall 1-1.5μ thick, verrucose. Uredinia mostly on adaxial leaf surface, yellowish brown; spores (18-)22-27(-30) x (16-)18-22 (-24)μ, mostly broadly ellipsoid or globoid, wall 1-2(-2.5)μ thick, golden brown, echinulate, germ pores 7-9, scattered. Telia amphigenous or mostly on abaxial surface, covered by epidermis, loculate with brown paraphyses; spores (35-)39-48 (-53) x (14-)18-35(-40)μ, extremely variable, (2)3-7(-9)-celled, usually muriformly septate, from oblong to globoid, wall 1-2μ thick at sides, (3-)4(5) in apex of apical cells, chestnut-brown, pedicels brown, very short.

Hosts and distribution: Bromus ciliatus L., B. latiglumis (Shear) Hitchc., B. purgans L.: the western Great Lakes region to Saskatchewan.

Type: Pammel, on Bromus ciliatus La Cross, Wisconsin (WIS; isotype PUR).

This strange species has commonly been treated as a synonym of P. recondita (P. rubigo-vera) to which it doubtless is related. But it is so aberrant that it cannot be "keyed" into the genus Puccinia.

Fraser (Mycologia 11:129-133. 1919) first demonstrated the life cycle experimentally. Kaufmann (Mycopathol. Mycol. Appl. 32:249-261. 1967) published photographs of teliospores showing the variability.

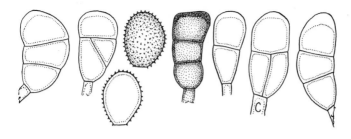

Figure 185

163. PUCCINIA AGROPYRICOLA Hirat. f. in Hiratsuka & Sato Bot.
Mag. Tokyo 64:221. 1951. Fig. 185.

Rostrupia miyabeana Ito J. Coll. Agr. Tohoku Imp. Univ.
3:243. 1909 not Miyabe 1906.

Aecia on Thalictrum thunbergii DC. var. hypoleuca Nakai but
description not published. Uredinia mostly on adaxial leaf
surface, "orange colored" (Ito); spores (18-)20-25(-28) x
17-20(-22)µ, mostly broadly ellipsoid, wall 1-1.5µ thick,
colorless or yellowish, echinulate, germ pores obscure, 7 or
8, scattered. Telia on abaxial surface, covered by the epidermis, blackish brown, tending to be loculate with brown paraphyses; spores (26-)30-46(-52) x (14-)16-22(-25)µ, variable
but mostly oblong, (1-)2 or 3(-4)-celled, the septa usually
horizontal but sometimes oblique or the lower one vertical in
3-celled spores, wall 1-1.5(-3)µ thick at sides (2.5-)3-6(-7)µ
apically, chestnut-brown, smooth; pedicels brown, to about 10µ
long.

Hosts and distribution: Agropyron ciliare (Trin.) Franch.,
A. tsukushiense (Honda) Ohwi, Brachypodium sylvaticum (Huds.)
Beauv.: Japan, Korea, and Manchuria.

Type: Yoshino, on Brachypodium sylvaticum (originally
listed as B. japonicum), Imizu-mura, Prov. Higo, Kiushu, Japan,
9 June 1904 (SAPA; isotype PUR). Ito listed 4 specimens (3
with telia) without designating a type. The above specimen was
received from Ito marked "Type collection", hence is cited as
such here.

Although Ito (loc. cit.) listed as hosts Brachypodium
japonicum and B. pinnatum, he later (Mycological Flora of Japan
Vol. II, No. 3, p. 345. 1950) lists only B. sylvaticum and
Agropyron ciliare.

The life cycle was first proved experimentally by Asuyama
(Ann. Phytopathol. Soc. Japan 4:108. 1934).

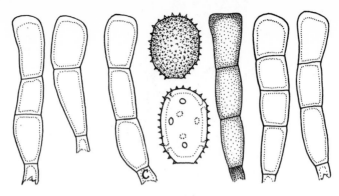

Figure 186

164. PUCCINIA ELYMI Westend. Bull. Acad. Roy. Belge 18:408.
1851 var. elymi. Fig. 186.

Rostrupia elymi (Westend.) Lagerh. J. Bot. Fr. 3:188. 1889.

Rostrupia elymi-sabulosi Tr. Savul. & O. Savul. Ann. Mycol.
35:118. 1937.

Rostrupia ammophilae Wilson Trans. Bot. Soc. Edinb. 33:iv.
1940, nomen nudem.

Aecia occur on Thalictrum, cupulate, in groups; spores
(Wilson & Henderson, British Rust Fungi) 14-28μ diam, angularly
ellipsoid or globoid, wall verrucose. Uredinia on the adaxial
leaf surface, pale cinnamon-brown; spores (25-)28-35(-38) x
(19-)22-25(-28)μ, ellipsoid, obovoid, or broadly ellipsoid,
wall (2-)2.5-3(-3.5)μ thick, yellowish to golden, echinulate,
germ pores 8-10(11), scattered or tending to be bizonate. Telia
mostly on the abaxial surface, covered by the epidermis, blackish, loculate with brown paraphyses; spores (45-)55-85(-100) x
(12-)14-18(-20)μ, 1-4- mostly 3-celled, mostly cylindrical, wall
1(-1.5)μ thick at sides, (2-)3-4(-6)μ at apex, chestnut-brown
apically, paler basally; pedicels brown, mostly less than 12μ
long.

Hosts and distribution: Ammophila arenaria (L.) Link (?),
Elymus arenarius L., E. sabulosus Marsch.-Bieb.: littoral areas
from Great Britain to the Black Sea and Omsk, U.S.S.R.

Type: Louis Landzweert, on Elymus arenarius, dunes d'
Ostende (BR; isotypes Westendorp and Wallays Pl. Crypt. Belge
No. 291. The specimen (No. 291) at BR is considered to be the
holotype).

Rostrup (Overs. Kgl. Danske Vidensk. Forh. 5:269-276. 1898)
proved the life cycle using aeciospores from Thalictrum. A
photograph of teliospores of the type was published by Cummins
and Caldwell (Phytopathology 46:81-82. 1956).

288

PUCCINIA ELYMI Westend. var. longispora var. nov.

Aeciis ignotis. Urediniis epiphyllis, flavidis; sporae (24-)27-34(-38;-43) x 20-24(-26)μ, plerumque obovoideae; membrana 1.5-2(-2.5)μ crassa, flavida, echinulata, poris germinationis 8-11, sparsis. Teliis hypophyllis, epidermide tectis, paraphysibus brunneis numerosis; sporae (52-)60-110(-128) x (12-)14-18(-20)μ, cylindraceae, (1-)3-4(-7)-septatae; membrana ad latere 1-1.5μ crassa, flavo-brunnea, ad apicem (2.5-)3-4(-6)μ crassa, pallide castaneo-brunnea; pedicello brunneo, brevissimo.

Hosts and distribution: Elymus mollis Trin., E. sibiricus L. (?): Kamchatka and Japan.

Type: S. Ito, on Elymus mollis, Prov. Shiribeshi, Japan (PUR F4122; isotypes Sydow Ured. No. 2583. The specimen (No. 2583) at PUR is considered to be the holotype).

The variety differs from the typical because of longer teliospores that have more cells and some tendency to have occasional vertical septa. The urediniospores are less pigmented and have thinner walls than in var. elymi.

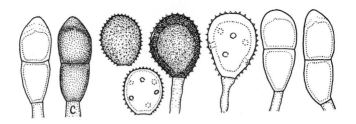

Figure 187

165. PUCCINIA SUBSTERILIS Ell. & Ev. Bull. Torrey Bot. Club
22:58. 1895 var. substerilis Fig. 187.

Uromyces scaber Ell. & Ev. J. Mycol. 6:119. 1891 (based
on amphispores).

Uredo luxurians Ell. & Ev. N. Amer. Fungi No. 3583. 1898.

Puccinia scaber (Ell. & Ev.) Barth. N. Amer. Ured. No.
2560. 1922.

Aecia unknown. Uredinia on adaxial leaf surface; spores
(20-)23-26(-30) x (16-)18-22(-25)μ, broadly ellipsoid or
globoid, wall 1-2μ thick, golden to cinnamon-brown, echinulate,
pores 6-8, scattered; amphispores (23-)25-30(-35) x (18-)20-25
(-28)μ, broadly ellipsoid or obovoid, wall (2.5-)3-3.5(-4)μ
thick or the apex to 5μ, chestnut-brown, closely echinulate,
pores (4-)5-7, scattered or tending to be equatorial when 4.
Telia on adaxial surface, early exposed, blackish, pulvinate;
spores (28-)32-40(-43) x (13-)16-19(-22)μ, mostly oblong-
ellipsoid, wall 1-1.5μ thick at sides, (4-)5-7(-9)μ apically,
golden brown, smooth; pedicels hyaline, thin-walled and
collapsing, to 80μ long but usually broken shorter; germination
occurs without dormancy.

Hosts and distribution: species of Stipa: Minnesota and
Alberta to New Mexico and Arizona.

Type: Baker, on Chrysopogon sp. (=error for Stipa viridula),
Fort Collins, Colorado (NY; isotypes, Ellis & Everh. N. Amer.
Fungi No. 3141).

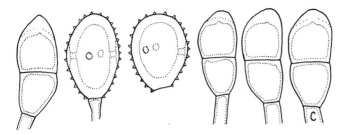

Figure 188

PUCCINIA SUBSTERILIS var. oryzopsidis H.C. Greene & Cumm. Mycologia 50:16. 1958. Fig. 188.

Similar to P. substerilis var. substerilis but urediniospores unknown; amphispores (25-)28-36(-43) x (18-)22-28(-30)μ, wall (3-)3.5-4.5(-5.5)μ or to 6μ apically, sparsely echinulate, germ pores 3 or 4(-5), equatorial; teliospores (32-)38-48(-56) x (17-)19-24(-26)μ, wall 1.5-2(-3)μ, (6-)7-9(-11)μ apically.

Hosts and distribution: Oryzopsis hymenoides (Roem. & Schult.) Ricker, Stipa arida M. E. Jones: Wyoming to Oregon southward to New Mexico and California.

Type: Bethel (Barth. Fungi Columb. No. 5075), Victorville, California (PUR).

Photographs of urediniospores and teliospores of the type were published by Greene and Cummins (loc. cit.).

PUCCINIA SUBSTERILIS var. scribneri H. C. Greene & Cumm. Mycologia 50:16. 1958.

Similar to P. *substerilis* var. *substerilis* but amphispores (23-)26-33(-38) x (19-)22-26(-30)μ, wall (3-)3.5-4(-5)μ thick or to 6μ apically, sparsely echinulate, pores (4-)5 or 6(-7), scattered; teliospores (31-)35-44(-50) x (16-)18-21(-23)μ, wall 1.5-2μ thick at sides, (5-)6-8(-9)μ, apically.

Hosts and distribution: Stipa scribneri Vasey: Colorado and New Mexico.

Type: Bethel, Manitou, Colo., April 11, 1921 (PUR).

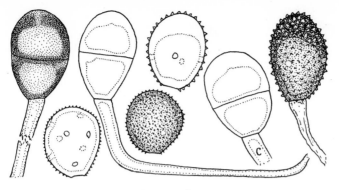

Figure 189

166. PUCCINIA VEXANS Farl. Proc. Am. Acad. 18:82. 1883.
Fig. 189.

Uromyces brandegei Pk. Bot. Gaz. 4:127. 1879 (based on amphispores).

Puccinia aristidicola P. Henn. Hedwigia 35:243. 1896.

The aecia (Aecidium cannonii Griff.) occur on Fouquieria splendens and Idria columnaris; spores 27-32(-34) x 23-27(-30)μ, wall 2.5-3.5μ thick, hyaline, verrucose. Uredinia amphigenous, cinnamon-brown; spores 26-30 x 23-29μ, globoid or broadly ellipsoid, wall (1.5-)2-3(-3.5)μ thick, cinnamon-brown, echinulate, pores 7 or 8, scattered; amphisori blackish, pulvinate; amphispores mostly obovoid, 34-42 x 26-35μ, wall 3-4μ thick laterally, 7-12μ apically, verrucose, chestnut-brown, pores 3 or 4, equatorial, pedicel usually persistent. Telia amphigenous, early exposed, pulvinate, blackish; spores 32-40 x (19-)23-29μ, mostly rather broadly ellipsoid, wall 2.5-3μ laterally, 6-8μ apically, chestnut-brown, smooth; pedicels hyaline, thick-walled, not collapsing, attaining a length of 95μ.

Hosts and distribution: Bouteloua breviseta Vasey (?), B. curtipendula (Michx.) Torr.: United States southward to Peru and Argentina.

Type: Holway (isotypes, Ellis N. Am. Fungi No. 1051), on Bouteloua curtipendula, Decorah, Iowa (FH).

Solheim and Cummins (Univ. Wyoming Publ. 23:37. 1959) proved the life cycle by successfully inoculating Fouquieria splendens.

A photograph of teliospores of the type was published by Hennen and Cummins (Mycologia 48:126-162. 1956).

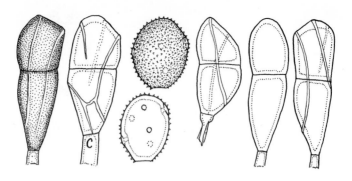

Figure 190

167. PUCCINIA PIPERI Ricker J. Mycol. 11:114. 1905. ssp.
piperi. Fig. 190.

Aecia unknown. Uredinia on adaxial leaf surface, yellowish
brown or perhaps orange when fresh; spores (23-)25-30(-32) x
(19-)22-25μ, mostly broadly ellipsoid, wall 1.5-2(-2.5)μ thick,
yellowish to pale golden, echinulate, pores 6-8, scattered.
Telia abaxial, usually in striae, covered by the epidermis,
with scanty marginal golden paraphyses; spores (33-)43-60(-68)
x 17-22(-24)μ, mostly oblong-ellipsoid or narrowly obovoid,
the apex obtuse or obtusely rounded, wall 1.5-2(-2.5)μ thick
at sides, (2-)2.5-3.5μ at apex, uniformly clear chestnut-brown;
with a few longitudinal ridges, these sometimes branched;
pedicels yellowish, collapsing, 20μ or less long.

Hosts and distribution: On Vulpia megalura (Nutt.) Rydb.,
V. microstachys (Nutt.) Munro ex Benth., V. pacifica (Piper)
Rydb., V. reflexa (Buckl.) Rydb.; Oregon and California, U.S.A.

Type: Piper 6502, on Festuca pacifica, 8 dollar Mt.,
Oregon, 12 June 1904 (WIS; isotype PUR).

There are no field indications as to aecial host but,
because of the morphological similarity to the following
variety, it will doubtless prove to be Liliaceae.

The Utah record on Festuca elatior L. and the Argentine
record (Jørstad, Ark. Bot. 4:63. 1959) on F. australis Nees
(=Vulpia australis (Nees) Blom) may not belong here.

PUCCINIA PIPERI Ricker spp. scillae-rubrae(P. Cruchet)
Cumm. ssp. nov.

Puccinia scillae-rubrae P. Cruchet Bull. Soc. Vaud. Sci. Nat.
51:625-627. 1919.

Aecia, Aecidium scillae Fckl., occur on Scilla bifolia L.;
spores mostly broadly ellipsoid, 21-28 x 17-23μ thick, finely
verrucose. Uredinia and spores as in ssp. piperi. Telia
abaxial without paraphyses; spores as in ssp. piperi except with
golden brown, slightly thinner walls.

Hosts and distribution: *Festuca* *rubra* L., *F.* sp.: Europe
from Switzerland east to the Black Sea.

Type: Cruchet, on *Festuca* *rubra* Montagny sur Yverdon,
Switzerland (Herb. Cruchet; isotype PUR).

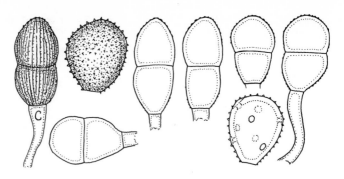

Figure 191

168. PUCCINIA PATTERSONIANA Arth. Bull. Torrey Bot. Club 33:29. 1906. Fig. 191.

Aecia occur on Brodiaea douglasii S. Wats. in local spots; spores (20-)23-28(-30) x (16-)19-22(-25)μ, broadly ellipsoid or globoid, wall (1.5-)2-2.5μ thick, colorless or pale yellowish, finely verrucose. Uredinia on adaxial leaf surface, yellowish brown, spores (26-)29-33(-36) x (19-)21-25(-28)μ, mostly broadly obovoid, wall 2.5-3.5(-4)μ thick, golden brown, echinulate, pores 6-8, rather obscure, scattered. Telia on adaxial surface, light chestnut-brown, exposed, pulvinate; spores (29-)32-38(-43) x (16-)18-21(-23)μ, mostly ellipsoid, wall uniformly 1-1.5(-2)μ thick, golden to clear chestnut-brown, closely striated with narrow, low, continuous or interrupted ridges; pedicels colorless, thin-walled, collapsing, to 85μ long but usually broken near the spore.

Hosts and distribution: species of Agropyron, Elymus, Sitanion: western Canada and western U.S.A.

Type: Anderson, on Agropyron divergens (=A. spicatum (Pursh) Scribn. & Smith, Sand Coulee, Montana, U.S.A. (PUR).

The relationship of the aecial and telial stages was proved by Mains (Indiana Acad. Sci. Proc. 1921: 133. 1922).

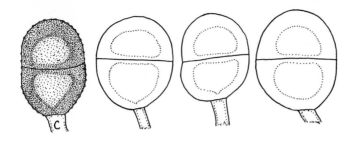

Figure 192

169. PUCCINIA WOLGENSIS Nawasch. in Sydow Monogr. Ured. 1:819.
1904. Fig. 192.

Aecia unknown. Uredinia not seen; spores 25-28(-30) x
23-25μ, broadly ellipsoid or globoid, wall 2-3μ thick, golden,
echinulate, germ pores 6 or 7, scattered. Telia on abaxial
leaf surface, early exposed, deeply pulvinate, brown; spores
(36-)40-63(-73) x (30-)34-50(-58)μ, broadly ellipsoid, wall
uniformly (5-)8-10(-16)μ, golden, rugose or appearing smooth;
pedicels hyaline, thin-walled, collapsing, at least 125μ long
but breaking near the spore.

Hosts and distribution: species of Stipa: Morocco to
Syria and southern U.S.S.R.

Type: Nawaschin, on Stipa pennata, Saratov, U.S.S.R. (S).

A photograph of teliospores of the type was published by
Greene and Cummins (Mycologia 50:6-36. 1958).

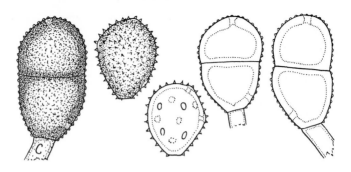

Figure 193

170. PUCCINIA PRATENSIS Blytt Christ. Vidensk-Selsk. For. 1896:52. 1896. Fig. 193.

Uredo avenae-pratensis Eriks. Ark. Bot. 18:17. 1923.

Puccinia versicoloris Semad. Centrlb. Bakteriol. II. 46:466. 1916.

Aecia unknown. Uredinia amphigenous or mostly on adaxial leaf surface, cinnamon-brown; spores (24-)28-35(-40) x (23-)26-30(-32)µ, mostly broadly ellipsoid or obovoid, wall (1.5-)2.5-3(-3.5)µ thick, golden or cinnamon-brown, echinulate, germ pores 8-14, mostly 10-12, large, scattered. Telia like the uredinia; spores (33-)42-60(-70) x (23-)28-35(-43)µ variable but mostly broadly ellipsoid or obovoid, wall uniformly (3-)4-5(-6)µ thick or slightly (-7µ) thicker apically, mostly golden brown, tending to be bilaminate, finely echinulate-verrucose, germ pore mostly apcial in upper cell, midway to hilum in lower cell; pedicels fragile and always broken near the spore.

Hosts and distribution: Avenochloa pratensis (L.) Holub, A. versicolor (Vill.) Holub: Europe.

Type: Aasen and Blytt, on Avena pratensis (=Avenochloa pratensis), Bygdo, Christiania, Norway (S).

298

171. PUCCINIA BROMOIDES Guyot Urediniana 3:67. 1951.

Aecia unknown. Uredinia amphigenous, cinnamon-brown; spores (23-)27-30(-32) x (22-)24-27(-30)μ, mostly globoid or broadly ellipsoid, wall (2-)2.5-3(-3.5)μ thick, about golden brown, echinulate, germ pores 4-10, mostly 6 or 7, scattered or when 4 or 5 tending to be or actually equatorial. Telia not seen; spores in the uredinia (34-)37-48(-54) x (24-)27-31 (-34)μ, wall uniformly (3-)3.5-4(-5.5)μ thick or slightly (-7)μ thicker apically, finely echinulate-verrucose, golden brown, germ pore apical in upper cell, midway to pedicel in lower cell; pedicels fragile and broken near the spore.

Hosts and distribution: Avenochloa bromoides (Gouan) Holub: France.

Type: Guyot, on Avena bromoides (=Avenochloa bromoides), pentes meridionales du col de Vence, France (Herb. Guyot; isotype PUR).

The species differs from P. pratensis only in the size of spores and the number of germ pores and there is overlap in these characters.

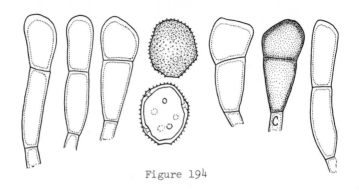

Figure 194

172. PUCCINIA EATONIAE Arth. J. Mycol. 10:18. 1904. Fig. 194.

Aecia systemic, on species of Ranunculus (Aecidium ranunculi Schw.) and on Myosotis virginica (Aecidium myosotidis Burr.); spores 15-24 x 12-20µ, globoid or broadly ellipsoid, wall 1.5µ thick, verrucose. Uredinia mostly on adaxial leaf surface, yellow; spores (19-)22-26 x (16-)18-21µ, mostly broadly ellipsoid or obovoid, wall 1-2µ thick, pale yellowish or colorless, echinulate, germ pores obscure 7 or 8. Telia mostly on abaxial leaf surface and sheaths, covered by epidermis, weakly loculate with brownish paraphyses, spores (28-)35-52(-58) x (10-)13-19(-21)µ, mostly oblong-clavate, wall 0.5-1µ thick at sides, 3-4(-6)µ apically, chestnut-brown apically, paler basally, smooth; pedicels mostly less than 15µ long, brownish; spores germinate without overwintering.

Hosts and distribution: Sphenopholis intermedia (Rydb.) Rydb., S. nitida (Bieler) Scribn., S. obtusata (Michx.) Scribn.: southern Canada, U.S.A., and the Dominican Republic.

Lectotype: Arthur, on Eatonia pennsylvanica (=Sphenopholis intermedia), from greenhouse inoculation, Lafayette, Ind. (PUR 23289). Lectotype designated here.

Arthur (loc. cit.) first proved the relationship of the aecia on Ranunculus and Mains (Mycologia 24:207-214. 1932) of the aecia on Myosotis to the grass rust. Mains (loc. cit.) recognized var. ranunculi and var. myosotidis.

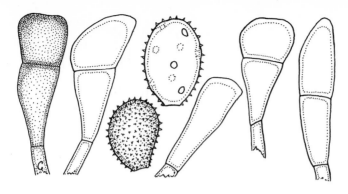

Figure 195

173. PUCCINIA HELICTOTRICHI Joerst. Ark. Bot. Ser. 2. 4:349.
1959. Fig. 195.

Aecia unknown. Uredinia amphigenous, probably bright yellow
or orange when fresh, nearly colorless when old and dry, spores
(24-)27-38(-42;-48) x (18-)20-26μ, variable in size and shape,
mostly broadly ellipsoid or obovoid, wall 1-1.5μ thick, echinulate,
germ pores. very obscure, at least 6-8 and in large spores probably 9-12. Telia amphigenous, blackish brown, covered by the
epidermis, loculate with abundant brown paraphyses; spores
(34-)40-60(-65) x (14-)17-22μ, variable but mostly clavate or
oblong-clavate, wall 1-1.5μ thick at sides, 2-4μ apically,
golden brown apically, nearly colorless basally, smooth; pedicels
yellowish, collapsing, mostly less than 15μ long; 1-celled
spores abundant.

Type: Smith No. 1122, on Avenochloa schelliana (Hack.) Holub
(as Helicotrichum schellianum), Chili Prov., China (UPS). Not
otherwise known.

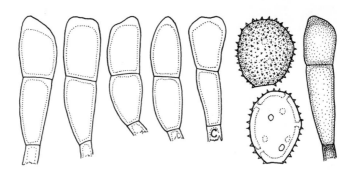

Figure 196

174. PUCCINIA AMMOPHILAE Guyot Rev. Pathol. Veg. et Entomol.
Agr. France 19:36. 1932. Fig. 196.

Uredo ammophilae H. Syd. & P. Syd. Bot. Notiser 1900:42. 1900.

Aecia unknown. Uredinia on adaxial leaf surface, cinnamon-brown; spores (26-)28-34(-38) x 20-25μ, broadly ellipsoid, obovoid, or ellipsoid, wall (1.5-)2.5-3.5(-4)μ thick, pale yellowish to golden, echinulate, germ pores (7)8 or 9, scattered or tending to be bizonate, rather obscure. Telia amphigenous, covered by the epidermis, blackish, loculate with brown paraphyses, spores (32-)38-60(-70) x (12-)15-19(-22)μ, mostly oblong or narrowly oblong-clavate, wall 1μ thick at sides, (3-)4-7(-10)μ apically, clear chestnut-brown apically, paler basally, smooth; pedicels brown, 12μ or less long.

Hosts and distribution: Ammophila arenaria (L.) Link: Europe.

Neotype: Guyot, Brighton près Cayeux-sur-mer, Somme, France 23 Sept. 1948 (Herb. Guyot; isotype PUR). Neotype designated here because Guyot (in litt.) has advised that the holotype no longer exists.

The species is similar to Puccinia procera but has paler urediniospores and no very conspicuous cuticular caps over the pores.

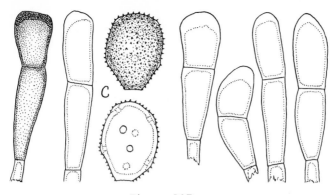

Figure 197

175. PUCCINIA PROCERA Diet. & Holw. in Dietel Erythea 1:249.
1893. Fig. 197.

Puccinia elymicola Constan. Ann. Mycol. 14:254. 1916.

Aecia occur on Phacelia; cupulate, in groups; spores 29-38 x
26-32μ, globoid, wall 1.5-2(-2.5)μ thick, verrucose. Uredinia
mostly on the adaxial leaf surface, cinnamon-brown; spores
(30-)32-44(-53) x (25-)28-34(-38)μ, broadly ellipsoid, obovoid,
or ellipsoid, wall (1.5-)2-2.5(-3)μ thick, golden to cinnamon-
brown, echinulate, germ pores (7)8 or 9(10), scattered or
tending to be bizonate. Telia on abaxial surface, covered
by the epidermis, blackish, tending to be loculate with golden
brown paraphyses; spores (40-)50-70(-80) x (14-)17-22(-25)μ,
mostly oblong or oblong-clavate, wall 1(-1.5)μ thick at sides,
(3-)4-6(-8)μ apically, chestnut-brown apically, paler basally,
smooth; pedicels brownish, mostly less than 15μ long.

Hosts and distribution: Elymus condensatus Presl, E. mollis
Trin., E. sabulosus Marsch.-Bieb.: coastal California, U.S.A.
and the Black Sea area of Eastern Europe.

Type: McClatchie, on Elymus condensatus, Pasadena, Calif.
(S; isotypes Bartholomew N. Amer. Ured. 658).

Mains (Papers Michigan Acad. Sci. Arts, Letters 17:289-394.
1932; publ. 1933) proved the life cycle by inoculation, using
Phacelia distans Benth. as the aecial host.

This species differs from Puccinia recondita mainly because
of the large urediniospores.

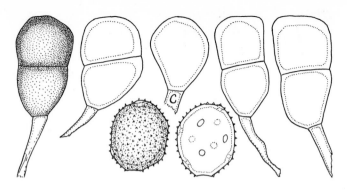

Figure 198

176. PUCCINIA CRYPTICA Arth. & Holw. in Arthur Proc. Amer. Phil.
Soc. 64:163-164. 1925 var. cryptica. Fig. 198.

Aecia unknown. Uredinia amphigenous, yellowish brown; spores
(18-)25-30(-32) x (18-)22-27(-28)μ, broadly ellipsoid or globoid,
wall (1-)1.5-2(-2.5)μ thick, yellowish brown, echinulate, germ
pores scattered, 7-10. Telia mostly on abaxial leaf surface,
covered by epidermis, blackish, with some brown paraphyses but
the sori not loculate; spores (25-)34-42(-48) x (16-)20-30(-37)μ,
variable but mostly oblong-obovoid or oblong, wall (1-)2-3(-5.5)μ
thick at sides, (2.5-)4-8(-9)μ apically, chestnut-brown, smooth;
pedicels brownish, mostly less than 20μ, 1-celled spores occur.

Hosts and distribution: Bromus coloratus Steud., B. trinii
Desv., Hordeum chilense Roem. & Schult., H. gussonianum Parl.:
Chile.

Type: Holway No. 40, on Bromus trinii, Papudo, Chile (PUR;
isotypes issued as No. 5 of Reliq. Holw.).

The rust on Hordeum differs only slightly and may belong to
this species. Kaufmann (Mycopathol. Mycol. Appl. 32:249-261.
1967) published a photograph of teliospores of the type.

PUCCINIA CRYPTICA Arth. & Holw. var. bromicola (Arth.& Holw.)
M. Kaufmann Mycopathol. Mycol. Appl. 32:260. 1967.

Uromyces bromicola Arth. & Holw. Proc. Amer. Phil. Soc.
64:210. 1925.

Differs from var. cryptica principally in having some 97-99%
1-celled teliospores (30-)33-34(-37) x (23-)27-31(-37)μ.

Hosts and distribution: Bromus coloratus Steud.: Chile.

Type: Holway No. 150, Concepcion, Chile (PUR F2353; isotypes
issued as No. 21 of Reliq. Holw.).

Kaufmann (loc. cit.) published a photograph of teliospores of
the type.

304

177. PUCCINIA AUSTROUSSURIENSIS Tranz. Conspectus Uredinalium URSS. p. 111. 1939.

Aecia unknown. Uredinia yellowish brown; spores 30-34 x 20-33μ (often 33 x 27μ), subglobose, wall brownish, loosely echinulate, germ pores 5 or 6, distinct, scattered. Telia epiphyllous, covered by the epidermis, blackish brown or black, loculate with brown paraphyses; spores 36 x 17μ, mostly clavate.

Type: Tranzschel, on Trisetum sibiricum Rupr., Primorskaja region, Far Eastern U.S.S.R. (LE; not seen).

The description is adapted from the original.

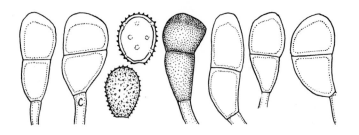

Figure 199

178. PUCCINIA PENNISETI-LANATI Ahmad Biologia 6:122. 1960.
Fig. 199.

Aecia unknown. Uredinia amphigenous or mostly on adaxial
surface of leaf, bright yellow or orange when fresh; spores
19-24 x 14-18µ, mostly broadly ellipsoid or obovoid, wall 1.5µ
thick, hyaline, echinulate, germ pores about 7 or 8, scattered.
Telia mostly on abaxial surface, blackish brown, covered by the
epidermis, without paraphyses; spores variable, (25-)33-42(-50)
x (15-)17-23(-25)µ, wall 1-1.5(-2.5)µ thick at sides, (2.5-)3-5
(-6.5)µ apically, golden or clear chestnut-brown, smooth; pedicels hyaline or yellowish, to 15µ long.

Hosts and distribution: Pennisetum lanatum Klotz.: West
Pakistan.

Type: Ahmad No. 2845, on Pennisetum lanatum, Batakundi,
Kagan Valley, West Pakistan (LAH; isotype PUR).

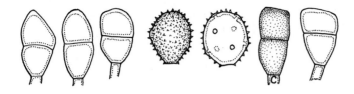

Figure 200

179. PUCCINIA LIMNODEAE Cumm. sp. nov. Fig. 200.

Aeciis ignotis. Urediniis epiphyllis, flavo-brunneis,
pulverulentis; paraphysibus nullis; sporae (18-)20-24(-30) x
17-20μ, plerumque late ellipsoideae vel obovoideae; membrana
1-1.5μ crassa, flavida, flavo-brunnea, vel fere hyalina, echinulata; poris germinationis 7 vel 8, sparsis, obscuris. Teliis
hypophyllis, epidermide tectis, loculatis, paraphysibus brunneis
numerosis; sporae (23-)26-34(-37) x (12-)14-18(-20)μ, oblongae
vel oblongo-ellipsoideae; membrana ad latere 1μ crassa, ad apicem
2-3(-4)μ crassa, pallide castaneo-brunnea vel aureo-brunnea,
deorsum pallidiore, levi; pedicello aureo-brunneo, brevi.

Type: B. C. Tharp, on Limnodea arkansana (Nutt.) L. H. Dewey,
Austin, Texas, U.S.A., 19 May 1922 (PUR 21471).

This fungus has been recorded previously as Puccinia schedonnardi from which it is separable because of the covered telia.
The type is the only collection known.

307

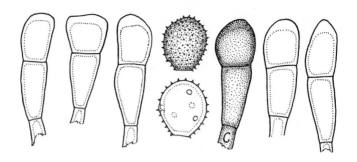

Figure 201

180. PUCCINIA ISHIKARIENSIS Ito J. Coll. Agr. Tohoku Imp.
Univ. 3:209. 1909. Fig. 201.

Aecia unknown. Uredinia on adaxial surface, about cinnamon-
brown; spores (23-)26-30(-33) x (19-)21-25µ, ellipsoid, broadly
ellipsoid or obovoid, wall (1.5-)2(-3)µ thick, golden to cinnamon-brown, echinulate, germ pores 4 to 6(-8?), scattered. Telia
amphigenous, rather tardily opening by a slit, with scant, brown,
stromatic paraphyses; spores (36-)42-54(-60) x (12-)14-20(-22)µ,
variable, cylindrical to clavate, wall 1-1.5µ thick at sides,
(3-)4-6(-7)µ apically, smooth; pedicel 10µ or less long, darker
brown than the base of the spore.

Hosts and distribution: Moliniopsis japonica (Hack.)
Hayata: Japan.

Type: Kasai, on Molinia japonica (=Moliniopsis japonica),
Tsuishikari, Prov. Ishikari, Japan (SAPA; isotype PUR).

Uredinial paraphyses were described by Ito but they are not
present in the isotype.

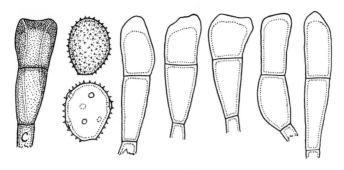

Figure 202

181. PUCCINIA GLYCERIAE Ito J. Coll. Agr. Tohoku Imp. Univ. 3:200. 1909. Fig. 202.

Aecia unknown. Uredinia mostly on adaxial leaf surface, yellowish brown (dry); spores (20-)23-27(-29) x 18-22(-24)μ, mostly obovoid, wall 1(-1.5)μ thick, pale yellowish or colorless, echinulate, germ pores scattered, obscure, 6-8. Telia amphigenous, covered or tardily exposed, blackish brown, without paraphyses; spores variable, both within and between sori, (30-)40-65(-80) x (11-)14-19(-22)μ, mostly clavate but sometimes cylindrical, wall 1(-1.5)μ thick at sides, (3-)4-6(-8)μ apically, clear chestnut-brown, smooth except sometimes with a few longitudinal ridges; pedicels persistent, brownish, mostly less than 12μ long.

Hosts and distribution: Glyceria alnasteretum Kom., G. ischyroneuron Steud., G. leptolepis Ohwi: Japan.

Type: K. Miyabe, on Glyceria aquatica Authors (=G. leptolepis), Prov. Ishikari: Jozankei, 19 Aug. 1898 (SAPA; isotype PUR). A type was not indicated originally but a portion of Miyabe's collection was received from Dr. Ito marked "Type collection".

This species has the general appearance of Puccinia recondita but the urediniospores have thin pale walls and the telia lack paraphyses.

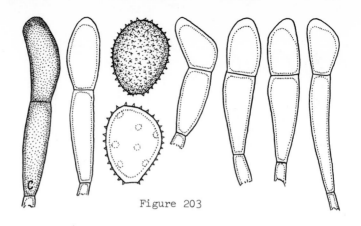

Figure 203

182. PUCCINIA COCKERELLIANA Bethel ex Arthur Bull. Torrey Bot. Club 46:113. 1919. Fig. 203.

Aecia occur on <u>Thalictrum</u> <u>fendleri</u> Engelm.; spores (20-) 23-29(-31) x (17-)19-23(-25)μ, wall 1.5-2(-2.5)μ thick, closely verrucose, yellowish or colorless. Uredinia on adaxial surface, yellowish brown; spores (24-)27-32(-36) x (19-)22-25(-27)μ, mostly ellipsoid or broadly ellipsoid, wall 1.5-2(-2.5)μ thick, yellowish to nearly colorless, echinulate, pores scattered, perhaps 8-10, very obscure. Telia adaxial, early exposed by a slit or broadly, blackish brown, without paraphyses; spores (40-)60-80(-90) x (12-)14-18(-22)μ, mostly cylindrical but slightly narrowed toward the base, rounded or obtusely rounded at apex, wall 1μ thick at sides, golden brown, (3-)4-6(-8)μ apically, clear chestnut-brown; pedicels persistent, brownish, 15μ or less long.

Hosts and distribution: <u>Festuca</u> <u>arundinacea</u> Schreb., <u>F.</u> <u>rubra</u> L., <u>F.</u> <u>scabrella</u> Torr., <u>F.</u> <u>thurberi</u> Vasey: Alaska south to New Mexico.

Type: Bethel, on <u>Festuca</u> <u>thurberi</u>, Lake Eldora, Colorado, 4 July 1911 (PUR).

The species differs from <u>P.</u> <u>recondita</u> because of nearly colorless urediniospores, early exposed, aparaphysate telia, and very long narrow teliospores.

Inoculations by Arthur (Mycologia 8:133. 1916) proved the life cycle.

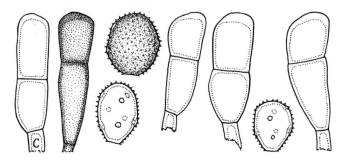

Figure 204

183. PUCCINIA SESSILIS W. G. Schneider in Schroeter Abh. Scles. Nat. Abth. 1869-72:19. 1870 var. sessilis. Fig. 204.

Puccinia linearis Peck Bull. Buffalo Soc. Nat. Sci. 1:67. 1873.

Puccinia striatula Peck Ann. Rept. New York State Mus. 33:38. 1880.

Puccinia phalaridis Plowr. J. Linn. Soc. London 24:88. 1888.

Puccinia digraphidis Soppitt J. Bot. London 28:213. 1890.

Puccinia paradis Plowr. J. Linn. Soc. London 30:43. 1893.

Puccinia schmidtiana Diet. Ber. Nat. Ges. Leipzig 1895-96: 195. 1896.

Puccinia festucina H. Syd. & P. Syd. Ann. Mycol. 10:217. 1912.

Puccinia angulosi-phalaridis Poev. in Poeverlein, Speyer & Schoenau Kryptog. Forsch. Bayer Bot. Ges. Erforsch. Heim. Flora 2:68. 1929.

Aecia, Aecidium majanthae Schum., occur on Araceae, Iridaceae, Liliaceae, and Orchidaceae; spores (16-)18-25(-27) x 15-20(-22)μ, globoid or more or less ellipsoid, wall 1(-1.5)μ thick, finely verrucose, colorless or yellowish. Uredinia amphigenous, about cinnamon-brown; spores (23-)27-32(-36) x (20-)22-26(-28)μ, broadly ellipsoid or obovoid, wall 1.5(-2)μ thick, golden or cinnamon-brown, echinulate, germ pores (4)5 or 6 scattered or tending to be equatorial (Japan), (5) 6 or 7(8) mostly 6, scattered (Europe), 7-9, mostly 8, tending to be bizonate (N. America). Telia amphigenous, blackish, covered by the epidermis, not or weakly loculate with scant brown paraphyses; spores (34-)40-56(-60) x (15-)18-23(26)μ, oblong or oblong-clavate, wall 1-1.5μ thick at sides, (2.5-)3-5(-6)μ apically, clear chestnut-brown, smooth; pedicels brownish, mostly less than 15μ long.

Hosts and distribution: species of Festuca and Phalaris: Europe to Turkey, the U.S.S.R., China, Japan, Canada, and the

U.S.A.

Type: Schneider, on _Phalaris arundinacea_ L., Neuhaus b. Pirscham (B).

Winter (Sitz.-Ber. Naturf. Ges. 1874:41-43; Hedwigia 14:113-115. 1875) first demonstrated an aecial host by inoculation, using _Allium ursinum_. Workers have confirmed Winter's results and demonstrated numerous other aecial hosts.

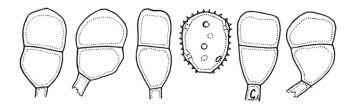

Figure 205

PUCCINIA SESSILIS Schneider var. minor var. nov. Fig. 205.

Aeciis ignotis; urediniosporis (22-)24-28(-30) x (18-)20-22µ,
membrana 1-1.5µ crassa, flavida, echinulata, poris germinationis plerumque 8, sparsis. Teliis valde loculatis, paraphysibus conspicuis, obscure brunneis, sporis (24-)28-38(-44)
x 16-20(-22)µ, oblongis vel oblongo-clavatis; mesosporis
numerosis.

Hosts and distribution: Phalaris angusta Nees ex Trin., P.
caroliniana Walt.: Kansas, Oklahoma, and Texas, U.S.A.; three
collections known.

Type: S. E. Wolf, on Phalaris caroliniana, Bell County,
Texas, 11 June 1931 (PUR 53511).

The variety differs from the typical because of small teliospores and abundant dark brown stromatic paraphyses that divide
the sorus into conspicuous locules.

313

184. PUCCINIA TSINLINGENSIS Wang   Acta Phytotax. Sinica
10:296. 1965.

Aecia unknown. Uredinia amphigenous or mostly on adaxial
leaf surface, yellowish brown; spores 21-25 x 20-23μ, globoid
or nearly so, orange color, wall 2-2.5μ thick, presumably
yellowish, echinulate, germ pores 6 or 7, scattered, with
"cuticular caps". Telia amphigenous, mostly on adaxial surface,
sometimes caulicolous, small, scattered, pale blackish, covered
by the epidermis, presumably without paraphyses; spores 41-58
x 17-23μ, clavate or oblong, apex conical or truncate, wall
1-2μ thick at sides, 3-5μ apically, smooth, chestnut-brown;
pedicels yellowish, to 18μ long.

Hosts and distribution: _Bromus japonicus_ Thunb., _B. tectorum_
L.: China (Wang cites 3 collections).

Type: Yang & Liu No. 1457 (Inst. Microbiol. Peking No.
17782) Not seen.

The description is adapted from the original. Wang published
a photograph of the spores.

314

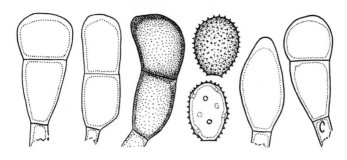

Figure 206

185. PUCCINIA POARUM Niels. Bot. Tidsskr. III. 2:34. 1877.
Fig. 206.

Puccinia poae-trivialis Bub. Ann. Mycol. 3:220. 1905.

Puccinia liatridis Bethel in Arthur Mycologia 9:301. 1917.
nom. nud.

Puccinia petasiti-pulchellae Luedi Centralbl. Bakt. II. 48:85.
1918.

Puccinia poae-alpinae Eriks. Ark. Bot. 18:1. 1923.

Puccinia conspicua Mains Mycologia 25:408. 1933.

Puccinia liatridis (Arth. & Fromme) Bethel ex Arthur Manual
Rusts U.S. and Canada. p. 146. 1934.

Puccinia petasiti-melicae Gaeum. Phytopathol. Z. 13:627. 1941.

Puccinia taminesis Gaeum. Phytopathol. Z. 13:629. 1941.

Puccinia kummeri Gaeum. Phytopathol. Z. 13:632. 1941.

Puccinia petasiti-poarum Gaeum. & Eich. Phytopathol. Z. 13:637.
1941.

Puccinia baldensis Gaeum. Ber. Schweiz. Bot. Ges. 61:48. 1951.

Puccinia paihuashanensis Wang Acta Phytotax. Sinica 10:292.
1965.

Aecia (Aecidium tussilaginis Pers. on species of Brickelia,
Helenium, Liatris, Ophryosporus, Petasites, Senecio, Tussilago,
as first demonstrated by Nielsen, loc. cit.); spores (18-)20-
27(-31) x (15-)18-24(-27)μ, wall (0.5-)1(-1.5)μ thick, incompletely
verrucose-echinulate. Uredinia mostly adaxial, bright orange-
yellow when fresh, usually without but occasionally with a few
short, capitate, peripheral paraphyses; spores (21-)23-30(-37) x
(14-)17-24(-26)μ, mostly obovoid or ellipsoid, wall 1.5μ thick,
colorless or pale yellowish, echinulate, pores scattered, (4-)5-8,
very obscure. Telia mostly abaxial, covered by the epidermis,
with variable development of colorless or brownish paraphyses
but the sori rarely loculate; spores (36-)40-58(-65;-77) x
(14-)17-25(-28)μ, mostly elongately obovoid or oblong-clavate,

315

wall 0.5–1.5μ thick at sides, (2–)3–6(–8)μ apically, chestnut-brown above, golden basally; pedicels colorless or yellowish, 15μ or less long.

Hosts and distribution: species of <u>Agrostis</u>, <u>Calamagrostis</u>, <u>Festuca</u>, <u>Koeleria</u>, <u>Melica</u>, <u>Peyritschia</u>, <u>Phleum</u>, <u>Poa</u>, <u>Trisetum</u>: Europe to China and in North and South America.

Lectotype: Nielsen, on <u>Poa trivialis</u>, Denmark (C); designated by Greene and Cummins (Mycologia 59:47-57. 1967).

Photographs of teliospores of the lectotype and from other specimens were published by Greene and Cummins (loc. cit.), who discussed this species complex in detail.

The species is difficult to distinguish from <u>P. recondita</u> but has paler uredinia and urediniospores and usually fewer telial paraphyses.

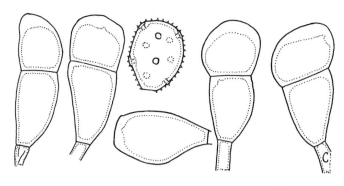

Figure 207

186. PUCCINIA HORDEI Otth Mitt. Naturf. Ges. Bern 1870:114.
1871. Fig. 207.

Puccinia straminis Fckl. var. simplex Koern. in Thuem. Herb.
Mycol. Oecon. 101. 1873.

Puccinia hordei Fckl. Jahrb. Nass. Ver. Nat. 15:16. 1873.

Uromyces hordei Niels. Ugeskr. Landm. IV, 9(1):567. 1875.

Puccinia koeleriae Bagnis Atti R. Acad. Lincei Ser. 2.
3:714. 1876, nom. nud.

Puccinia anomala Rostr. in Thuem. Flora 61:92. 1878.

Puccinia simplex (Koern.) Eriks. & Henn. Z. Pflanzenkr.
4:260. 1894, not Peck, 1881.

Puccinia triseti Eriks. Ann. Sci. Nat. 8 ser. 9:277. 1899.

Puccinia holcina Eriks. Ann. Sci. Nat. 8 ser. 9:274. 1899.

Puccinia pseudo-myuri Kleb. Kryptogam. Fl. Mark Brandenb.
5:618. 1913.

Uredo elymi capitis-medusae Gz. Frag. Bol. R. Soc. Espan.
Hist. Nat. 1913:197. 1913.

Puccinia schismi Bub. Ann. Naturhist. Hofmus. Wien 28:193.
1914.

Puccinia fragosoi Bub. Hedwigia 57:2. 1915.

Puccinia laguri Jaap Ann. Mycol. 14:23. 1916.

Puccinia laguri-chamaemoly Maire Bull. Soc. Hist. Nat.
Afr. Nord. 10:139. 1919.

Puccinia paraphysata Reichert Bot. Jahrb. 56:690. 1921.

Puccinia loliina H. Syd. Ann. Mycol. 19:147. 1921.

Puccinia brachypus Speg. var. loliiphila Speg. Rev. Argent.
Bot. 1:109. 1925.

Puccinia baudysii Picb. Inst. Jard. Bot. Bull. Univ. Belgrade
1:63. 1928.

Puccinia koeleriae Politis Pragmat. Acad. Athenes 3(4):12. 1935.

Puccinia loliicola V.-Bourgin Rev. Pathol. Entomol. Agr. France 24:78. 1937.

Puccinia hordei-murini Buchw. Ann. Mycol. 41:308. 1943.

Puccinia tetuanensis Guyot & Malen. Trav. Inst. Sci. Cherif. ser. Bot. 28:114. 1963.

Puccinia vulpiana Guyot Uredineana 2:53. 1946.

Puccinia gaudiniana Guyot Uredineana 2:56. 1946.

Puccinia vulpiae-myuri Mayor & V.-Bourgin Rev. Mycol. 15:103. 1950.

Puccinia holcicola Guyot Uredineana 3:63. 1951.

Puccinia ifraniani Guyot & Malen. Trav. Inst. Sci. Cherif. ser. Bot. 11:99. 1957.

Puccinia cutandiae Guyot Uredineana 5:368. 1958.

Aecia, _A_. ornithogaleum Bub., occur on _Allium, Orinthogalum_ and _Sedum_, cupulate, in groups; spores (18-)20-26(-29) x (15-)18-21(-22)μ, wall 1.5(-2)μ thick, colorless, finely verrucose. Uredinia mostly on adaxial surface, yellow or brownish yellow; spores (18-)21-30(-32) x (15-)18-25(-28)μ, ellipsoid, or obovoid, wall (1-)1.5-2(-2.5)μ thick, yellowish to very pale brownish, echinulate, pores obscure, scattered, 7-9. Telia amphigenous or mostly abaxial, covered by the epidermis, blackish, loculate with abundant brown paraphyses; spores (36-)45-63(-74) x (15-)19-25(-32)μ, mostly elongate obovoid or oblong-clavate, often angular, wall 1-1.5(-2)μ thick in lower cell, side wall of upper cell (1-)1.5-2.5(-3.5)μ thick, usually gradually thickened toward apex, (3-)4-7(-10)μ thick at apex, deep golden brown or clear chestnut-brown, often paler basally, commonly with surface ridges, otherwise smooth, 1-celled spores common, 3-celled spores occasional; pedicels yellowish, 20μ or less long.

Hosts and distribution: On species of _Aegilops, Arrhenatherum, Avellinia, Avena, Boissiera, Bromus, Cutandia, Deschampsia, Deyeuxia, Echinaria, Gaudinia, Holcus, Hordeum, Koeleria, Lagurus, Lolium, Psilurus, Schismus, Taeniatherum, Trisetum_, and _Vulpia_: circumglobal, especially in littoral climates.

Neotype: Eriksson, on _Hordeum_ vulgare L., Stockholm, Sweden (PUR F4222; isotypes, Eriksson Fungi Paras. Scand. No. 431). Neotype designated here, there being no holotype in BERN.

Tranzschel (Mycol. Centralbl. 4:70-71. 1914), using _Hordeum_ vulgare L. and _Ornithogalum_ umbellatum L., Maire (Bull. Soc. Mycol. France 61:XIV-XXIV. 1914), using _Lagurus_ ovatus L. and _Allium_ chamaemoly L., and Dupias (Compt. Rend. Acad. Sci. Paris 236:962-963. 1953) using _Trisetum_ flavescens (L.) Beauv. and _Sedum_ nicaeensis All. first proved the life cycles. Inoculations have not established the aecial-telial host relationship between the rust fungus on _Arrhenatherum, Deschampsia, Echinaria, Holcus_,

318

Lolium, Psilurus, and Taeniatherum.

The species has often been confused with P. recondita but differs because of paler urediniospores and broader teliospores. P. triseti Eriks., although many specimens have been referred to it, probably is not synonymus. Eriksson's specimens apparently are not extant. Dupias (Uredineana 5:303-312. 1956) suggested relationship with P. fragosoi. P. hordei obviously is a "complex" more or less like the P. recondita complex. Puccinia blasdalei Diet. & Holw., on Allium, is similar morphologically.

Photomicrographs of the teliospores from various hosts were published, as P. holcina, by Greene and Cummins (Mycologia 59: 47-57. 1967).

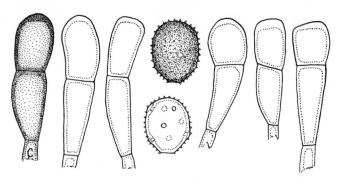

Figure 208

187. PUCCINIA RECONDITA Roberge ex Desmazieres Bull. Soc. Bot. France 4:798. 1857. Fig. 208.

Uredo rubigo-vera DC. Fl. France 5:83. 1815.

Puccinia rubigo-vera Wint. Rabh. Kryptog.-Fl. Ed. 2. I (1): 217-218. 1882.

Puccinia perplexans Plowr. Quart. J. Microscop. Sci. 25:164. 1885.

Puccinia persistens Plowr. Monogr. Brit. Ured. Ustil. 180. 1889.

Puccinia apocrypta Ell. & Tracy J. Mycol. 6:76. 1890.

Puccinia agrostidis Plowr. Gard. Chron. 3 ser. 8:139. 1890.

Puccinia piptatheri Lagh. Bol. Soc. Bot. 8:136. 1890.

Puccinia agropyri Ell. & Ev. J. Mycol. 7:131. 1892.

Puccinia dispersa Eriks. & Henn. Z. Pflanzenkr. 4:17. 1894.

Puccinia borealis Juel Oefvers. K. Ventensk.-Akad. Foerh. 51:411. 1894.

Puccinia adspersa Diet. & Holw. Erythea 3:81. 1895.

Puccinia agropyrina Eriks. Ann. Sci. Nat. 8 ser. 9:273. 1899.

Puccinia bromina Eriks. Ann. Sci. Nat. 8 ser. 9:271. 1899.

Puccinia triticina Eriks. Ann. Sci. Nat. 8 ser. 9:270. 1899.

Puccinia actaeae-agropyri E. Fisch. Ber. Schweiz. Bot. Ges. 11:8. 1901.

Puccinia symphyti-bromorum F. Muell. Bot. Centralbl. Beih. 10:201. 1901.

Puccinia brachypus Speg. An. Mus. Nac. B. Aires 3:61. 1902.

Puccinia brachysora Diet. Bot. Jahrb. 32:49. 1902.

320

Puccinia triticorum Speg. An. Mus. Nac. B. Aires 3:65. 1902.

Puccinia dactylidina Bub. Ann. Mycol. 3:219. 1905.

Puccinia cerinthes-agropyrina Tranz. Trav. Mus. Bot. Acad.
Imp. Sci. St. Petersb. 3:52-53. 1907.

Puccinia cinerea Arth. Bull. Torrey Bot. Club 34:583. 1907.

Puccinia perminuta Arth. Bull Torrey Bot. Club 34:584. 1907.

Puccinia dietrichiana Tranz. Ann. Mycol. 5:418. 1907.

Puccinia subalpina Lagh. ex Liro Bidr. Kaenned. Finl. Nat.
Folk 65:144. 1908.

Puccinia thulensis Lagh. ex Liro Bidr. Kaenned. Finl. Nat.
Folk 65:139. 1908.

Puccinia alternans Arth. Mycologia 1:248. 1909.

Puccinia obliterata Arth. Mycologia 1:250. 1909.

Puccinia bromi-japonicae Ito J. Coll. Agr. Tohoku Imp. Univ.
3:205-206. 1909.

Puccinia elymi-sibericae Ito J. Coll. Agr. Tohoku Imp. Univ.
3:202-203. 1909.

Puccinia fujiensis Ito J. Coll. Agr. Tohoku Imp. Univ. 3:210.
1909.

Puccinia actaeae-elymi Mayor Ann. Mycol. 9:361. 1911.

Puccinia secalina Grove The Brit. Rust Fungi 261. 1913.

Puccinia agropyri-juncei Kleb. Kryptog.-fl. Mark Brandenb.
5(1):618. 1914.

Puccinia hierochloina Kleb. Kryptog.-fl. Mark Brandenb.
5(1):622. 1914.

Puccinia aconiti-rubri Luedi Mitt. Naturf. Ges. Bern
1918:200-211. 1919.

Puccinia madritensis Maire Bol. Soc. Hist. Nat. Afr. Nord
10:145. 1919.

Puccinia arrhenathericola E. Fisch. Mitt. Naturf. Ges. Bern.
1920:XLII. 1921.

Puccinia thalictri-distichophylli E. Fisch. & Mayor Mitt.
Naturf. Ges. Bern 3:7. 1924.

Puccinia scarlensis Gaeum. Ber. Schweiz. Bot. Ges. 46:245.
1936.

Puccinia thalictri-koeleriae Gaeum. Ber. Schweiz. Bot. Ges.
46:241. 1936.

Puccinia hordei-secalini V.-Bourgin Ann. Ecole Natl. Agr.
Grignon 2:156. 1941.

Puccinia tritici-duri V.-Bourgin Ann. Ecole Natl. Agr.
Grignon 2:146. 1941.

Puccinia sardonensis Gaeum. Ber. Schweiz. Bot. Ges. 55:72. 1945.

Puccinia milii-effusi Dupias Bull. Soc. Mycol. France 61:61. 1945.

Puccinia bromi-maximi Guyot Uredineana 2:50. 1946.

Puccinia bromicola Guyot Uredineana 2:52. 1946.

Puccinia clematidis-secalis Dupias Bull. Soc. Mycol. France 64:182. 1948.

Puccinia haynaldiae Mayor & V.-Bourgin Rev. Mycol. 15:96. 1950.

Puccinia hordei-maritimi Guyot Uredineana 3:62. 1951.

Puccinia aneurolepidii Korbon. Trud. Inst. Bot. Acad. Sci. Tadzhik S.S.R. 30:61. 1954 (nomen nudum).

Puccinia dasypyri Guyot & Malen. Trav. Inst. Sci. Cherif. ser. Bot. 28:62. 1963.

Aecia (Aecidium clematidis DC.) occur on the Balsaminaceae, Boraginaceae, Hydrophyllaceae, and Ranunculaceae; localized, cupulate; spores (18-)21-26(-28) x (14-)17-22(-24)µ, globoid or broadly ellipsoid, wall 1-2µ thick, hyaline, verrucose. Uredinia on the adaxial leaf surface, or the abaxial surface, or often amphigenous, mostly about cinnamon-brown; spores (20-)24-32(-36) x (17-)20-25(-28)µ, mostly broadly ellipsoid or obovoid, wall 1-2µ thick, yellowish brown to cinnamon-brown, echinulate, germ pores 6-10, scattered. Telia mostly on abaxial surface but commonly on the adaxial surface and the sheaths, covered by the epidermis, blackish-brown, brown paraphyses present, the sori usually loculate; spores variable in size and shape, (32-)40-60(-75) x (12-)15-22(-25)µ, mostly oblong-clavate, wall 1-1.5µ thick at sides, 3-5(-7)µ apically, chestnut-brown, smooth; pedicels usually less than 20µ long, brown or brownish.

Hosts and distribution: species of Aegilops, Agropyron, Agrostis, Alopecurus, Anthoxanthum, Arrhenatherum, Avena, Boissiera, Brachypodium, Briza, Bromus, Calamagrostis, Cinna, Colpodium, Dactylis, Deschampsia, Deyeuxia, Elymus, Festuca, Gaudinia, Glyceria, Haynaldia, Hierochloe, Hordeum, Hystrix, Koeleria, Leersia, Lolium, Milium, Oryzopsis, Poa, Scolochloa, Secale, Sitanion, Trisetum, Triticum, and Vulpia: circumglobal; especially common in temperate climates.

Type: Roberge, in Secale, France (isotypes, Desmazieres Plantes Cryptog France No. 252).

Puccinia recondita is treated here as a "species complex". This is not unique nor is it particularly satisfactory. But on a world basis, the variability in morphological features is continuous from extreme to extreme. Distinctive segments of the population may exist regionally and will, undoubtedly, receive separate names. Fifty-one such names are listed above as synonymous; there can hardly be need for more.

A photograph of teliospores of the type was published by Cummins and Caldwell (Phytopathology 46:81-82. 1956).

188. PUCCINIA KOELERIICOLA Tranz. Conspectus Uredinalium URSS. p. 111. 1939.

Aecia unknown. Uredinia not described except aparaphysate; spores globose or ovate, germ pores (4-5?) indistinct. Telia with abundant brown paraphyses; spores 52-75 x 12-15µ, elongate-clavate, the apex slightly thickened and darker; 1-celled spores lacking.

Type: on Koeleria gracilis, Transbaicalia, Burjato-Mongolia, U.S.S.R. (LE; not seen).

Tranzschel notes "Sequenti speciei videtur". This refers to Puccinia fragosoi (see P. hordei) but the teliospore width, as given, does not indicate this species.

The description is adapted from the original.

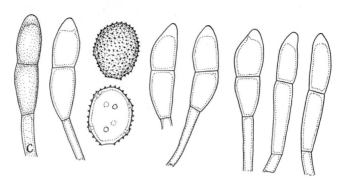

Figure 209

189. PUCCINIA AGROPYRI-CILIARIS Tai & Wei Sinensia 4:111. 1933.
Fig. 209.

Aecia unknown. Uredinia mostly on adaxial leaf surface,
yellowish brown; spores (19-)22-26(-28) x (17-)18-22μ, mostly
broadly ellipsoid, wall 1-1.5μ thick, yellowish, germ pores 7 or
8, scattered. Telia mostly on adaxial surface, early exposed,
compact, waxy in appearance; spores (32-)40-60(-70) x (7-)9-12
(-14)μ, cylindrical or fusiform-cylindrical, wall 0.5-1μ thick
at sides, (3-)4-6(-7)μ apically, pale yellowish, smooth; pedicels
colorless, fragile, to 25μ but usually broken near the spore.
The spores germinate without a dormant period.

Hosts and distribution: Agropyron ciliare (Trin.) Franch.,
A. tsukushiense (Honda) Ohwi, Poa achroleuca Steud.: China, Japan,
and Korea.

Type: Tai No. 4020, on Agropyron ciliare, Ting-kia-chao,
Nanking, China (N?; not seen).

The delicate, long, and narrow teliospores are unlike those
of most grass rust fungi. Because they germinate immediately
one would anticipate a systemic aecial stage.

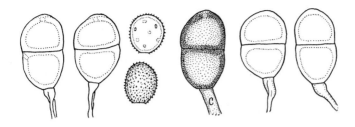

Figure 210

190. PUCCINIA KANSENSIS Ell. & Barth. Erythea 4:1. 1896.
Fig. 210.

Aecia occur on Physalis, systemic; spores 16-24 x 14-19µ,
wall 1-2µ thick, hyaline. Uredinia mostly on abaxial leaf
surface, yellowish; spores 17-22 x 15-18µ, mostly globoid, wall
1.5-2µ thick, hyaline, echinulate, pores obscure, 6 or more,
scattered. Telia mostly on abaxial surface, early exposed,
blackish, pulvinate; spores 23-30(-32) x 17-22µ, oblong-ellipsoid,
wall uniformly 1.5-2.5µ thick, chestnut-brown, smooth; pedicels
colorless, thin-walled, attaining a length of 30µ but usually
broken short.

Hosts and distribution: Buchloë dactyloides Engelm.: U.S.A.,
Kansas and Nebraska, and in Mexico (on Physalis).

Type: Bartholomew, on Buchloë dactyloides, Rockport, Kansas
(NY; isotype PUR).

Baxter and Cummins (Plant Dis. Reptr. 47:1040. 1963) proved
the life cycle by inoculation. The aecial stage corresponds to
Aecidium physalidis Burr. but Parmelee (Res. Branch Can. Dept.
Agr. Publ. 1080:3-4. 1960.) has suggested that A. physalidis
may be an Endophyllum. If so, there is a similar aecial form
that is associated with P. kansensis.

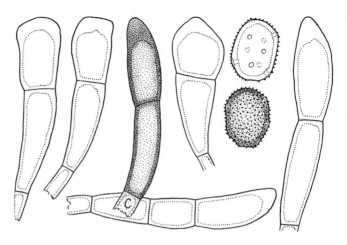

Figure 211

191. PUCCINIA LONGISSIMA Schroet. Beitr. Biol. Pfl. 3:70. 1879.
Fig. 211.

Aecia (Aecidium sedi (DC.) Schroet.) occur on species of
Sedum, Rhodiola, systemic, opening by a pore; spores (22-)24-27
(-31) x (18-)21-24(-26)μ, from globoid to oblong, wall (1-)2-3
(-4)μ thick, colorless or yellowish, verrucose-echinulate.
Uredinia on adaxial leaf surface, yellowish brown; spores (23-)
25-29(-34) x (21-)23-26(-30)μ, globoid, broadly ellipsoid, or
obovoid, wall 1.5-2(-3)μ thick, yellowish or golden, echinulate,
germ pores (7?-)9-12, scattered. Telia on adaxial surface,
blackish brown, early exposed, compact; spores (54-)70-100(-125)
x (13-)17-22(-30)μ, mostly cylindrical to elongately clavate,
wall 1-2(-2.5)μ thick at sides, (5-)7-12(-18)μ apically, mostly
golden brown, smooth; pedicels persistent, brownish, less than
25μ long.

Hosts and distribution: species of Koeleria: Europe and
northern Africa.

Lectotype: Gerhardt, on Koeleria cristata (L.) Pers., "durch
H. Gerhardt in Liegnitz erhielt." Schroeter (in Cohn Krytog.
Flora Scles. III. 1, p. 339. 1887) lists the locality as Jauer:
Hesseberge am Rehbock. In B, there is a specimen in "Herb. G.
Winter" collected by Gerhardt 19. 9. 78, which doubtless is a
part of the original. The lectotype designation is mine.

Bubák (Centrlbl. Bakt. II. 9:126. 1902) first demonstrated
the life cycle experimentally.

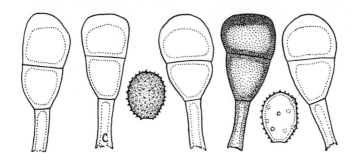

Figure 212

192. PUCCINIA MEXICENSIS H. C. Greene & Cumm. Mycologia 50:24.
1958. Fig. 212.

Aecia not known. Uredinia on abaxial leaf surface, spores
(17-)19-23(-24) x (16-)17-20µ, broadly ellipsoid or globoid,
wall 1-1.5µ thick, hyaline or yellowish, echinulate, pores 7 or
8, scattered. Telia on abaxial surface, early exposed, blackish,
pulvinate; spores (31-)35-45(-53) x (16-)20-27(-30)µ, clavate
or oblong-clavate, wall 1.5-2(-3)µ thick at sides, (4-)6-9(-10)µ
apically, chestnut-brown, smooth; pedicels brownish, mostly
25-40µ long.

Hosts and distribution: Stipa constricta Hitchc., S. eminens
Cav., S. lettermani Vasey: central and south central Mexico
and southern New Mexico, U.S.A.

Type: Lyonnet No. 1957, on S. constricta, Lomas de Michoac,
Dist. Fed., Mexico (PUR).

Greene and Cummins (loc. cit.) published a photograph of
teliospores of the type.

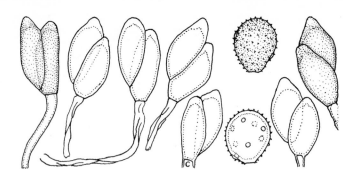

Figure 213

193. PUCCINIA ABNORMIS P. Henn. Hedwigia 35:243. 1896. Fig. 213.

Diorchidium flaccidum Lagh. Tromso Mus. Aarsh. 17:80. 1895, non Puccinia flaccida Berk. & Br. 1873.

Puccinia subdiorchidioides P. Henn. Hedwigia 35:244. 1896.

Aecial stage unknown. Uredinia amphigenous, cinnamon-brown; spores 18-20(-24) x (13-)17-19µ, mostly broadly ellipsoid or obovoid, wall 1.5-2.5µ thick, cinnamon-brown, echinulate, germ pores 4-6, scattered. Telia amphigenous, early exposed, chestnut-brown, pulvinate; spores (26-)30-35(-48) x 12-15(-27)µ, mostly ellipsoid or oblong-ellipsoid, usually variously diorchidioid, wall 1-1.5µ thick at sides, 2-4µ apically, golden, smooth; pedicels colorless, thin-walled and collapsing, fragile, to 50µ long.

Hosts and distribution: Echinochloa crus-galli (L.) Beauv., E. zelayensis (H.B.K.) Schult., E. holciformis (H.B.K.) Chase: The United States southward to Mexico, Chile, and Argentina.

Type: Galander, on Gymnothrix sp. (error for Echinochloa sp.), Rio Tercero, Prov. de Cordoba, Argentina (B; isotype PUR).

328

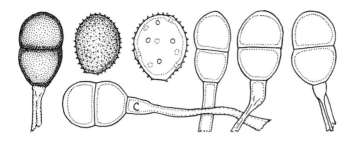

Figure 214

194. PUCCINIA TORNATA Arth. & Holw. in Arthur Proc. Amer. Phil.
Soc. 64:186. 1925. Fig. 214.

Aecia unknown. Uredinia mostly on adaxial leaf surface,
yellowish brown; spores (23-)26-30(-33) x (18-)20-24(-25)μ,
mostly ellipsoid or obovoid, wall 1.5-2(-2.5)μ thick, yellowish,
echinulate, germ pores 9-12 scattered. Telia mostly on adaxial
surface, early exposed, rather pulverulent, chocolate-brown;
spores (28-)32-38(-41) x (18-)20-23(-25)μ, ellipsoid, wall
(1-)1.5-2(-2.5)μ thick at sides, 2-4μ apically, smooth, chestnut-
brown, germ pores very obscure but the lower one often depressed
half way to the hilum; pedicels colorless or yellowish, collapsing, to 50μ long but usually broken near spore.

Hosts and distribution: Hordeum andinum Trin.: Bolivia.

Type: Holway No. 474, La Paz, Bolivia (PUR); isotypes
Reliq. Holw. No. 73). Arthur (loc. cit.) published a photograph
of teliospores of the type.

329

195. PUCCINIA AGROSTIDICOLA Tai Farlowia 3:115-116. 1947.

Aecia unknown. Uredinia hypophyllous or sometimes on sheaths, elongate or linear, yellowish brown; spores 24-33 x 23-30µ, globoid or rarely ovoid, wall 1.5-2µ thick, yellowish, germ pores 8-10, scattered. Telia like the uredinia but pulvinate, blackish; spores 36-56 x 17-27µ, ellipsoid or ellipsoid-oblong, conically attenuate or rarely rounded apically, slightly constricted at the septum, wall 1.5-2µ thick at sides, 8.5-11µ apically, chestnut-brown, smooth; pedicels brownish, to 46µ long; 1-celled spores occasional.

Type: W. L. Hsian, on Agrostis sp., Yungdun, Kansu, China, 23 Aug. 1943 (Pl. Pathol. Herb. No. 8404, Tsing Hua Univ., Kunming - not seen).

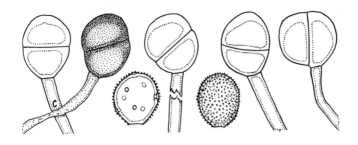

Figure 215

196. PUCCINIA AEGOPOGONIS Arth. & Holw. in Arthur Amer. J.
Bot. 5:467. 1918. Fig. 215.

Aecia occur on species of Eupatorium; peridium short-
cylindric; spores 19-36 x 15-26µ, mostly angularly globoid
or ellipsoid, wall 1-2µ thick at sides, to 7µ apically,
colorless, finely verrucose. Uredinia mostly on abaxial leaf
surface, yellowish brown; spores 24-29 x 22-25µ, broadly
ellipsoid, wall 1-1.5µ thick, yellowish to golden, echinulate,
germ pores 7 or 8, scattered. Telia abaxial and on sheaths,
early exposed, small, blackish; spores 27-31(-33) x (21-)23-28µ,
mostly broadly ellipsoid, commonly diorchidioid, wall 2-3µ
thick at sides, (4-)5-7µ apically, chestnut-brown, smooth;
pedicels thick-walled, mostly not collapsing, yellowish, to
55µ long.

Hosts and distribution: Aegopogon cenchroides Humb. &
Bonpl., A. tenellus (DC.) Trin.: Mexico, Guatemala, Bolivia,
and Ecuador.

Type: Holway No. 54, on A. cenchroides, San Rafael, Dept.
Guatemala, Guatemala (PUR).

Uromyces aegopogonis Diet. & Holw. is similar, except for
the teliospores. It is doubtful if the aecia are distinguishable.

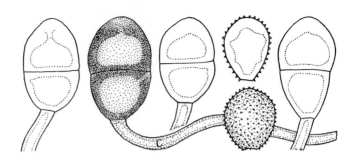

Figure 216

197. PUCCINIA VERSICOLOR Diet. & Holw. in Holway Bot. Gaz.
24:28. 1897. Fig. 216.

Uredo themeda Diet. Ann. Mycol. 6:228. 1908.

Puccinia trachypogonis Speg. Mus. Nac. Buenos Aires Anal.
19:301. 1909.

Puccinia calchakiana Speg. Rev. Argent. Bot. 1:110. 1925.

Puccinia variospora Arth. & Holw. in Arthur Amer. Philos.
Soc. Proc. 64:182. 1925.

Puccinia filipodia Cumm. Ann. Mycol. 35:98. 1937.

Puccinia themedae Hirat. f. Bot. Mag. Tokyo 56:279. 1942.

The aecia, Aecidium plectroniae Cke., occur on species of
Canthium and Lantana; spores 23-25 x 19-21µ, wall 1.5-2µ thick
at sides, to 5µ apically, verrucose. Uredinia mostly on abaxial
leaf surface, yellow; spores (22-)25-33(-38) x (19-)21-28(-30)µ,
mostly broadly ellipsoid, wall 3-6(-8)µ thick, the inner surface
irregular and giving a stellate appearance to the lumen, colorless, moderately echinulate, germ pores 8-11, scattered, very
obscure; occasional collections have cinnamon-brown spores
with a uniformly 2-2.5µ thick wall. Telia mostly on abaxial
surface, to 4 mm long, early exposed, pulvinate, blackish brown;
spores (33-)35-46(-50) x (22-)25-32(-35)µ, mostly broadly
ellipsoid or oblong-ellipsoid, wall (2.5-)3-4(-5)µ thick at
sides, 4-8(-12)µ apically, deep golden or clear chestnut-brown,
smooth; pedicels colorless, mostly thin-walled and collapsing
at least in the lower part, to 130µ long.

Hosts and distribution: Andropogon, Bothriochloa, Capillipedium, Cymbopogon, Heteropogon, Hyparrhenia, Ischaemum,
Monocymbium, Themeda, Trachypogon: Mexico southward to
Argentina and eastward to Africa, India, New Guinea, Japan and
the Hawaiian Islands.

Type: E. W. D. Holway, on Heteropogon melanocarpus, Guadalajara, Mexico (S; isotype MIN, PUR).

332

Cummins (Uredineana 4: Plate IX. 1953) published photographs of teliospores of the species and of most of the synonyms.

Inoculations proving the aecial stage were made on Canthium (Plectronia) parviflorum by Thirumalachar and Narasimhan (Current Sci. 18:252-253, 1949) and on Lantana indica by Patil and Thirumalachar (Current Sci. 33:253. 1964).

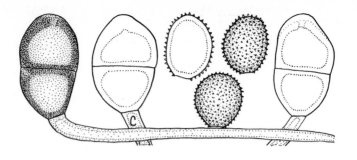

Figure 217

198. PUCCINIA CHRYSOPOGI Barcl. Asiatic Soc. Bengal J. 58:247. 1889. Fig. 217.

Puccinia jasmini-chrysopogonis Barcl. Linn. Soc. London Trans. Ser. II. 3:241. 1891.

Puccinia hookeri Syd. Monogr. Ured. 1:723. 1904.

Puccinia burmanica Syd. & Butl. Ann. Mycol. 10:261. 1912.

Aecia, Aecidium jasmini Barcl., occur on species of Jasminum; spores 23-28μ diam, nearly globoid, wall 1.5μ thick, verrucose. Uredinia on abaxial leaf surface, often confluent, yellow; spores (20-)24-30(-32) x (18-)20-23(25)μ, globoid, broadly ellipsoid, or ellipsoid, wall 2-3(-3.5)μ thick, the lumen tending to be stellate, finely echinulate, colorless or yellow-ish, germ pores scattered, very obscure. Telia amphigenous, early exposed, pulvinate, blackish brown; spores (38-)42-52(-57) x 24-32(-35)μ, mostly broadly ellipsoid or oblong-ellipsoid, wall (2.5-)3-4μ thick at sides, 6-10μ apically, clear chestnut-brown or golden brown, smooth; pedicels colorless, moderately thick-walled, usually collapsing only in lower part if at all, to 140μ long.

Hosts and distribution: Chrysopogon echinulatus (Steud.) W. Wats., C. gryllus (L.) Trin., Themeda anathera (Nees) Hack., T. quadrivalvis (L.) Kuntze: Burma and India.

Neotype: Hooker and Thompson, on Andropogon echinulatus (=Chrysopogon echinulatus), Himalaya bor. or. (type of P. hookeri), (S.).

Barclay (loc. cit., 1891) proved the life cycle by inocula-tion, using Jasminum humile as the aecial host. Cummins (Uredineana 4:5-89. 1953) published a photograph of teliospores of the neotype.

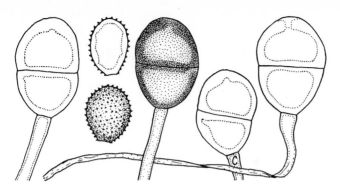

Figure 218

199. PUCCINIA ARTHRAXONIS Syd. & Butl. Ann. Mycol. 5:499. 1907.
Fig. 218.

Uromyces arthraxonis P. Henn. Bot. Jahrb. 14:370. (based on uredia).

Spermogonia and aecia unknown. Uredinia amphigenous, to 2 mm long, yellow; (20-)28-30(-33) x 18-25μ, globoid, ellipsoid, or obovoid, wall (2.5-)3(-3.5)μ thick, the lumen tending to be stellate, colorless, echinulate, germ pores 7-9, very obscure. Telia on abaxial surface, to 3 mm long, pulvinate, blackish brown; spores (32_)35-42(-47) x (25-)27-33(-35)μ, mostly broadly ellipsoid, wall 3-4μ thick at sides, 4-5(-6)μ apically, rather clear chestnut-brown, smooth; pedicels colorless, thin-walled, collapsing, to 100μ long but usually broken shorter.

Hosts and distribution: Arthraxon lanceolatus (Roxb.) Hochst., A. meeboldii Stapf, A. serrulatus (Link) Hochst.: Eritrea, India.

Type: E. J. Butler No. 764, on A. lanceolatus, Dehra Dun, India. 23 Nov. 1902 (S).

Cummins (Uredineana 4:1-89. 1953) published a photograph of teliospores of the type.

335

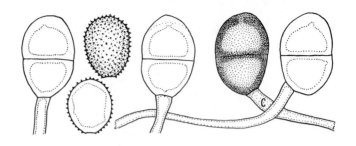

Figure 219

200. PUCCINIA AGROPHILA H. Syd. Ann. Mycol. 35:246. 1937.
Fig. 219.

Aecia (Aecidium habunguense P. Henn.) on Solanum incanum
L., S. indicum L., spores 18-22 x 15-18μ, angularly globoid,
wall 1μ thick, finely verrucose, colorless or pale yellowish.
Uredinia on abaxial surface, yellow; spores (18-)23-27(-29) x
(16-)18-23μ, mostly globoid or broadly ellipsoid, wall
uniformly (1.5-)2-3(-3.5)μ thick, or the inner surface
invaginated at the pores to give a slightly stellate appearance
to the lumen, echinulate with low spines, germ pores 6-8,
obscure. Telia on abaxial surface, early exposed, blackish
brown; spores (28-)33-40(-43) x (19-)21-26(-28)μ, broadly
ellipsoid or broadly obovoid, wall (2-)3(-4)μ thick at sides,
4.5-6(-8)μ apically, deep golden or clear chestnut-brown, smooth;
pedicel thin-walled, commonly collapsing, colorless, 60-135μ
long.

Hosts and distribution: Andropogon appendiculatus Nees, A.
gabonensis Stapf, A. gayanus Kunth, A. tectorum Schum. & Thonn.,
Capillipedium hugelii (Hack.) Stapf: Africa and India.

Type: Deighton 692, on Andropogon tectorum, Rokupr, Sierra
Leone (Isotypes IMI, PUR).

The life cycle was proved by reciprocal inoculations, using
Solanum indicum and Capillipedium hugelii, by Patil and
Thirumalachar.

A photograph of teliospores of the type was published by
Cummins (Uredineana 4: Pl. VIII, Fig. 45. 1953).

336

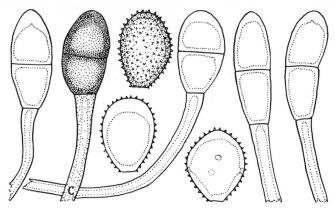

Figure 220

201. PUCCINIA ARUNDINELLAE-ANOMALAE Diet. Bot. Jahrb. 37:100.
1905. Fig. 220.

Uredo yoshinagai Diet. Bot. Jahrb. 37:109. 1905.

Aecia on Stachys japonica Mig. var. intermedia Ohwi; spores
21-28(-30) x 17-23(-27)μ, wall yellowish, 1.5(-2.5)μ thick,
verrucose. Uredinia amphigenous or mostly on adaxial surface,
pale yellowish when dry, probably bright yellow when fresh;
spores (24-)28-36(-39) x (18-)22-28(-30)μ, mostly ellipsoid,
or obovoid, wall 2-3(-4)μ thick at sides, apical wall the same
or often 4-8μ thick (or to 12μ in type of Uredo yoshinagai),
colorless, echinulate, germ pores obscure, about 6-8, scattered
but tending to be in the equatorial region. Telia amphigenous,
blackish brown, compact, early erumpent; spores (32-)38-54 x
(16-)19-24μ, spores ellipsoid or obovoid, tending to be
dimorphic with the elongate spores paler than the robust spores,
wall (1.5-)2-3(-3.5)μ thick at sides, 4-7(-9)μ at apex,
chestnut-brown, smooth; pedicels colorless or yellowish, thick-
walled, persistent, to 100μ long but usually about 80μ long.

Hosts and distribution: Arundinella anomala Steud., A. sp.:
China and Japan.

Type: Kusano, on Arundinella anomala, Tokyo, Japan (S).

This species differs from P. arundinellae in having narrower
teliospores with thinner walls and urediospores that are larger
and have thicker walls. The apical thickening of the urediniospore wall is variable in both magnitude and frequency.

Cummins and Greene (Trans. Mycol. Soc. Japan 7:52-57. 1966)
published photographs of spores of the type. Hiratsuka and Sato
(Bot. Mag. Tokyo 64:219-222. 1951) proved the life cycle.

337

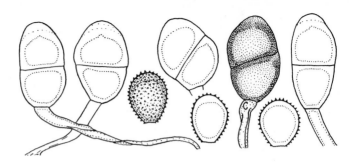

Figure 221

202. PUCCINIA DIETELII Sacc. & Syd. in Saccardo Syll. Fung. 14:358. 1899. Fig. 221.

Puccinia chloridis Diet. Hedwigia 31:290. 1892, not Speg. 1891.

Puccinia chloridina Bacc. Annali Bot. 4:269. 1906.

Puccinia chloridicola P. Henn. Flora Bas. Moy.-Congo Ann. Mus. Congo 2(2):90. 1907.

Puccinia dactyloctenii Pat. & Har. Bull. Soc. Mycol. France 24:13. 1908.

Aecia on Acalypha; spores (13-)15-18(-20) x (10-)12-15(-17)μ, wall 1μ thick, colorless, verrucose. Uredinia amphigenous, yellow or pale brownish; spores 17-26 x 15-21μ, ellipsoid or obovoid, wall 1.5-2μ thick laterally, 3-10μ apically, hyaline to golden, echinulate, pores obscure, probably scattered. Telia amphigenous, early exposed, blackish, pulvinate; spores 24-35 x 17-24μ, mostly broadly ellipsoid, wall 2-3μ thick at sides, 5-7μ apically, dark chestnut-brown, smooth; pedicels golden, thin-walled, collapsing, to 75μ long.

Hosts and distribution: species of Chloris, Dactyloctenium aegypticum (L.) Beauv.: southern U.S.A. to Argentina and in Africa.

Type: Bartholomew No. 526, on Chloris verticillata, Rooks County, Kansas (S; isotype PUR).

Cummins proved the life cycle by inoculation (Mycologia 55:73-78. 1963).

338

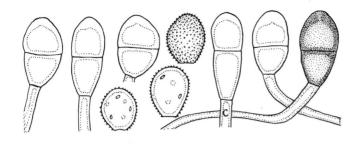

Figure 222

203. PUCCINIA ZOYSIAE Diet. Bot. Jahrb. 32:48. 1902. Fig. 222.

Puccinia ischaemi Diet. Ann. Mycol. 15:493. 1917.

Aecia (Aecidium paederiae Diet.) occur on species of Paederia, peridia short; spores 16-21 x 12-17μ, globoid or broadly obovoid, wall 1-1.5μ thick at sides, 3-8μ apically, hyaline, finely verrucose. Uredinia on adaxial leaf surface, bright yellow when fresh, nearly colorless when dry; spores 17-22 x (14-)15-18μ, mostly obovoid or ellipsoid, wall uniformly 1.5-2.5μ thick or thickened apically to 8μ, the thick-walled spores common in some collections, rare in others, echinulate, yellowish or colorless, germ pores very obscure, probably about 6, scattered. Telia amphigenous, early exposed, blackish brown; spores (28-)30-40(-42) x (15-)16-22(-24)μ, mostly ellipsoid, wall 1.5-2.5μ thick at sides, (3-)4-6(-7)μ apically, chestnut-brown except a usually pale differentiated area at the apex, smooth; pedicels mostly thick-walled and not collapsing, yellowish, to 100μ long.

Hosts and distribution: species of Zoysia: Manchuria, China, Japan, and the United States.

Type: Kusano No. 249, on Zoysia pungens Willd. (= Z. matrella) Komaba in Tokyo, Sept. 1899 (S). The specimen is "Ex Herb. Dietel" and Puccinia zoysiae Diet. is in Dietel's handwriting, hence is to be taken as the holotype.

The presence and proportion of urediniospores with a thickened apical wall varies greatly. Short, colorless, thin-walled paraphyses occur in some collections but apparently not in all.

Asuyama (Ann. Phytopathol. Soc. Japan 5:23-29. 1935) proved the life cycle by inoculation of Paederia chinensis.

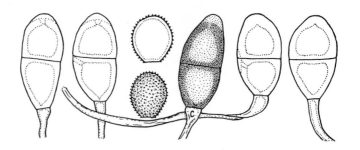

Figure 223

204. PUCCINIA GYMNOPOGONICOLA Hennen, in Hennen & Cumm.,
Mycologia 48:134. 1956. Fig. 223.

Aecia unknown. Uredinia on adaxial leaf surface, yellow;
spores 16-22 x 13-16µ, oval or nearly globoid, wall 1.5-2µ thick,
hyaline or yellowish, echinulate, pores obscure, probably scattered. Telia amphigenous, blackish, early exposed, pulvinate;
spores (26-)28-41 x (16-)18-22µ, ellipsoid or oblong-ellipsoid,
wall 3-5µ thick laterally, 4-6µ apically, golden or clear
chestnut-brown, bilaminate, smooth; pedicels hyaline or brownish, thin-walled, collapsing, attaining a length of 80µ.

Hosts and distribution: Gymnopogon burchellii (Munro) Ekman,
G. spicatus (Spreng.) Kuntze: Brazil, Argentina.

Type: Holway No. 1888, (Isotypes, Reliq. Holw. No. 146 as
Puccinia gymnopogonis Syd.), on Gymnopogon burchellii, Mandaqui,
Sao Paulo, Brazil (PUR).

Hennen and Cummins (loc. cit.) published a photograph of
teliospores of the type.

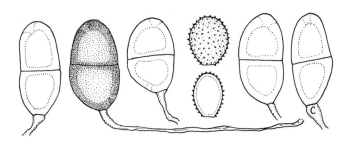

Figure 224

205. PUCCINIA NYASALANDICA Cumm. Torrey Bot. Club Bull. 83:228. 1956. Fig. 224.

Aecia unknown. Urediniospores in the telia 19-21 x 14-18μ, ovate or globoid, wall 1μ thick, colorless, finely echinulate, germ pores obscure, probably scattered. Telia epiphyllous, to 2 mm long and commonly confluent, early exposed, pulvinate, blackish brown; spores (25-)30-38 x 17-20(-22)μ, ellipsoid or oblong-ellipsoid, wall 3-4μ thick at sides, 4-5(-6)μ apically, golden brown, smooth, germ pore near pedicel in lower cell; pedicels colorless, thin-walled, collapsing, to at least 85μ long but usually broken short.

Hosts and distribution: <u>Brachiaria</u> <u>decumbens</u> Stapf: Nyasaland.

Type: P. O. Wiehe No. 752, Muso, Kirk Range, Nyasaland, June 13, 1950 (PUR; isotype IMI).

Cummins (loc. cit.) published a photograph of teliospores of the type.

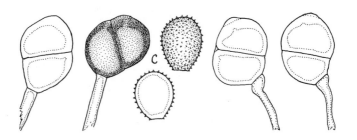

Figure 225

206. PUCCINIA BOUTELOUAE (Jennings) Holw. Ann. Mycol. 3:20.
1905. Fig. 225.

Diorchidium boutelouae Jennings Bull. Texas Exp. Sta. 9:25.
1890.

Puccinia gymnopogonis Syd. Monogr. Ured. 1:755. 1903.

Uredo chardonii Kern in Seaver et al. Sci. Surv. Puerto Rico
& Virgin Isl. 8:140. 1932.

Aecia unknown. Uredinia amphigenous, orange when fresh;
spores (14-)16-23 x (12-)15-19μ, spores globoid or obovoid,
wall hyaline or yellowish; (1.5)2-3μ thick, echinulate, germ
pores obscure, probably 6-8, scattered. Telia amphigenous,
blackish, pulvinate; spores (21-)25-33 x (18-)20-27(-29)μ,
mostly broadly ellipsoid, mostly diorchidioid, wall 2.5-3μ
thick at sides, 5-7μ apically, chestnut-brown, smooth; pedicel
hyaline or golden, thin-walled and collapsing, to 120μ long.

Hosts and distribution: species of Bouteloua, Cathestecum
erectum Vasey & Hack., Gymnopogon foliosus (Willd.) Nees: southwestern U.S.A. south to Panama, Puerto Rico and Brazil.

Type: Jennings, on Bouteloua curtipendula, College Station,
Texas, (BPI; isotype PUR).

Hennen and Cummins (Mycologia 48:126-162. 1965) published
a photograph of teliospores of the type.

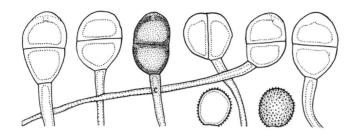

Figure 226

207. PUCCINIA SUBTILIPES Speg. An. Mus. Nac. Hist. Nat. Buenos Aires 31:386. 1922. Fig. 226.

Aecia unknown. Uredinia amphigenous, yellow; spores 16-18 x 13-15μ, obovoid or globoid, wall 1-1.5μ thick, hyaline or yellowish, echinulate, pores obscure, probably scattered. Telia amphigenous, blackish, early exposed, pulvinate; spores 23-31 x 18-22μ, oblong-ellipsoid or broadly ellipsoid, wall 2-3μ thick at sides, 3-4μ apically, chestnut-brown, smooth; pedicels thin-walled, collapsing, yellowish or colorless, attaining a length of 130μ.

Hosts and distribution: Leptochloa scabra Nees, L. virgata Beauv.: Mexico and the Dominican Republic southward to Argentina.

Type: Spegazzini, on Leptochloa virgata, Ascuncion, Paraguay, (LPS; isotype PUR).

Hennen and Cummins (Mycologia 48:126-162. 1956) published a photograph of teliospores of the type.

208. PUCCINIA SINICA H. Syd. Ann. Mycol. 27:419. 1929.

Aecia unknown. Uredinia mainly on abaxial leaf surface,
yellowish; spores 14-19 x 12-13µ subgloboid, globoid, or ovoid,
wall 1.5µ thick, colorless, finely echinulate, pores obscure.
Telia abaxial, blackish, early exposed, pulvinate; spores 26-38
x 12-17µ ellipsoid, ovoid, or oblong, wall 1.5-2µ thick at
sides, 3-3.5µ apically; pedicels colorless, to 70µ long, occasionally inserted obliquely.

Hosts and distribution: Muhlenbergia longistolon Ohwi (M.
huegelii Auth. not Trin.): China.

Type: Sydow No. 2254, Kiangsu, Nanking, 24 Sept. 1928.

No material of this species has been available.

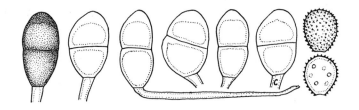

Figure 227

209. PUCCINIA SCLEROPOGONIS Cumm. Southw. Nat. 8:190. 1964.
Fig. 227.

Aecia on Chamaesaracha sordida (Dunal) Gray; spores (16-)
17-21(-23) x (14-)15-18(-20)μ, wall 2(-2.5)μ thick, hyaline or
yellowish, verrucose. Uredinia mostly on adaxial surface,
yellow; spores 16-19 x 13-16μ, broadly ellipsoid or obovoid,
wall (0.5-)1-1.5μ thick, colorless, echinulate, germ pores
obscure, scattered, probably 7 or 8. Telia amphigenous or
mostly adaxial, blackish brown, compact, early erumpent; spores
(26-)28-35(-40) x (15-)17-20(-23)μ, mostly ellipsoid, wall 2-3μ
thick at sides, 4-7μ at apex, nearly uniformly chestnut-brown,
smooth; pedicels colorless or yellowish, persistent, to 100μ
long.

Hosts and distribution: Scleropogon brevifolius Philippi:
New Mexico and Texas, U.S.A., and San Luis Potosi, Mexico.

Type: Cummins No. 62-423, on Scleropogon brevifolius, Texas
(PUR).

Puccinia diplachnicola Diet. is similar morphologically.

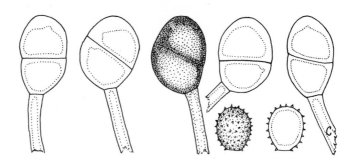

Figure 228

210. PUCCINIA HILARIAE Cumm. Southw. Nat. 12:78. 1967. Fig. 228.

Aecia unknown. Uredinia not seen; spores 17-19 x 14-18μ, broadly obovoid or globoid, wall 2-2.5μ thick, hyaline, echinulate, pores obscure, doubtless scattered. Telia amphigenous or mostly on adaxial leaf surface, early exposed, pulvinate, blackish brown; spores (25-)28-35(-39) x 22-25(-29)μ, broadly ellipsoid, sometimes diorchidioid, wall (2-)2.5-3.5(-5)μ thick at sides, (4-)5-8(-9)μ at apex, chestnut-brown, smooth; pedicel hyaline, or often brownish apically, usually not collapsing, to 110μ long.

Hosts and distribution: Hilaria hintonii Sohns: Mexico.

Type: Pringle 11225 (=PUR 59559), Yautepec, Morelos (PUR).

The species is generally similar to P. scleropogonis but has broader teliospores and urediniospores with thicker walls. A photograph of teliospores of the type was published with the original description.

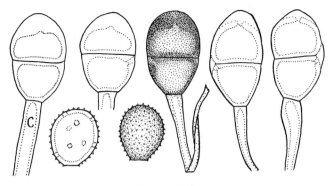

Figure 229

211. PUCCINIA AUSTRALIS Koern. in Thuemen Fungi Austr. No. 842.
1873. Fig. 229.

Aecia (Aecidium erectum Diet.) on Sedum spp; spores 18-20
x 16-18μ, wall 1μ thick, finely verrucose, hyaline. Uredinia
mostly on adaxial surface, nearly colorless when dry, doubtless
bright orange when fresh; spores (15-)17-22 x (14-)16-18(-20)μ,
mostly broadly ellipsoid or obovoid, wall pale yellowish or
hyaline (1.5)2-3μ thick, echinulate, pores obscure, about 8.
Telia mostly on abaxial surface, early exposed, pulvinate,
blackish; spores (27-)30-40(-42) x (17-)21-24(-26)μ, mostly
ellipsoid or broadly obovoid, wall 2-3(-4)μ thick at sides,
(5-)7-10(-12)μ at apex, mostly uniformly chestnut-brown or
deep golden-brown, smooth; pedicels hyaline or pale yellowish,
rather thick-walled and mostly not collapsing, to 100μ long
but usually shorter.

Hosts and distribution: Cleistogenes serotina (Lk.) Keng,
C. squarrosa (Trin.) Keng: Europe to U.S.S.R. and China.

Type: Körnicke, on Molinia serotina, near Bozen, Austria
(B; isotypes No. 842 Thuem. Fungi Austriaci).

Pazschke (Hedwigia 33:84-85. 1894) first demonstrated the
aecial stage.

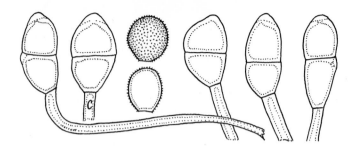

Figure 230

212. PUCCINIA DIPLACHNICOLA Diet. Ann. Mycol. 8:308. 1910.
Fig. 230.

Puccinia moliniicola Cumm. Mycologia 43:92. 1951.

Aecia unknown. Uredinia mostly abaxial, colorless with age, doubtless bright yellow when fresh; spores (14-)15-18(-19) x (10-)12-14(-15)μ, obovoid, ellipsoid, or nearly globoid, wall 1-1.5μ thick, colorless, finely echinulate, pores obscure, 5-7 (?), scattered. Telia mostly abaxial, early exposed, pulvinate, blackish brown; spores (23-)27-38(-40) x (12-)14-19 (-20)μ mostly ellipsoid, sometimes broadly so, tending to be dimorphic, wall 1.5-2μ thick at sides or to 3μ in robust spores, 3-5(-6)μ at apex, the apex often with a pale, outer area, clear chestnut-brown, smooth; pedicels thick-walled, not collapsing, hyaline, to 90μ long.

Hosts and distribution: Cleistogenes hackelii Honda, C. nakaii Keng, C. serotina (Lk.) Keng: China, Japan, and Korea.

Type: Yoshinaga, on Cleistogenes serotina (as Diplachne serotina var. aristata), Tosa, Japan (S).

The species is characterized by very small urediniospores and dimorphic teliospores, i.e. narrow and robust teliospores in varying proportions. P. moliniicola Cumm. is based on a predominantly narrow-spored collection.

Jørstad (Ark. Bot. Ser. 2. 4(8):333-370. 1959) suggests that this species is synonymous with P. australis but the differences are constant and of recognizeable magnitude.

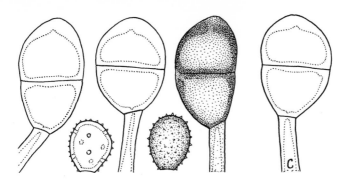

Figure 231

213. PUCCINIA PERMIXTA P. Syd. & H. Syd. Ann. Mycol. 10:216.
1912. Fig. 231.

Aecia on species of Allium as first proved by Treboux (Ann.
Mycol. 10:303-306. 1912); spores 16-22 x 11-16μ, wall about
1μ thick, hyaline, verrucose. Uredinia mostly abaxial, yellowish brown; spores (17-)19-22(-24) x (14-)16-19(-21)μ, mostly
broadly obovoid to nearly globoid, wall (1.5-)2-3(-4)μ, yellowish to pale golden, echinulate, pores obscure, scattered, 8-10.
Telia mostly abaxial, early exposed, pulvinate, blackish brown;
spores (32-)36-43(-46) x (20-)24-27(-32)μ, mostly broadly
ellipsoid or obovoid, wall 2-3(-4)μ thick at sides, (4-)5-8μ at
apex, mostly uniformly chestnut-brown, smooth or sometimes
minutely reticulate-rugose; pedicels thick-walled, mostly not
collapsing, hyaline or pale yellowish, to 90μ long.

Hosts and distribution: Cleistogenes serotina (Lk.) Keng,
C. squarrosa (Trin.) Keng: U.S.S.R. and Afghanistan to China.

Type: Treboux, on Cleistogenes serotina (as Diplachne
serotina), Nowotscherkask, U.S.S.R. (S).

P. permixta is similar to P. australis, differing mainly in
the size of the spores and in having more pigment in the urediniospore walls.

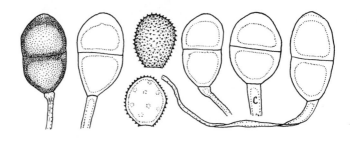

Figure 232

214. PUCCINIA CHLORIDIS Speg., Rev. Argent. Hist. Nat. Buenos
Aires 1:172. 1891. Fig. 232.

Puccinia bartholomaei Diet., Hedwigia 31:290. 1892.

Puccinia jamesiana Arth., Bot. Gaz. 35:18. 1903.

Puccinia trichloridis Speg., An. Mus. Nac. Buenos Aires 19:298.
1909.

Aecia, Aecidium brandegei Pk., occur on species of Asclepias,
Matelea, and Sarcostoma; spores 18-26 x 16-23μ, wall colorless,
2-3μ thick at sides, 7-10μ at apex. Uredinia mostly on adaxial
surface, orange when fresh; spores 18-23 x 16-22μ, mostly broadly
obovoid or globoid, wall 1.5-2.5μ, hyaline or yellowish, echinulate, pores obscure, 5-8, scattered. Telia mostly on adaxial
surface, blackish, pulvinate; spores 26-40 x 16-25μ mostly oblong-
ellipsoid, wall 1.5-2.5μ at sides, 5-9μ apically, chestnut,
smooth; pedicels golden brown, thin-walled, usually collapsing,
attaining a length of 100μ.

Hosts and distribution: Bouteloua curtipendula (Michx.)
Torr., B. gracilis (H.B.K.) Lag., B. hirsuta Lag., Chloris
distichophylla Lag., C. ciliata Swartz, C. venusta Lag.,
Trichloris mendocino (Phil.) Kurtz., T. pluriflora Fourn.:
southern U.S. to Mexico, Bolivia, Brazil, and Argentina.

Type: Balansa, on Chloris sp., Paraguari, Paraguay (LPS).

Arthur (Bot. Gaz. 35:18. 1903) first demonstrated the life
cycle by inoculation.

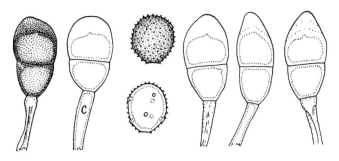

Figure 233

215. PUCCINIA MICRANTHA Griffiths Bull. Torrey Bot. Club
29:299. 1902. Fig. 233.

Aecia occur on species of <u>Ribes</u>; spores (17-)20-25(-28) x
16-22μ, ellipsoid or globoid, wall 1.5-2μ thick, colorless or
pale yellowish, verrucose. Uredinia on adaxial leaf surface,
cinnamon-brown; spores 18-23(-25) x (15-)17-20(-22)μ, mostly
broadly ellipsoid or globoid, wall 1.5(-2)μ thick, cinnamon-
brown or often dark cinnamon-brown, echinulate, germ pores (5)6
or 7(8), scattered. Telia on adaxial surface, rarely amphigenous,
early exposed, blackish brown, compact; spores (30-)36-48(-60)
x (16-)20-26(-30)μ, mostly obovoid when deeply pigmented and
ellipsoid and longer when lightly pigmented, wall 1-1.5(-2.5)μ
thick at sides, (6-)8-12(-16)μ apically, the thicker apex usually
associated with the longer paler spores, mostly chestnut-brown,
the apical thickening progressively paler externally, smooth;
pedicels persistent, yellowish, usually collapsing, to 90μ long.

Hosts and distribution: <u>Oryzopsis micrantha</u> (Trin. & Rupr.)
Thurb.: Nebraska and South Dakota to Montana and New Mexico, U.S.A.

Type: Williams and Griffiths, Billings, Montana (WIS; isotypes
Griffiths W. Amer. Fungi No. 386).

Mains (Mycologia 25:407-417. 1933) published proof of the
life history but Bethel had previously successfully inoculated
<u>Ribes</u> in garden and greenhouse.

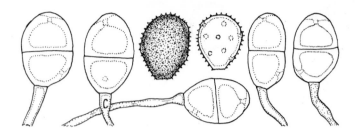

Figure 234

216. PUCCINIA POGONARTHRIAE Hopkins Trans. Rhodesian Sci. Assoc. 35:126. 1938. Fig. 234.

Uredo pogonarthriae H. Syd. & P. Syd. Ann. Mycol. 10:35. 1912.

Puccinia eragrostidis-chalcanthae Doidge Bothalia 3:499. 1939.

Aecia unknown. Uredinia on adaxial leaf surface, cinnamon-brown; spores (23-)25-28(-30) x (19-)21-24(-25)μ, mostly broadly ellipsoid or broadly obovoid, wall 1.5-2.5μ thick, golden brown, echinulate, germ pores 6-9, mostly 7 or 8, scattered. Telia on adaxial surface, early exposed, pulverulent, chestnut- or chocolate-brown; spores (25-)30-37(-40) x (19-)21-25(-27)μ ellipsoid, wall 3-4(-4.5)μ thick at sides, 4-5(-6)μ apically, golden brown or clear chestnut-brown, smooth, germ pore apical in upper cell, midway or lower in lower cell; pedicels to at least 100μ long but fragile, collapsing and often broken short, colorless.

Hosts and distribution: Eragrostis chalcantha Trin., Pogonarthria squarrosa (Licht.) Pilger: Nyasaland, S. Rhodesia, and South Africa.

Type: Hopkins No. 2163, on Pogonarthria squarrosa, Maandellas, S. Rhodesia (IMI; isotype PUR).

The teliospores of P. eragrostidis-chalcanthae are not distinctive and unless it proves to have distinguishing urediniospore features the species surely is synonymous. The length of teliospores published by Doidge is incorrect (probably a typographical error) because they commonly are 36μ long and attain 40μ.

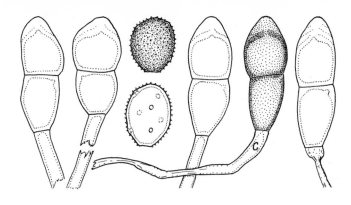

Figure 235

217. PUCCINIA MONOICA Arth. Mycologia 4:61. 1912. Fig. 235.

Aecia (Aecidium parryi Pk.) occur on several genera, especially Arabis, of the Cruciferae, systemic; spores 17-30 x 15-24μ, wall 1.5-2.5μ thick, verrucose, hyaline. Uredinia on adaxial leaf surface, cinnamon-brown; spores (22-)26-30(-35) x (18-)22-26(-28)μ, mostly broadly ellipsoid, wall mostly 1.5-2.5μ thick, golden to cinnamon-brown, echinulate, pores mostly 5-8, scattered or tending to be 3 or 4 and equatorial on Stipa. Telia adaxial, early exposed, pulvinate, blackish brown, early cinereous from germination; spores (33-)40-51(-63) x (16-)19-23(-27)μ, mostly oblong-ellipsoid, wall mostly 1-1.5μ thick at sides, 5-10(-14)μ apically, golden or clear chestnut-brown, smooth; pedicels color- less, thin-walled and collapsing, attaining 120μ in length but usually 100μ or less; germination occurs without dormancy.

Hosts and distribution: Koeleria cristata (L.) Pers., Oryzopsis hymenoides (Roem. & Schult.) Ricker, Poa secunda Presl, Stipa californica Merr. & Davy, S. elmeri Piper & Brodie, S. occidentalis Thurb., Trisetum spicatum (L.) Richter: Wisconsin to British Columbia southward to New Mexico and California.

Type: Garrett, on Trisetum spicatum (T. subspicatum (L.) Beauv.), Big Cottonwood Canyon, Salt Lake County, Utah (PUR; isotypes Fungi Utahensis No. 194.).

Greene and Cummins (Mycologia 50:6-36. 1958) published a photograph of teliospores of the type. Arthur (Mycologia 4:59-61. 1912) first demonstrated the life cycle by inoculation.

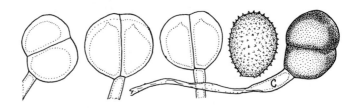

Figure 236

218. PUCCINIA SIERRENSIS Cumm. Southw. Nat. 12:81. 1967.
Fig. 236.

Aecia unknown. Uredinia on abaxial surface, yellow; spores
(20-)22-27(-30) x (17-)19-22(-24)μ, broadly ellipsoid or broadly
obovoid, wall (1.5-)2(-2.5)μ thick, hyaline, echinulate, pores
obscure, scattered, about 7 or 8. Telia usually abaxial, early
exposed, pulvinate, blackish brown; spores (27-)29-35(-39) x
(21-)23-26(-28)μ, broadly ellipsoid, frequently diorchidioid,
wall (1.5-)2-3(-4)μ thick at sides, (3-)4-6(-7)μ at apex,
chestnut-brown, smooth; pedicel hyaline, or brownish next the
spore, mostly not collapsing, to 125μ long.

Hosts and distribution: Muhlenbergia speciosa Vasey: Mexico.

Type: Cummins 63-580, Durango (State), Mexico (PUR).

A photograph of teliospores of the type was published with
the original description.

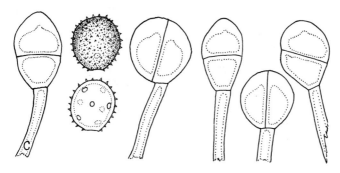

Figure 237

219. PUCCINIA EXASPERANS Holw. Ann. Mycol. 3:21. 1905. Fig. 237.

Aecia unknown. Uredinia amphigenous, cinnamon-brown; spores 22-29 x 17-25μ, broadly ellipsoid or globoid, wall 1.5-2μ thick, cinnamon-brown, echinulate, pores 6-8, scattered. Telia amphigenous, blackish, pulvinate, early exposed; spores 24-31 x 17-26μ globoid or broadly ellipsoid, often diorchidioid, wall 1.5-3μ thick laterally, 4-10μ apically, chestnut-brown, smooth; pedicels yellowish, thick-walled, not collapsing, attaining a length of 125μ.

Hosts and distribution: Bouteloua disticha (H.B.K.) Benth., B. curtipendula (Michx.) Torr., B. pringlei Scribn., B. triathera Benth.: southern United States and Mexico.

Type: Holway No. 5280, on Bouteloua curtipendula, Cuernavaca, Morelos, Mexico (MIN; isotype PUR).

Hennen and Cummins (Mycologia 48:126-162. 1956) published a photograph of teliospores of the type.

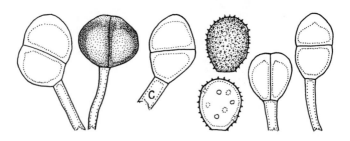

Figure 238

220. PUCCINIA DOCHMIA Berk. & Curt. Amer. Acad. Sci. Proc. 4:126. 1885. Fig. 238.

Puccinia windsoriae Schw. var. australis F. W. Anderson J. Mycol. 6:123. 1891.

Aecia unknown. Uredinia amphigenous, cinnamon-brown or fading to yellowish; spores (19-)22-26(-31) x (16-)18-23(-28)μ, broadly ellipsoid or globoid, wall 1-1.5(-2)μ thick, pale cinnamon or golden, finely echinulate, pores 6-8, scattered. Telia amphigenous and on stems and inflorescence, blackish, early exposed, pulvinate, often confluent in lines; spores (22-)26-30(-38) x (19-)22-25(-29)μ, globoid or broadly ellipsoid, often diorchidioid, wall (1.5-)2-3(-3.5)μ thick at sides, (3-)4-7μ apically, clear chestnut-brown, smooth; pedicels colorless or yellow, mostly thick-walled but tending to collapse, to 125μ long but usually 100μ or less.

Hosts and distribution: species of Muhlenbergia, Pereilema crinitum Presl: Mexico and Central America.

Type: Wright, on Muhlenbergia sp., Nicaragua, before 1858 (K; isotype PUR).

A photograph of teliospores of the type was published by Cummins and Greene (Brittonia 13:271-285. 1961). From field observations, Cummins (Southw. Nat. 12:70-86. 1967) suggested that Abutilon might be the aecial host.

356

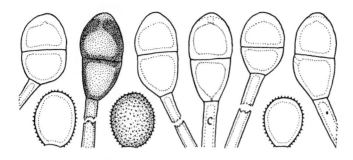

Figure 239

221. PUCCINIA DIPLACHNIS Arth. Bull. Torrey Bot. Club 31:4.
1904. Fig. 239.

Aecia (Aecidium bouvardiae Diet. & Holw.) occur on species
of Bouvardia; spores 20-26(-28) x (17-)19-23μ, ellipsoid,
obovoid, or globoid, wall 1μ thick, finely verrucose. Uredinia
mostly on adaxial leaf surface, orange when fresh, colorless
when old and dry; spores (20-)22-26(-28) x (18-)20-24(-26)μ,
mostly broadly ellipsoid or broadly obovoid, wall (1.5-)2-2.5μ
thick, colorless or pale yellowish, finely echinulate, germ
pores very obscure, scattered, probably 7 or 8. Telia mostly
on abaxial surface and on sheaths, early exposed, blackish brown,
compact; spores (28-)32-40(-44) x (16-)19-25(-28)μ, mostly
broadly ellipsoid or broadly obovoid, wall 1.5-2(-3.5)μ thick
at sides (3.5-)4-6(-7)μ apically, chestnut-brown, smooth;
pedicels rather thin-walled and collapsing, brown next to the
spore, to 125μ long.

Hosts and distribution: Bouteloua gracilis (H.B.K.) Lag.,
Leptochloa dubia (H.B.K.) Nees: Arizona and Texas, U.S.A. south
to Mexico City, Mexico.

Type: Tracy No. 8270, on Diplachne dubia (=Leptochloa dubia)
Big Springs, Texas (PUR 22975).

Cummins (Mycologia 55:73-78. 1963) proved the life cycle
by inoculation.

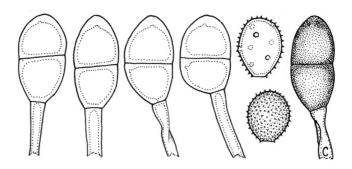

Figure 240

222. PUCCINIA ERAGROSTIDIS Petch Ann. Roy. Bot. Gard. Peradeniya 6:209. 1917. Fig. 240.

Uredo kigeziensis Cumm. Bull. Torrey Bot. Club 70:527. 1943.

Puccinia eragrostidis-ferrugineae Tai Farlowia 3:116. 1947.

Aecia unknown. Uredinia amphigenous or mostly on abaxial leaf surface, orange or yellow (colorless when dry); spores (18-)20-25(-27) x (16-)18-20(-22)µ, mostly broadly ellipsoid or broadly obovoid, wall 1-1.5µ thick, pale yellow or colorless, echinulate, germ pores very obscure, scattered, about 7 or 8. Telia amphigenous, exposed, blackish brown, compact; spores (26-)30-38(-42) x (16-)19-22(-24)µ, mostly ellipsoid or broadly ellipsoid, wall (1.5-)2(3)µ thick at sides, (3-)4-5(-7)µ apically, chestnut-brown, smooth; pedicels colorless or pale yellowish, mostly thin-walled, collapsing or not, to 130µ long, mostly about 100µ.

Hosts and distribution: Eragrostis barrelieri Daveau, E. ferruginea (Thunb.) Beauv., E. nigra Nees: Ceylon and India to China.

Neotype: Petch, on Eragrostis nigra, Hakgala, Ceylon, Apr. 1917 (K), designated here.

The record of E. barrelieri was reported by Joerstad (Ark. Bot. Ser. 2. 4:333-370. 1959). The only Petch collection with telia at Kew is the neotype. Also courtesy of Kew, 7 specimens of E. nigra from the grass herbarium, all rusted, and 4 with telia, from India, Assam, Nepal, Tehri, and N. Burma.

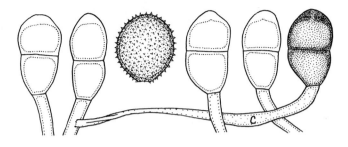

Figure 241

223. PUCCINIA MALALHUENSIS Lindq. Rev. Fac. Agron. Univ. Nac.
La Plata 38:85-86. 1962. Fig. 241.

Aecia unknown. Urediniospores in telia 26-33 x 22-26, broadly
ellipsoid or globoid, wall 1.5-2μ thick, pale yellowish, echinulate, germ pores obscure, scattered. Telia on adaxial leaf
surface, early exposed, blackish brown, compact; spores (28-)32-
40(-45) x (15-)18-22(-24)μ, mostly ellipsoid or obovoid, wall
(1.5-)2-2.5μ thick at sides, (2.5-)3.5-6(-7)μ apically, chestnut-
brown, smooth; pedicels persistent, yellowish to brownish, not
collapsing, to 120μ long; brown sporogenous basal cells conspicuous.

Type: Ruiz Leal No. 21. 547, on Stipa gynerioides Phil.,
Malahue, Mendoza, Argentina (LPS 30:707; isotype PUR). Not
otherwise known.

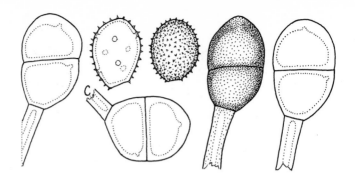

Figure 242

224. PUCCINIA NEYRAUDIAE H. Syd. & P. Syd. in Sydow and Butler
Ann. Mycol. 10:260. 1912. Fig. 242.

Aecia unknown. Uredinia amphigenous, colorless when dry,
doubtless yellow or orange fresh; spores (23-)25-30(-32) x (17-)
19-22(-23)μ, mostly broadly ellipsoid or obovoid, wall 1.5μ
thick, colorless or very pale yellowish, echinulate, germ pores
6-8, scattered, very obscure. Telia amphigenous, exposed, loosely
pulvinate, chocolate-brown; spores (28-)32-40(-45) x (20-)23-
28(-32)μ, mostly broadly ellipsoid or broadly obovoid, wall
(2-)2.5-3.5(-4)μ thick at sides, (4-)4.5-6(-7)μ apically, clear
chestnut-brown, smooth; pedicels yellowish or colorless, thick-
walled, not collapsing, to 160μ long.

Type: Kawakami (Butler No. 1610), on Neyraudia madagascarensis
Hook. f., Mungpoo, Darjeeling, India (S). Not otherwise reported.

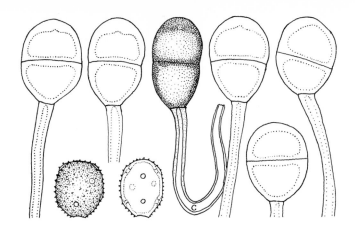

Figure 243

225. PUCCINIA SCHEDONNARDI Kell. & Swing. J. Mycol. 4:95. 1888.
Fig. 243.

Puccinia windsoriae Burr. Ill. Lab. Nat. Hist. Bull. 2:197.
1885, not Schweinitz 1832.

Puccinia triodiae Ell. & Barth. Erythea 4:3. 1896.

Puccinia epicampis Arth. Bull. Torrey Bot. Club. 28:662.
1901.

Puccinia muhlenbergiae Arth. & Holw. Univ. Iowa Lab. Nat.
Hist. Bull. 5:317. 1902.

Puccinia tosta Arth. Bull. Torrey Bot. Club 29:228. 1902.

Puccinia subglobosa Speg. Mus. Nac. Buenos Aires Anal. 19:300.
1909.

Puccinia spegazziniella Sacc. & Trott. in Sacc. Syll. Fung.
20:627. 1911.

Puccinia melicina Arth. & Holw. Am. Philos. Soc. Proc. 64:191.
1925.

Aecia, Aecidium hibisciatum Schw., occur on Hibiscus and other
genera of Malvaceae, spores (16-)20-24(-28) x (12-)16-19(-23)μ,
ellipsoid or globoid, wall (0.5-)1-1.5(-2.5)μ thick, colorless,
finely verrucose. Uredinia amphigenous, pale cinnamon-brown;
spores (18-)21-26(-30) x (15-)18-24(-28)μ, wall 1-2μ thick, pale
cinnamon-brown, echinulate, pores (5-)6-8(-10), scattered. Telia
amphigenous, blackish, early exposed, pulvinate; spores (24-)28-
36(-45) x (16-)18-25(-29)μ, mostly ellipsoid or oblong-ellipsoid,
rarely diorchidioid, wall (1-)1.5-2(-3)μ thick at sides, 3-7(-10)μ
apically, chestnut-brown, smooth; pedicels mostly colorless,
mostly thick-walled but sometimes collapsing, to 125μ long but
usually less than 100μ.

Hosts and distribution: Lycurus, Melica, Muhlenbergia, Schedonnardus, Sporobolus, and Triplasis: U.S.A. from New York to Washington and southward to the Gulf of Mexico, Mexico and southward to Peru and Argentina; and in the Philippines and Japan.

Type: Kellerman & Swingle, on Schedonnardus paniculatus (as S. texanus), Manhattan, Kansas (KSC; isotype PUR). Apparently the same specimen issued as No. 2246 in Ellis & Everhart N. Am. Fungi.

The first successful inoculations proving the life cycle were by Kellerman (J. Mycol. 9:225-238. 1903) using Muhlenbergia and Hibiscus.

Greene and Cummins (Brittonia 13:271-285. 1961) published photographs of teliospores of the type of P. schedonnardi and P. epicampis.

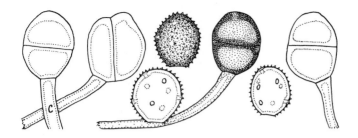

Figure 244

226. PUCCINIA LEPTOCHLOAE-UNIFLORAE Cumm. n. sp. Fig. 244.

Aeciis ignotis. Urediniis plerumque epiphyllis, cinnamomeo-brunneis; sporae 20-24(-26) x (18-)19-22µ, globoideae vel late ellipsoideae, membrana 1.5(-2)µ crassa, pallide cinnamomeo-brunnea, echinulata, poris germinationis 6-8, sparsis. Teliis amphigenis, pulvinatis, atro-brunneis; sporae (25-)28-34 x (21-)22-26(-28)µ, membrana ad latere (2-)2.5-3.5(-4.5)µ crassa, ad apicem (3-)4-6(-7)µ crassa, lucide castaneo-brunnea, levi; pedicello pallide flavido, tenue tunicati, usque ad 60µ longo, persistenti; sporis unicellularibus frequens.

Type: Newbold and Harley No. 4398, on <u>Leptochloa</u> <u>uniflora</u> Hochst., Mt. Kasoje, Kiza Distr., Western Prov., Tanganyika (K; isotype PUR). Not otherwise known.

363

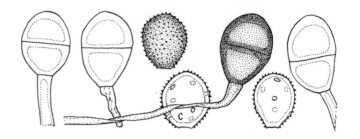

Figure 245

227. PUCCINIA PEROTIDIS Cumm. Torrey Bot. Club Bull. 83:229.
1956. Fig. 245.

Aecia unknown. Uredinia amphigenous, to 1.5 mm long, cinnamon-brown; spores (22-)24-27(-30) x 18-22(-24)μ, ellipsoid or obovoid, wall (1.5-)2-2.5μ thick, cinnamon-brown, echinulate, germ pores 5-7, scattered. Telia on the sheaths (but few seen), early exposed, pulvinate, blackish brown; spores (27-)30-36(-38) x 19-25μ, ellipsoid or clavate-ellipsoid, wall 2.5-3.5(-4)μ thick at sides, 4-6μ apically, dark chestnut-brown, smooth; pedicels yellowish brown, moderately thick-walled, mostly collapsing, to 65μ long.

Hosts and distribution: Perotis indica (L.) O. Kuntze; Sierra Leone.

Type: F. C. Deighton No. 3464, Newton, Sierra Leone (PUR; isotype IMI).

A photograph of spores of the type was published with the original description.

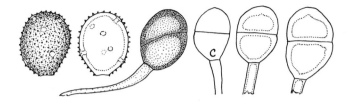

Figure 246

228. PUCCINIA LEPTURI Hirat. f. Trans. Sapporo Nat. Hist. Soc.
17:28. 1941. Fig. 246.

Aecia unknown. Uredinia amphigenous, cinnamon-brown; spores
(20-)23-28(-30) x (17-)19-23(-25)μ, mostly broadly obovoid, wall
1.5-2.5μ thick, pale cinnamon-brown, echinulate, germ pores 6-8,
scattered. Telia amphigenous, early exposed, blackish brown;
spores (22-)24-28(-32) x (18-)22-24(-26)μ, wall (2.5-)3-3.5(-4)μ
thick at sides, 3.5-5(-7)μ apically, chestnut-brown, smooth;
pedicels persistent, mostly collapsing, colorless to brownish,
to 80μ long but usually broken shorter.

Hosts and distribution: Lepturus repens (G. Forst.) R. Br.:
Japan.

Type: Hiratsuka No. 277, Okinawa Island (Herb. Hirat.).
Known only from the Ryuku Islands.

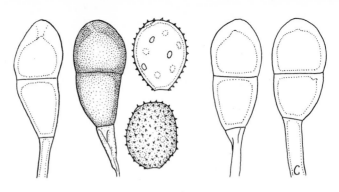

Figure 247

229. PUCCINIA MACRA Arth. & Holw. in Arthur Am. J. Bot. 5:465.
1918. Fig. 247.

Aecia unknown. Uredinia mainly on abaxial leaf surface,
orange or yellow; spores 27-35 x (19-)23-25μ, ellipsoid or
broadly ellipsoid, wall pale yellowish, 1.1.5μ thick, echinulate,
germ pores 4-6(-8), scattered. Telia abaxial and on sheaths,
early exposed, pulvinate, blackish brown; spores (40-)44-53(-63)
x (20-)24-30μ, mostly clavate, wall 1.5-2μ thick at sides, 4-9μ
apically, golden or clear chestnut-brown, smooth; pedicels golden,
thin-walled, collapsing, to 65μ long.

Hosts and distribution: Paspalum candidum (H. B. K.) Kunth,
P. prostratum Scribn. & Merr.: Central America and in northern
South America.

Type: E. W. D. Holway No. 168, on P. candidum, Solola,
Guatemala (PUR).

366

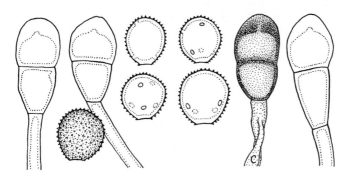

Figure 248

230. PUCCINIA ANDROPOGONIS Schw. Am. Philos. Soc. Trans. II.
4:295. 1832. Fig. 248.

Puccinia pustulata Arth. J. Mycol. 10:18. 1904.

Aecia (Aecidium pentastemonis Schw.) occur on Leguminosae,
Oxalidaceae, Polygalaceae, Rutaceae, Santalaceae, and Scrophulariaceae; spores 16-30 x 15-24μ, wall 1-2μ thick, colorless,
verrucose. Uredinia mostly on abaxial surface, small, to 0.5 mm
long, cinnamon-brown; spores (19-)21-25(-30) x (17-)20-23(-26)μ,
oblate sphaeroid, globoid, or broadly ellipsoid, wall (1-)1.5-
2μ thick, finely echinulate, cinnamon-brown, germ pores various,
5 or 6, scattered, 3 or 4 equatorial, 3 or 4 near the hilum, 2
or 3 near the hilum and 1 or 2 near the apex, or 3 near the
hilum and 3 near the apex. Telia mostly on abaxial surface, to
2 mm long, often confluent, pulvinate, chestnut-brown; spores
(26-)30-44(-50) x (14-)16-21(-24)μ, ellipsoid, oblong-ellipsoid,
or clavate, wall 1.5-2.5(-3)μ thick at sides, (4-)5-8(-9)μ
apically, chestnut-brown, smooth; pedicels yellow or golden,
rather thick-walled, mostly not collapsing, to 60(-70)μ long.

Hosts and distribution: species of Andropogon: Canada
southward to Guatemala.

Type: von Schweinitz, on A. sp. (probably A. scoparius Michx.),
Bethlehem (probably), Pennsylvania, U.S.A. (PHIL; isotype (PUR).

Arthur (Bot. Gaz. 29:272-273. 1900) first proved the life
cycle by inoculating Penstemon as the aecial host. Cummins
(Uredineana 4:1-89. 1953) described, without providing names,
4 variants based mostly on the various arrangements of germ
pores but to some extent on the sizes of spores. He also published a photograph of teliospores of the type.

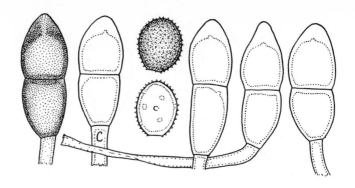

Figure 249

231. PUCCINIA STIPAE Arth. Iowa Agr. Coll. Dept. Bot. Bull.
1884:160. 1884 var. stipae. Fig. 249.

Aecia, Aecidium bigeloviae Peck, occur on several genera of
the Compositae; spores (17-)23-28(-36) x (15-)22-26(-33)μ, wall
(1-)2.5-3.5(-5)μ thick, mostly yellowish to golden, verrucose.
Uredinia on adaxial leaf surface, cinnamon-brown; spores (19-)23-
26(-36) x (16-)20-23(-30)μ, wall mostly 1.5-2.5μ thick, golden
or cinnamon-brown; echinulate, germ pores (4-)6 or 8(-10),
scattered. Telia adaxial, exposed, blackish brown, compact;
spores (33-)43-53(-82) x (17-)20-25(-33)μ, ellipsoid or oblong-
ellipsoid, wall 1.5-2.5μ thick at sides, (4-)5-10(-14)μ apically,
chestnut-brown, smooth; pedicels yellowish, thin-walled and mostly
collapsing, to 175μ long, usually more than 100μ.

Hosts and distribution: species of Stipa: Indiana and Alberta
southward to Mexico and Bolivia.

Type: Bessey, on Stipa spartea Trin., Ames, Iowa (PUR).

Arthur (J. Mycol. 11:63-64. 1905) first proved the life
cycle by inoculation, producing aecia on 4 species of Aster. A
photograph of teliospores of the type was published by Greene
amd Cummins (Mycologia 50:6-36. 1958).

368

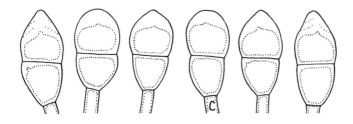

Figure 250

PUCCINIA STIPAE Arth. var. stipae-sibiricae (Ito) H. C. Greene & Cumm. Mycologia 50:22. 1958. Fig. 250.

Puccinia stipae-sibiricae Ito J. Coll. Agr. Tohoku Imp. Univ. 3:228. 1909.

The aecia (Aecidium libanotidis Thuem., A. sedi-aizoontis Tranz.) occur on genera of the Umbelliferae and on Sedum of the Crassulaceae; spores (16-)18-26(-30) x (15-)17-21(-22)μ, wall 1.5-2(-4)μ thick, golden or pale cinnamon-brown, verrucose; urediniospores (16-)18-23(-26) x (15-)16-20(-22)μ; teliospores (34-)36-50(-59) x (15-)18-23(-25)μ, wall 1.5-2μ thick at sides, 6-10(-14)μ apically, golden to clear chestnut-brown.

Hosts and distribution: Stipa effusa Nakai, S. extremiorientalis Hara, S. sibirica Lam.: central Sibiria to Manchuria, and Japan.

Type: Miyabe, on S. effusa, Sapporo, Japan (SAPA; isotype PUR).

Tranzschel (Mycol. Centralbl. 4:70. 1914) first proved the life cycle by inoculation, using Sedum aizoon as the aecial host. Greene and Cummins (Mycologia 50:6-36. 1958) published a photograph of teliospores of the isotype.

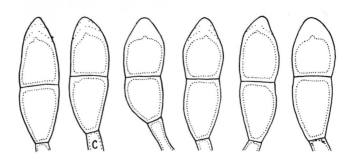

Figure 251

PUCCINIA STIPAE Arth. var. stipina (Tranz.) H. C. Greene & Cumm. Mycologia 50:21. 1958. Fig. 251.

Puccinia graminis foliorum stipae Opiz Seznam Rost Kvet. Ceske p. 138. 1852, nom. nud.

Puccinia stipae (Opiz) Hora Sydow Uredineen No. 28. 1888, nom. nud.

Puccinia stipina Tranz. Trav. Mus. Bot. Acad. Imp. Sci. St. Petersb. 7:114. 1909, nom. nud.

Puccinia stipina Tranz. ex Klebahn Kryptogfl. Mark Brandenburg 5a:477. 1913.

The aecia (Aecidium thymi Fckl.) occur on several genera of the Labiatae; spores (17-)21-23(-30) x (15-)18-21(-25)μ, wall 1-1.5(-3)μ thick, golden or cinnamon-brown; urediniospores (18-)21-24(-30) x (16-)19-22(-25)μ; teliospores (36-)45-56(-67) x (17-)20-24(-27)μ, wall 1.5-2μ thick at sides, 5-10μ apically, golden to clear chestnut-brown.

Hosts and distribution: Stipa capillata L., S. dasyphylla Czern., S. pennata L., S. pulcherrima C. Koch, S. szovitsiana Trin.: Switzerland and France eastward to south central Siberia and perhaps Manchuria and China.

Type: Diedicke, on S. capillata, Schwellenburg bei Erfurt, Thuringen, Germany (isotypes Sydow Mycoth. Germ. No. 563 as Puccinia stipae.)

Bubák (Centrlbl. Bakt. II. 9:917. 1902) used aeciospores from Thymus ovatus to produce uredinia on Stipa capillata, the first proof of the life cycle.

232. PUCCINIA CHANGTUENSIS Wang Acta Phytotax. Sinica 10:291-292. 1965.

Aecia unknown. Uredinia epiphyllous and sometimes on sheaths, yellowish brown; spores 27-31 x 27-30, globoid or broadly ellipsoid, wall 2.5-5µ thick, pale golden yellow, verrucose-echinulate, germ pores 7-8, scattered, conspicuous. Telia epiphyllous, blackish brown, exposed; spores 18-68 x 18-25µ, elongate-ellipsoid, often conically narrowed apically, wall 2-3µ thick at sides, 13-22µ apically, brown or yellow-brown, smooth; pedicels yellowish, deciduous, 100µ long.

Type: Kia No. 283, on Deyeuxia sp., Changtu, (Inst. Microbiol., Peking No. 34718; not seen). Not otherwise reported.

The description is adapted from the original.

Wang (loc. cit.) published a photograph of spores of the type.

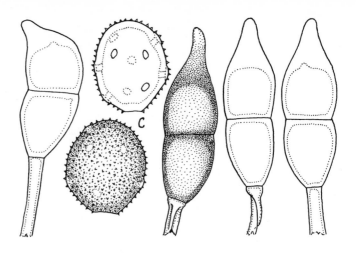

Figure 252

233. PUCCINIA HARRYANA Joerst. Ark. Bot. Ser. 2. 4:348. 1959.
Fig. 252.

Aecia unknown. Uredinia on adaxial leaf surface, cinnamon-brown; spores (25-)30-36(-40) x (25-)27-31(-33)μ, broadly ellipsoid, broadly obovoid, or globoid, wall 2.5-3μ thick, golden to near cinnamon-brown, finely echinulate, germ pores (7)8-10(11), scattered. Telia on adaxial surface, early exposed, blackish brown, compact; spores (40-)50-70(-75) x (18-)20-26μ, mostly ellipsoid or fusiform-ellipsoid, wall 1-1.5(-2.5)μ thick at sides, (8-)12-20μ apically, the apex usually narrowly elongate and pale, clear chestnut-brown, smooth; pedicels colorless, thin-walled, collapsing, to 100μ long but usually about 60μ.

Type: Smith No. 4023, on Lasiagrostis pappiformis (Keng) Handel-Maz. (=Stipa pappiformis Keng), Sze-ch'uan Prov., China (UPS). Not otherwise known.

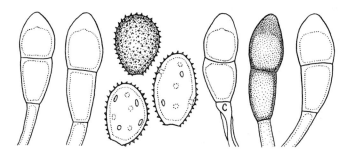

Figure 253

234. PUCCINIA PHAENOSPERMAE Hino & Katumoto Fac. Agr. Yamaguti
Univ. Bull. 7:265. 1956. Fig. 253.

Puccinia takikibicola Y. Morimoto Japan. J. Bot. 34:187.
1959.

Puccinia phaenospermae Wang Acta Phytotax. Sinica 10:293.
1965.

Aecia unknown. Uredinia mostly on adaxial leaf surface;
spores (22-)24-30(-33) x (18-)20-24μ, ellipsoid or obovoid,
wall 1.5-2(-2.5)μ thick, pale cinnamon-brown, echinulate, germ
pores 5-8, scattered, rather obscure. Telia on adaxial surface,
exposed, blackish; spores (35-)40-50(-56) x (13-)16-20(-23)μ,
mostly ellipsoid or narrowly obovoid, wall 1.5-2(-2.5)μ thick
at sides, 4-9(-12)μ apically, uniformly golden brown, or clear
chestnut-brown, or the apex paler; pedicels golden brown, mostly
collapsing, to 60μ long.

Hosts and distribution: Phaenosperma globosum Munro: China
and Japan.

Type: Katumoto, Koiwai Isl., Kamimoseki, Yamaguti Pref.,
Japan (YAM; isotype PUR).

I have not seen Wang's species but, despite his description
of the germ pores as equatorial, there is little doubt that it
is synonymous.

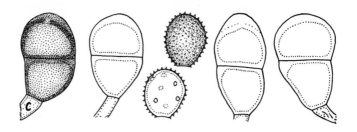

Figure 254

235. PUCCINIA FLAVESCENS McAlp. Proc. Linn. Soc. New S. Wales
28:558. 1903. Fig. 254.

Aecia unknown. Uredinia on adaxial leaf surface; spores
mostly globoid, (22-)24-28(-33) x (20-)22-25μ, wall 1-1.5
(-2.5)μ thick, golden to cinnamon-brown, echinulate, pores 4-7,
scattered. Telia on adaxial surface, early exposed, pulvinate,
blackish; spores (35-)38-46(-50) x (19-)22-25(-29)μ, mostly
ellipsoid, wall 1-1.5(-2)μ thick at sides, (4-)5-7(-8)μ
apically, dark chestnut-brown, smooth; pedicels thin-walled,
yellowish or brownish, attaining a length of 85μ but usually
broken shorter.

Hosts and distribution: Stipa flavescens Lobell, S.
semibarbata R. Br.: Australia.

Type: McAlpine, on S. flavescens, Hampton, Victoria (MEL;
isotype PUR).

Greene and Cummins (Mycologia 50:6-36. 1958) published a
photograph of teliospores of the type.

374

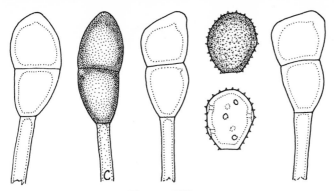

Figure 255

236. PUCCINIA POLYPOGONIS Speg. An. Mus. Nac. B. Aires 19:300. 1909. Fig. 255.

Uredo polypogonis Speg. An. Mus. Nac. B. Aires 6:240. 1899.

Aecia unknown. Uredinia on abaxial leaf surface, cinnamon-brown; spores (22-)24-28(-30) x (20-)22-26µ, broadly ellipsoid or obovoid, wall (1.5-)2-2.5(-3)µ thick, golden or cinnamon-brown, echinulate, germ pores 5-7(-8), with conspicuous caps, scattered. Telia mostly on abaxial surface, exposed, compact, blackish brown; spores (35-)40-55(-60) x (17-)20-24(-30)µ, varying from broadly clavate to ellipsoid, wall (2.5)3-4µ thick at sides, 4-6(-8)µ apically, deep golden brown, smooth; pedicels colorless to brownish, mostly collapsing, to 70µ long but usually broken shorter.

Hosts and distribution: Polypogon chilensis (Kunth) Pilger, P. interruptus H.B.K., P. monspeliensis (L.) Desf.: South America and South Africa.

Type: Spegazzini, on Polypogon monspeliensis, near Lake Muster, Patagonia (LPS; isotype PUR).

Arthur's report of the species on Polypogon elongatus (Proc. Amer. Phil. Soc. 64:183. 1925) is erroneous. The identity of the fungus (uredinia only) is uncertain.

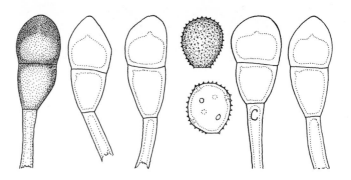

Figure 256

237. PUCCINIA AMPHIGENA Diet. Hedwigia 34:291. 1895. Fig.
256.

Aecia on Smilax spp.; spores (16-)18-22(-24) x (14-)16-19
(-21)µ, wall 1µ thick, colorless, finely verrucose. Uredinia
amphigenous, cinnamon-brown; spores (21-)23-29(-32) x (17-)
19-23(-25)µ, mostly broadly ellipsoid, wall cinnamon-brown,
1.5-2µ thick, echinulate, pores 6-8, scattered. Telia amphigenous, blackish brown, exposed, compact; spores (33-)40-54
(-62) x (14-)18-23(-25)µ, mostly clavate or oblong-clavate,
wall 1.5-2(-3)µ thick at sides, (4-)7-10(-15)µ apically,
chestnut-brown, smooth; pedicels golden, thin-walled and collapsing, to 80µ long.

Hosts and distribution: Calamovilfa longifolia (Hook.)
Scribn.: Canada and the United States from Indiana and Michigan
to Alberta and Oklahoma.

Type: Arthur, Chicago, Illinois (PUR).

Arthur proved the life cycle by inoculation (Bot. Gaz.
35:20. 1903). A photograph of teliospores of the type was
published by Cummins and Greene (Brittonia 13:271-285. 1961).

376

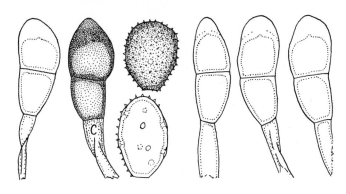

Figure 257

238. PUCCINIA CRANDALLII Pam. & Hume in Hume Proc. Davenport (Iowa) Acad. 7:250. 1899. Fig. 257.

Puccinia kreageri Ricker J. Mycol. 11:114. 1905.

Aecia (Aecidium abundans Peck) occur on species of Symphoricarpos; spores 21-33 x 18-26μ, wall 1.5-2μ thick, verrucose, colorless. Uredinia on adaxial surface of leaves, about cinnamon-brown; spores (27-)30-37(-42) x 24-28(-30)μ, ellipsoid to nearly globoid, wall 1.5-2(-2.5)μ thick, cinnamon-brown, echinulate, pores large, 7-10, scattered. Telia on adaxial surface, early exposed, compact, blackish brown; spores (34-)40-50(-53) x (16-)20-26(-28)μ, mostly ellipsoid or narrowly obovoid, wall 1.5-2(-2.5)μ thick at sides, 5-10(-12)μ apically, clear chestnut-brown, smooth; pedicels rather thin-walled, mostly collapsing, yellow or brownish, to about 70μ long.

Hosts and distribution: species of Festuca, Hesperochloa, kingii (S. Wats.) Rydb., species of Poa: The western United States.

Type: Pammel No. 69, on Festuca kingii (=Hesperochloa kingii), Larimer County, Colorado (ISC; isotype PUR).

Arthur (Mycologia 4:27. 1912) proved the life cycle by inoculation.

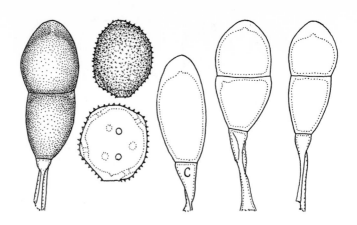

Figure 258

239. PUCCINIA MOYANOI Speg. An. Mus. Nac. Buenos Aires 19:299.
1909. Fig. 258.

Aecia unknown. Uredinia not seen; spores in the telia
(28-)30-32(-34) x (23-)26-30μ, globoid or nearly so, (1.5-)2μ
thick, yellowish to golden, echinulate, germ pores 8-10, scattered,
with conspicuous "caps". Telia on adaxial leaf surface, cinnamon-
brown, early exposed; 2-celled spores (35-)42-60(-64) x (20-)22-
28(-30)μ, mostly ellipsoid, wall 1-1.5(-2)μ thick at sides,
4-6(-8)μ apically, pale golden brown, smooth; pedicels colorless
to yellowish, mostly collapsing, to 65μ long; 1-celled spores
common.

Hosts and distribution: Agrostis moyanoi Speg.: Argentina;
known from the type only.

Type: Spegazzini, near Lago San Martin, Patagonia (LPS;
isotype PUR).

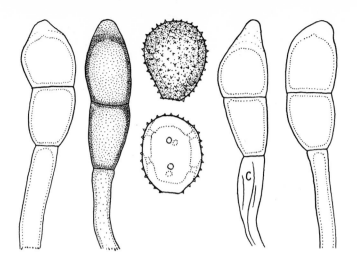

Figure 259

240. PUCCINIA DISTICHLIDIS Ell. & Ev. Proc. Acad. Phila. 1893: 152. 1893. Fig. 259.

Puccinia kelseyi P. Syd. & H. Syd. Monogr. Ured. 1:806. 1904.

Aecia occur on species of Glaux and Steironema; spores 18-27 x 15-24μ, wall 1.5-2μ thick, finely verrucose, hyaline. Uredinia on adaxial leaf surface, yellow; spores 26-33(-35) x 23-28μ, globoid or broadly ellipsoid, wall 3-4μ thick, echinulate, pale yellowish, pores 6-8, scattered, very obscure. Telia on adaxial surface, blackish, early exposed; spores 42-64 x 21-27μ, mostly lanceolate-oblong, wall 1.5-2.5μ thick at sides, 8-13μ apically, clear chestnut, smooth; pedicels golden or paler, mostly thin-walled, attaining a length of 115μ.

Hosts and distribution: Spartina gracilis Trin., S. pectinata Link: New York to Colorado, Montana, and Saskatchewan.

Type: Kelsey, on Distichlis spicata (=error for Spartina gracilis), Helena, Montana (NY; isotype PUR).

Arthur (Mycologia 8:136. 1916) first demonstrated the life history by inoculation of Steironema.

Hennen and Cummins (Mycologia 48:126-162. 1956) published a photograph of teliospores of the type.

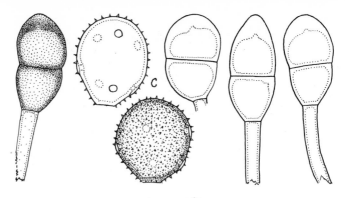

Figure 260

241. PUCCINIA DURANGENSIS Cumm. sp. nov. Fig. 260.

Aecia ignotis. Urediniis epiphyllis, cinnamomeo-brunneis; sporae (29-)32-39 x 29-36μ, globoideae, late ellipsoideae vel late obovoideae, membrana 1.5-2(-2.5)μ crassa, cinnamomeo-brunnea, echinulata, poris germinationis 6-10, plerumque 7 vel 8, sparsis. Teliis epiphyllis, atro-brunneis, pulvinatis, compactis; sporae (32-)34-42(-46) x (18-)20-26(-28)μ, ellipsoideae vel late obovoideae, membrana ad latere 1.5-2.5(-3.5)μ crassa, ad apicem (5-)6-8(-10)μ crassa, lucide castaneo-brunnea, levi; pedicello tenui tunicati, hyalino, persistenti, usque ad 110μ longo sed plerumque breviori.

Type: Hennen 69-203 (=PUR 62782), on Piptochaetium fimbriatum (H.B.K.) Hitchc., 39 miles west of Durango, Dgo., along highway Mex 40, 22 Oct. 1969. Not otherwise known.

The species differs from P. stipae because of larger urediniospores and smaller teliospores.

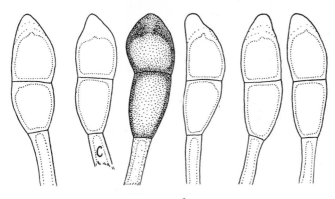

Figure 261

242. PUCCINIA LASIAGROSTIS Tranz. Consp. Ured. URSS p. 96.
1939. Fig. 261.

Puccinia lasiagrostis Tranz. Notulae Syst. Inst. Cryptog.
Hort. Bot. Petrop. II. 6:83. 1923, nom. nud.

Aecia occur on species of Artemisia and probably other
Compositae, cylindrical; spores 22-29.5 x 17-22.5μ thick,
verrucose. Uredinia on abaxial leaf surface, spores 20-35μ
diam, globoid or ellipsoid, wall yellowish, echinulate, pores
several, presumably scattered. Telia abaxial and on sheaths,
early exposed, pulvinate, blackish, attaining a length of 5 mm;
teliospores (40-)50-70(-76) x (15-)19-27(-30)μ, cylindrical,
fusiform or long-clavate, wall 1.5-2(-3)μ thick at sides,
5-12(-14)μ apically and progressively paler externally, smooth;
pedicels thick-walled, hyaline or yellowish, attaining a length
of 175μ.

Hosts and distribution: Stipa splendens Trin.: southern
U.S.S.R. from Kirgiz region to Buryat-Mongol'skaya and China.

Type: Tranzschel, on S. splendens, Buryat-Mongolia near
Kiachta (LE).

Greene and Cummins (Mycologia 50:6-36. 1958) published a
photograph of teliospores purported to be from the type. The
specimen was sent to PUR by Tranzschel but is not identifiable
with certainty as isotype material.

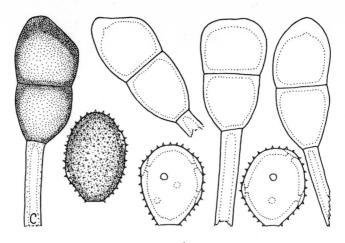

Figure 262

243. PUCCINIA TREBOUXII H. Syd. & P. Syd. Ann. Mycol. 10:215.
1912. Fig. 262.

Puccinia heimerliana Bub. var. melicae-cupani Magn. Hedwigia
51:282-283. 1912.

Aecia unknown. Uredinia on sheaths and the adaxial leaf
surface, cinnamon-brown; spores (26-)28-35(-38) x (22-)24-28
(-30)μ, mostly broadly ellipsoid, wall (2-)2.5-3.5(-4)μ thick,
yellowish brown or golden brown, echinulate, germ pores 5-7(8),
scattered or occasionally equatorial. Telia mostly on adaxial
surface, exposed, blackish brown, compact; spores (40-)45-58
(-60) x (20-)24-28(-30;-36)μ, ellipsoid, oblong-ellipsoid, or
obovoid, wall (1.5-)2-2.5(-3.5)μ thick at sides, (4-)6-8(-10)μ
apically, chestnut-brown, smooth; pedicels persistent, thick-
walled, only occasionally collapsing, colorless or pale yellowish, to 110μ long but usually less than 100μ.

Hosts and distribution: Melica canescens (Regel) Lavr.,
M. cupanii Guss.: southern U.S.S.R. to Iran and Afghanistan (S).

Type: Treboux, on Melica ciliata (now considered to be
M. cupanii), Samarkand, U.S.S.R. (S).

This species is readily distinguished from, although
commonly treated as a synonym of, P. heimerliana (=P. graminis),
as pointed out by the Sydows (loc. cit.).

382

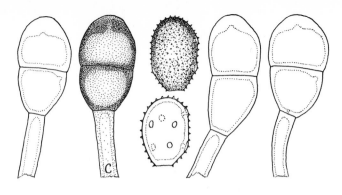

Figure 263

244. PUCCINIA CRYPTANDRI Ell. & Barth. var. luxurians (Arth).
Cumm. & H. C. Greene Brittonia 13:274. 1961. Fig. 263.

Puccinia tosta luxurians Arth. Bull. Torrey Bot. Club
29:229. 1902.

Puccinia luxuriosa Syd. Monogr. Ured. 1:812. 1904.

Aecia on Sarcobatus; spores (21-)24-30(-33) x (14-)20-25
(-28)μ; wall colorless (1-)1.5-2.5(-3)μ thick, finely verrucose.
Uredinia on adaxial surface, cinnamon; spores (23-)26-33(-36)
x (18-)22-26(-31)μ; broadly ellipsoid or oblong-ellipsoid, wall
cinnamon or paler, 1.5-2.5(-3)μ thick, pores (4-)5-7(-9),
scattered or equatorial in occasional spores. Telia on adaxial
surface, early exposed, blackish, pulvinate, compact, to 1 cm
long; spores (35-)42-54(-60) x (21-)25-30(-37)μ, mostly
broadly ellipsoid or oblong-ellipsoid, wall chestnut, (1-)1.5-
2.5(-3)μ thick at sides, 4-11μ apically, smooth; pedicels
colorless or tinted, thick-walled, not collapsing, to 150μ
long; 1-celled teliospores occasional.

Hosts and distribution: Sporobolus airoides Torr., S.
wrightii Munro: U.S.A. from Nebraska to Montana and Oregon
south to New Mexico, California, and northern Mexico.

Type: D. Griffiths (West Am. Fungi No. 304), on Sporobolus
airoides, Andrews, Oregon (PUR).

Bethel (Phytopathology 7:93. 1917) demonstrated the life
cycle with reciprocal inoculations. Cummins & Greene (loc.
cit.) published a photograph of teliospores of the type. See
p.270 for var. cryptandri.

245. PUCCINIA PSAMMOCHLOAE Wang Acta Phytotax. Sinica 10:293. 1965.

Aecia unknown. Uredinia epiphyllis, yellowish brown; spores 26-28µ diam, globoid, wall 2µ thick, golden yellow, densely verrucose-echinulate, germ pores 8 or 9, scattered. Telia epiphyllous, blackish brown, exposed; spores 50-58 x 20-23µ, ellipsoid or obovoid, wall 1.5-2.5µ thick at sides, 5-9µ apically, chestnut-brown, smooth; pedicels colored, firm, to 120µ long.

Type: Lee No. 304, on Psammochloa villosa (Trin.) Bor, Ku-nei-meo, Interior Mongolia (Inst. Microbiol., Peking No. 34716; not seen). One other collection is reported.

The description is adapted from the original. Wang published a photograph of the teliospores. Schmiedeknecht and Puncag (Feddes Repert. 74:177-199. 1967) treat this as Puccinia magnusiana, or presumably they had the same fungus and on the same host. They did not cite Wang's publication.

246. PUCCINIA CAGAYANENSIS H. Syd. in Sydow & Petrak Ann. Mycol. 29:148. 1931.

Aecia unknown. Uredinia amphigenous, cinnamon-brown; spores 19-24 x 15-18μ, ellipsoid, ovoid, or almost globoid, wall 1.5-2μ thick, yellowish brown to golden brown, finely verrucose, germ pores 3 or 4, equatorial. Teliospores in the uredinia 28-38 x 14-18μ, oblong-ellipsoid, oblong, or subclavate, rounded apically and basally or often narrowing basally, wall uniformly 1.5-2μ thick or the apex to 2.5μ, finely and closely punctate or appearing almost smooth, brown; pedicels brownish, fragile, probably thin-walled and collapsing, to 35μ long.

Type: Clemens, on Phragmites vulgaris (=P. communis Trin.), Aparri, Prov. Cagayan, Philippine Islands, Jan. 1924. Not seen; probably not extant. Not otherwise known.

The punctate teliospores should make the species easy to recognize.

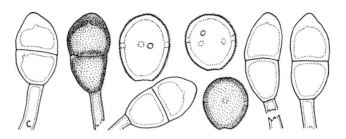

Figure 264

247. PUCCINIA INFUSCANS Arth. & Holw. in Arthur Amer. J. Bot. 5:463. 1918. Fig. 264.

Puccinia meridensis Kern Mycologia 30:547. 1938.

Aecia unknown. Uredinia on abaxial leaf surface, cinnamon-brown or paler; spores 25-29(-32) x (19-)21-24(-26)μ, mostly globoid or broadly ellipsoid, wall 2.5-3(-3.5)μ thick, golden or cinnamon-brown, finely and closely verrucose, germ pores 3-5, equatorial or sometimes with 1 or 2 extra-equatorial. Telia on abaxial surface, exposed, pulvinate, chocolate-brown; spores (26-)30-40(-46) x (16-)18-21(-23)μ, mostly ellipsoid or oblong-ellipsoid, wall 1.5-2(-3)μ thick at sides, (3-)5-7(-9)μ apically, golden or clear chestnut-brown, smooth; pedicels colorless or yellowish, thin-walled and collapsing, to 60μ long; 1-celled spores relatively common.

Hosts and distribution: species of Bothriochloa: Mexico, Guatemala, and Venezuela.

Type: Holway No. 15, on Imperata brasiliensis (=error for Bothriochloa saccharoides, Guatemala City (PUR).

A photograph of teliospores of the type was published by Cummins (Uredineana 4: Plate XI, Fig. 61. 1953).

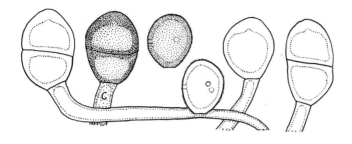

Figure 265

248. PUCCINIA ANTHEPHORAE Arth. & J. R. Johnst. Mem. Torrey Bot. Club 17:137. 1918. Fig. 265.

Uredo anthephorae H. Syd. & P. Syd. Ann. Mycol. 1:22. 1903.

Aecia unknown. Uredinia amphigenous, yellowish brown, rather compact; spores (26-)28-31 x (22-)24-27(-29)μ, mostly broadly ellipsoid or broadly obovoid, wall (2.5-)3-4μ thick, or 4-6μ apically, golden to cinnamon-brown, rugosely verrucose in a labyrinthiform pattern, germ pores 3 or 4, equatorial. Telia amphigenous, early exposed, blackish; spores (30-)33-40 (-42) x (21-)25-30(-32)μ, broadly ellipsoid or broadly obovoid, wall 2.5-3.5(-4)μ at sides, (5-)6-8(-9)μ apically, chestnut-brown, smooth; pedicels thin-walled, collapsing, yellowish, to 100μ long; 1-celled teliospores are common.

Hosts and distribution: Anthephora hermaphrodita (L.) Kuntze: the West Indies to Guatemala and Colombia.

Type: Britton No. 1917, on Anthephora hermaphrodita, Jamaica, 5 Mar. 1908 (PUR 18337).

The species is similar to Puccinia aristidae var. chaetariae.

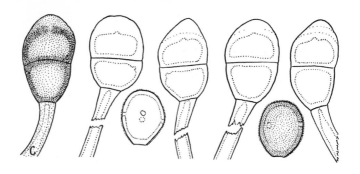

Figure 266

249. PUCCINIA MIYOSHIANA Diet. Bot. Jahrb. 27:569. 1899. Fig. 266.

Puccinia kozukensis Diet. Bot. Jahrb. 32:48. 1902.

The aecia (Aecidium bupleuri-sachalinensis Miyaki), occur on species of Bupleurum, spores 22-25μ diam, hyaline. Uredinia on abaxial surface, pale cinnamon-brown, to 0.8 mm long; (20-) 22-26 x 19-23(-25)μ, mostly globoid or broadly oval, wall 2.5-3(-3.5)μ thick, golden or pale cinnamon-brown, finely verrucose, germ pores 3 or 4, equatorial. Telia abaxial, early exposed, pulvinate and blackish brown; spores (29-)30-43(-48) x (16-)19-26(-28)μ, mostly broadly ellipsoid or oblong-ellipsoid, wall 2-3μ thick at sides, 6-10(-14)μ apically, chestnut-brown, smooth; pedicels colorless or yellowish, moderately thick-walled, seldom collapsing, to 100μ long; 1-celled spores sometimes abundant.

Hosts and distribution: Capillipedium parviflorum (R. Br.) Stapf, Eccoilopus cotulifer (Thunb.) A. Camus, Spodiopogon sibiricus Trin.: China, Japan, western U.R.S.S.

Type: Miyoshi, on Eulalia cotulifer (=Eccoilopus cotulifer), Tokyo Japan, 31 Oct., 1898 (S; isotypes, Sydow Ured. No. 1317).

Cummins (Uredineana 4: Plate X, Figs. 57, 58. 1953) published photos of the types of both P. miyoshiana and P. kozukensis.

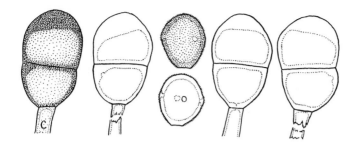

Figure 267

250. PUCCINIA CYMBOPOGONIS Mass. Kew Bull. Misc. Inform. 1911: 224. 1911. Fig. 267.

Uredinia on abaxial surface, yellowish brown or cinnamon-brown, to 1.5 mm long; spores 23-29(-33) x 19-24(-26)μ, globoid or oval, wall 3-4(-4.5)μ thick, golden to dark cinnamon-brown, finely verrucose, germ pores 3-5, equatorial or rarely scattered in some spores. Telia abaxial, early exposed, pulvinate, blackish brown; spores 35-42(-44) x 24-30μ, mostly broadly ellipsoid, wall 2-3(-4)μ thick at sides, 7-9(-10)μ apically, clear chestnut-brown, smooth; pedicels colorless or yellowish, thin-walled and collapsing laterally, to 80μ long but usually broken shorter.

Hosts and distribution: Cymbopogon citratus (DC.) Stapf: central and southern Africa.

Type: Fyffe, Entebbe, Uganda, 1911 (K; isotype PUR).

A photograph of teliospores of the type was published by Cummins (Uredineana 4: Plate X, Fig. 59. 1953).

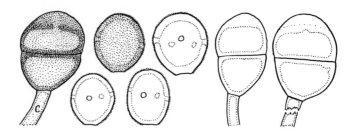

Figure 268

251. PUCCINIA CESATII Schroet. Cohn Beitr. Biol. Pflanzen
3:70. 1879. Fig. 268.

Uredo (Podocystis) andropogonis Ces. Rabenh. Herb. Myc. I.
No. 1997.

Uredo andropogonis Fckl. Nass. Naturw. Jahrb. 15:16. 1861.

Uredo andropogonis Cast. Cat. Pl. Mars. Supp. p. 89. 1851.

Puccinia andropogonis Fckl. Symbol. Mycol. p. 59. 1870, not
Schw. 1832.

Puccinia andropogonis Otth Naturf. Ges. Bern Mitth. 1873:
86. 1873.

Puccinia propinqua Syd. & Butl. Ann. Mycol. 5:499. 1907.

Aecia unknown. Uredinia on abaxial surface, yellowish brown,
or chocolate-brown when amphisporic; urediniospores (19-)23-28
(-30) x 19-24(-26)µ, mostly globoid or broadly oval, wall
(2.5-)3-4µ thick, golden, closely and finely verrucose, germ
pores (3-)4 or 5(-6); amphispores like the urediniospores but
24-30(-32) x 23-26µ, wall chestnut, 3-5µ thick. Telia abaxial,
early exposed, pulvinate, chestnut-brown; spores (30-)32-38(-40)
x (22-)24-27(-29)µ, mostly broadly ellipsoid, wall 1.5-3µ thick
at sides, (3-)4-7(-8)µ apically, clear chestnut-brown, smooth;
pedicels colorless, or brownish next the spore, thin-walled,
collapsing, to 80µ long but often broken short.

Hosts and distribution: Bothriochloa insculpta (Hochst.)
A. Camus, B. ischaemum (L.) Keng, Capillipedium glaucopsis
(Steud.) Stapf, C. parviflorum (R. Br.) Stapf, C. spicigerum
(Benth.) S. T. Blake, Dichanthium annulatum (Forsk.) Stapf:
France and Italy eastward to Egypt, Iran, India and China
and in the southwestern United States and Mexico.

Type: Fuckel, on Andropogon ischaemum (=B. ischaemum),
Beibrich, Germany, autumn (isotypes Fuckel, Fungi Rhenani exs.
No. 2223).

Cummins published a photograph of teliospores of the type
(Uredineana 4: Plate XI, Fig. 60. 1953).

390

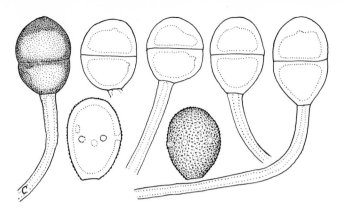

Figure 269

252. PUCCINIA ESCLAVENSIS Diet. & Holw. in Holway Bot. Gaz.
24:29. 1897 var. esclavensis Fig. 269.

Aecia (Aecidium mirabilis Diet. & Holw.) occur on Mirabilis;
spores 21-28 x 16-19µ, wall 1-1.5µ thick, hyaline, verrucose.
Uredinia mainly on abaxial surface, cinnamon-brown or darker;
spores 26-33(-39) x 19-23(-25)µ, wall (2-)2.5-3.5µ thick,
golden or darker, closely verrucose, germ pores 4-6, equatorial
or scattered in occasional spores. Telia amphigenous, early
exposed, pulvinate, blackish brown; teliospores 28-36(-41)
x 22-27(-31)µ, mostly ellipsoid, wall 2.5-3.5µ thick at sides,
4-8µ apically, deep chestnut-brown, smooth; pedicels thick-
walled and not collapsing, yellowish, to 80µ long.

Hosts and distribution: Panicum bulbosum H. B. K., P.
plenum Hitchc. & Chase, Pennisetum chilense (Desv.) Jackson,
P. bambusiforme Hemsl., P. peruvianum Trin.: The southwestern
United States southward to Honduras, the West Indies, Ecuador,
and Argentina.

Type: E. W. D. Holway, on Panicum bulbosum, Eslava, Mexico,
3 Oct. 1896 (S; isotype PUR).

Cummins and Baxter (Madroño 16:201-203. 1962) proved the
life cycle by inoculation.

A photograph of teliospores of the type was published by
Cummins (Mycologia 34:669-695. 1942).

The urediniospores resulting from infection by aeciospores
have thinner and paler walls, perhaps indicating that spores
produced later in the season tend to be amphisporic.

PUCCINIA ESCLAVENSIS Diet. & Holw. var. panicophila (Speg.) Ramachar & Cumm. Mycopathol. Mycol. Appl. 25:55. 1965.

Puccinia atra Diet. & Holw. in Holway Bot. Gaz. 24:29. 1897, not Spreng. 1827.

Puccinia panicophila Speg. An. Mus. Nac. Buenos Aires 19:300. 1909.

Uredo panicophila Speg. Bol. Acad. Nac. Cien. Rep. Argentina 29:149. 1926.

Aecia unknown. Urediniospores (24-)26-30(-35) x 20-25(-27)μ, mostly broadly ellipsoid or globoid, wall (2-)2.5-3.5μ thick, usually rugose with wartlets fused in a labyrinthiform pattern; teliospores not distinctive.

Hosts and distribution: Digitaria californica (Benth.) Henrard, D. cognata (Schultes) Pilger, D. insularis (L.) Mez, Paspalum laxum Lam., Setaria grisebachii Fourn., S. scheelei (Steud.) Hitchc.: southwestern United States to Guatemala, Puerto Rico, and Argentina.

Type: Spegazzini, on Digitaria insularis (as Panicum insulare), near Cacheuta, Argentina (LPS; isotype PUR).

PUCCINIA ESCLAVENSIS Diet. & Holw. var. unicellula Ramachar & Cumm. Mycopathol. Mycol. Appl. 25:56. 1965.

Aecia unknown. Uredinia unknown; teliospores mostly 1-celled, 25-33 x (18-)21-26μ, mostly broadly ellipsoid or broadly obovoid, wall 2.5-3.5μ thick at sides, 5-9μ apically, chestnut-brown, pedicel colorless, long.

Hosts and distribution: Digitaria californica (Benth.) Henrard: Tamaulipas State, Mexico.

Type: Swallen No. 1710, Chamal, Tamps., Mexico (PUR 58725).

253. PUCCINIA ERAGROSTIS-ARUNDINACEAE Tranz. & Erem. in
Tranzschel Conspectus Uredinalium URSS. p. 100. 1939.

Aecia unknown. Uredinia not described; spores 24-35 x
24-35µ, subglobose, wall 3.5µ thick, densely verruculose,
brown, germ pores 2 or 3 (equatorial?). Telia not described
but doubtless exposed; spores 35-48 x 21-32µ, rounded at the
ends, to 6µ thick apically, chestnut-brown, smooth; pedicels
persistent.

Type: Eremeeva (?), on Eragrostis arundinacea (L.) Rosh.
in Rynpeski sand, Kazachstan, USSR (LE; not seen). Paratype
near Lake Zajsan.

The description is adapted from the original.

The species apparently is similar to P. aeluropodis.

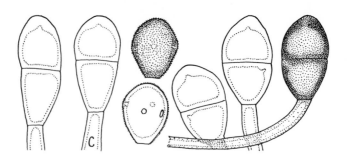

Figure 270

254. PUCCINIA REDFIELDIAE Tracy J. Mycol. 7:281. 1893. Fig. 270.

Aecia (Aecidium anograe Arth.) occur on Oenothera nuttallii Sweet; spores (19-)22-27(-29) x (17-)18-23μ, globoid or ellipsoid, wall 1.5(-2)μ thick, finely verrucose. Uredinia mostly on adaxial leaf surface, cinnamon-brown; spores (23-)26-31(-34) x (19-)21-25(-28)μ, mostly obovoid, wall 2-2.5(-3)μ thick, finely verrucose-rugose, the wartlets tending to unite in a reticulate pattern, cinnamon-brown, germ pores (3)4(5), equatorial, large. Telia mostly on adaxial surface, early exposed, compact, blackish brown; spores (36-)40-50(-52) x (21-)23-30(-34)μ, mostly ellipsoid, wall 1.5-2.5(-3.5)μ thick at sides, (4-)5-8(-9)μ apically, chestnut-brown, smooth; pedicel thick-walled, colorless, mostly not collapsing, to 80μ long.

Hosts and distribution: Redfieldia flexuosa (Thurb.) Vasey: North Dakota to Kansas and Colorado, U.S.A.

Type: Vasey, Kansas, Sept. 1889 (NY; isotype PUR).

The life cycle was proved by inoculation by Solheim and Cummins (Univ. Wyo. Publ. 23:35. 1959).

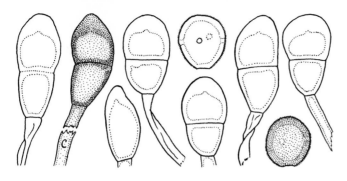

Figure 271

255. PUCCINIA ELLISIANA Thuem. Bull. Torrey Bot. Club 6:215.
1878. Fig. 271.

Puccinia americana Lagh. Tromso Mus. Aarsh. 17:45. 1895.

Puccinia sagittata Long Phytopathology 2:167. 1912.

Puccinia mariae-wilsonii Barth. N. Am. Ured. No. 204. 1922,
not G. W. Clint. 1873.

Aecia (Aecidium mariae-wilsoni Pk.) occur on Viola; spores
12-19µ diam, wall 1-1.5µ thick, yellowish. Uredinia on abaxial
surface, pale cinnamon-brown; spores (17-)19-22(-24) x (16-)
18-20(-21)µ, globoid or broadly ellipsoid, wall golden or pale
cinnamon-brown, finely and closely verrucose, germ pores 3 or
4, equatorial, 2.5-4µ thick. Telia abaxial and on the sheaths,
early exposed, pulvinate, blackish brown; spores (28-)31-45(-55)
x (14-)18-23(-25)µ mostly clavate or oblong-ellipsoid, wall
(1.5-)2-3(-4)µ thick at sides, (5-)7-9(-10)µ apically, chestnut-
brown or sometimes golden, smooth; pedicels yellow or brownish,
moderately thick-walled and collapsing partially, to 85µ long.

Hosts and distribution: species of Andropogon: Canada
southward to Mexico east of the Continental Divide.

Type: Ellis, on Andropogon virginicus, Newfield, New Jersey
(BPI; isotype Thuemen Mycotheca univers. 1336).

Arthur (Mycologia 7:230-231. 1915) first proved the life
cycle by inoculation. A photograph of teliospores of the type
was published by Cummins (Uredineana 4: Plate XI, Fig. 62.
1953).

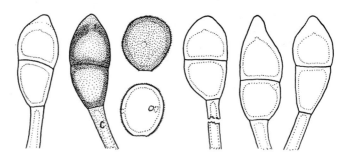

Figure 272

256. PUCCINIA CYNODONTIS Lacroix ex Desm. Pl. Crypt. Ser. III,
No. 655. 1859. Fig. 272.

Puccinia cynodontis Fckl. Symb. Mycol. Nachr. 2:16. 1875.

Puccinia varians Diet. Ann. Mycol. 6:224. 1908.

Uredo elusine-indicae Saw. J. Taihoku Soc. Agr. For. 7:41.
1943.

The aecia (Aecidium plantaginis Ces.) occur on Euphorbiaceae,
Plantaginaceae, Ranunculaceae, Saxifragaceae, Scrophulariaceae,
Valerianaceae, and Violaceae; spores 15-24 x 16-29μ, wall 1.5-2μ
thick, colorless, verrucose. Uredinia mostly on abaxial surface,
cinnamon-brown; spores globoid, 20-26 x 19-23μ, wall 2-3μ thick,
cinnamon-brown, verrucose, pores 2 or 3, equatorial. Telia
mostly abaxial, early exposed, blackish, pulvinate; spores 30-55
x 16-22μ, mostly ellipsoid, often acuminate apically, wall
1.5-2.5μ thick at sides, 6-12μ apically, chestnut-brown; pedicels
yellow or colorless, thin-walled, to 80μ long.

Hosts and distribution: Cynodon dactylon (Pers.) L.: circumglobal in temperate and warmer regions.

Type: De Lacroix, on Cynodon dactylon, St. Romain-sur-Vienne,
Arroundissement de Chatellerault, 1857, (isotypes, Desmaz. Pl.
Crypt. III, No. 655).

Hennen and Cummins (Mycologia 48:126-162. 1956) published
a photograph of teliospores of the type.

Tranzschel (Trav. Mus. Bot. Acad. Imp. Sci. St. Petersb.
3:39-40. 1906) first proved the life cycle by successfully
inoculating species of Plantago.

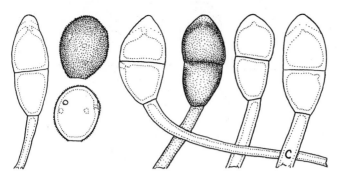

Figure 273

257. PUCCINIA WINDSORIAE Schw. Trans. Amer. Phil. Soc. II.
4:295. 1832. Fig. 273.

Puccinia omnivora Ellis & Ev. Bull. Torrey Bot. Club 22:59.
1895.

Aecia (Aecidium pteleae Berk. & Curt.) occur on Ptelea tri-
foliata L.; spores 16-23 x 15-18μ, globoid or ellipsoid, wall
colorless 1-1.5μ thick, finely verrucose. Uredinia amphigenous,
cinnamon-brown; spores (22-)24-30(-34) x (18-)21-24(-26)μ,
globoid or ellipsoid, wall 1.5-2μ thick, uniformly golden or
cinnamon-brown or slightly darker apically, finely verrucose
with discrete wartlets or these sometimes striately arranged,
germ pores 3-5, mostly 3 or 4, equatorial. Telia mostly on
abaxial leaf surface and on stems, early exposed, blackish
brown; spores (28-)32-42(-52) x (15-)17-22(-24)μ, mostly ellipsoid or obovoid, wall 1.5-2(-2.5)μ thick at sides, 5-8(-10)μ
apically, chestnut-brown, smooth; pedicels thin-walled and mostly
collapsing, golden, to 60μ long.

Hosts and distribution: Tridens flavus (L.) Hitchc: New
York and Georgia west to Nebraska and Texas, U.S.A..

Type: Schweinitz, on Poa quinquedentata (error for Tridens
flavus), Bethlehem, Pennsylvania (PH; isotype PUR).

Arthur (Bot. Gaz. 29:273. 1900) first proved the life cycle
by inoculation.

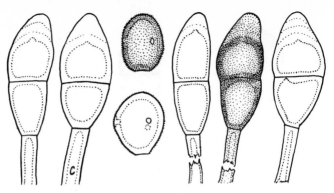

Figure 274

258. PUCCINIA CRASSAPICALIS Bub. Naturh. Hofmus. Wien Ann.
28:192. 1914. Fig. 274.

Uredinia not seen; urediniospores in the telia 23-30(-34)
x 19-25(-28)μ, globoid or oval, wall 3-3.5(-4.5)μ thick,
verrucose, golden or cinnamon-brown, germ pores 2 or 3 equatorial. Telia on abaxial surface, early exposed, pulvinate,
blackish brown; spores (35-)40-56(-78 according to Bubak) x
(17-)19-27μ, ellipsoid, oblong-ellipsoid or almost fusiform,
wall (1.5-)2-3(-4)μ thick at sides, (8-)10-16(-20)μ apically,
chestnut-brown, smooth; pedicels colorless or yellowish, thick-
walled, seldom collapsing, to 100μ long.

Hosts and distribution: Spodiopogon pogonatherus (Boiss.)
Benth.: Turkey.

Type: Handel-Mazzetti, Kutmis, Kurdistan region, 17 Aug.
1910 (BPI).

The teliospores differ from those of P. daniloi and P.
pseudocesatii in being longer and having a thicker more conically elongated apex.

A photograph of teliospores of the type was published by
Cummins (Uredineana 4: Plate X, Fig. 55. 1953).

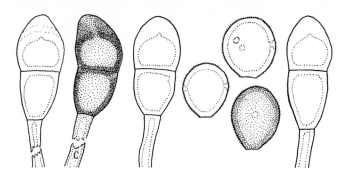

Figure 275

259. PUCCINIA DANILOI Bub. Ann. Mycol. 3:219. 1905. Fig. 275.

Aecia unknown. Uredinia on abaxial surface, yellowish, to 1 mm or by confluence to at least 3 mm long; spores 25-30(-33) x 19-25(-27)µ, globoid, oval, or ellipsoid, wall 3-3.5µ thick, pale cinnamon-brown or golden, finely verrucose, germ pores 2 (or 3), equatorial. Telia abaxial, early exposed, pulvinate and blackish brown; spores (33-)36-50(-55) x (18-)20-24(-27)µ, mostly oblong-ellipsoid or clavate, wall 2-3µ thick at sides, 6-12(-16)µ apically, chestnut-brown, smooth; pedicels yellowish or colorless, thick-walled, seldom collapsing, to 80µ long.

Hosts and distribution: _Erianthus hostii_ Griseb.: Yugoslavia.

Type: F. Bubak, between Spuz and Danilov Grad, Yugoslavia, 6 Aug. 1904 (BPI).

The species produces very long infections which are only 1 sorus wide and reminiscent of stripe smut lesions.

A photograph of teliospores of the type was published by Cummins (Uredineana 4: Plate X, Fig. 54. 1953).

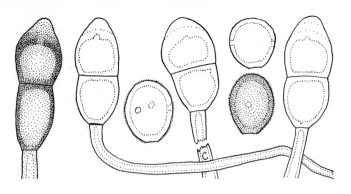

Figure 276

260. PUCCINIA PSEUDOCESATII Cumm. Uredineana 4:71. 1953. Fig. 276.

Aecia unknown. Uredinia mostly on abaxial surface, cinnamon-brown, spores (21-)23-28(-32) x 19-24(-26)μ, mostly globoid or ovate, wall (2-)2.5-3.5μ thick, golden or cinnamon-brown, finely verrucose, germ pores 2 or 3, equatorial. Telia mostly abaxial, early exposed, pulvinate, and blackish brown; spores (36-)40-48 (-52) x (16-)18-23(26)μ oblong-ellipsoid, ellipsoid, or clavate, wall (1.5-)2-3(-4)μ thick at sides, (5-)8-12μ apically, chestnut-brown, smooth; pedicels yellowish brown, thick-walled, seldom collapsing, to 80(-105)μ long.

Hosts and distribution: Bothriochloa ischaemum (L.) Keng, Chrysopogon gryllus (L.) Trin.: southern Europe.

Type: F. Petrak, on C. gryllus, Niederdonau, Braunsberg bei Hainberg, Austria, Oct., 1940 (PUR; isotypes, Petrak Mycotheca gen. No. 2026, issued as P. cesatii Schroet.).

A photograph of teliospores of the type was published by Cummins (Uredineana 4: Plate X, Fig. 55. 1953).

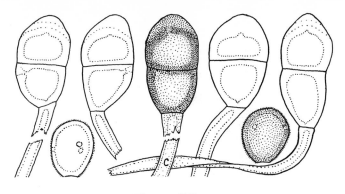

Figure 277

261. PUCCINIA SCHOENANTHI Cumm. & Guyot sp. nov. Fig. 277.

Aeciis ignotis. Urediniis hypophyllis, cinnamomeo-brunneis; sporae (21-)23-26 x (19-)20-24µ, late ellipsoideae vel obovoideae vel globoideae, membrana (2-)2.5-3.5µ crassa, plus minusve castaneo-brunnea, dense verruculosa, poris germinationis 2 vel 3, aequatorialibus, obscuris. Teliis hypophyllis, seriatim dispositis, pulvinatis, compactis, atro-brunneis; sporae (36-) 42-48(-53) x (23-)26-30(-34)µ, ellipsoideae vel obovoideae, membrana ad latere (2-)2.5-4(-6)µ crassa, ad apicem (6-)8-10(-13)µ, castaneo-brunnea vel lucide castaneo-brunnea, minutissime punctato-rugosa vel levi; pedicello hyalino, persistenti, plus minus crasse tunicati, usque ad 130µ longo.

Hosts and distribution: Cymbopogon oliviera (Boiss.) Bor, C. schoenanthus (L.) Spreng.: Iran.

Type: Pasquier, on Cymbopogon schoenanthus, west of Kermanchah, Iran, 1957 (PUR F16543; isotype herb. Guyot).

It is probable that Urban's report (Uredineana 6:5-58. 1966) of Puccinia crassapicalis from Iraq refers to this fungus.

262. PUCCINIA DANTHONIAE Korbon. Akad. Nauk Tadzhik SSR. 22:30. 1957.

Aecia unknown. Urediniospores in the telia rare, 26-28μ diam, globoid, verruculose, color not stated. Telia mainly hypophyllous, exposed, compact, grouped or in lines; spores 37-53 x 18-32μ, broadly ellipsoid or clavate, rounded at the ends or subattenuate basally, wall 3-5μ thick, clavate spores to 9μ (apically?), smooth, color not stated; pedicels firm, thick, to 155μ long.

Type: Nikitin, on Danthonia forsskalii R. Br. (Asthenatherum forsskalii Nevski), in Kurdzhala-Kum sand, southern Tadzhik SSR (TAD?, not seen).

The description is adapted from the original.

Germ pores of the urediniospores were not described but I assume that they are equatorial and that the fungus is similar to P. aeluropodis.

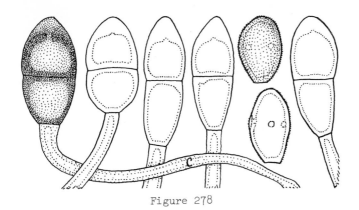

Figure 278

263. PUCCINIA ARISTIDAE Tracy J. Mycol. 7:281. 1893 var.
aristidae. Fig. 278.

Uredo aristidae-acutiflorae Maire Botaniste 34:308. 1949.

Aecia, Aecidium caspicum Jacz., occur on Heliotropium
europaeum L; spores 18-21 x 15-18µ, wall 1.5µ thick, colorless
or nearly so. Uredinia on adaxial leaf surface, in lines,
cinnamon-brown; spores (22-)25-33(-36) x (16-)18-23(-24)µ,
ellipsoid or broadly ellipsoid, wall (2-)2.5-3(-3.5)µ thick,
mostly golden brown, closely verrucose, often in an obscurely
striolate pattern, germ pores equatorial, usually 2 or 3 in
elongate spores, 3 or 4 in robust spores. Telia amphigenous
or mostly on adaxial surface, linear and often confluent to
5 cm, blackish brown, exposed, pulvinate; spores (34-)40-58
(-65) x (17-)20-27(-30)µ, oblong, ellipsoid, or broadly ellipsoid, wall (2-)2.5-3µ thick at sides, (5-)6-10(-12)µ apically,
clear chestnut-brown, the apical thickening usually paler
externally, smooth; pedicels colorless, thick-walled, to 175µ
long.

Hosts and distribution: species of Aristida: northern Africa
eastward through the Transcaspian region to Afghanistan.

Type: Regel, on Aristida pungens, Turkestan (NY).

Macroscopically, this species is distinctive because of the
strikingly seriate telia. It is of much more limited distribution than var. chaetariae.

A photograph of teliospores of the type was published by
Cummins and Husain (Bull. Torrey Bot. Club 93:56-67. 1966).

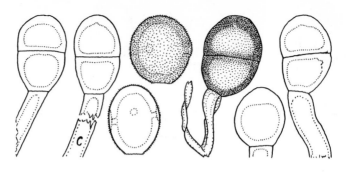

Figure 279

PUCCINIA ARISTIDAE Tracy var. chaetariae Cumm. & Husain
Bull. Torrey Bot. Club 93:63. 1966. Fig. 279.

Aecia (Aecidium pupaliae Prasad, Sharma & Singh) occur on
species of Pupalia of the Amaranthaceae and Boerhaavia of the
Nyctaginaceae; spores 15-22(-28) x (11-)13-19μ, wall 1.5(-2)μ
thick, hyaline, verrucose. Uredinia adaxial, scattered or
grouped, cinnamon-brown; spores 23-30(-32) x (19-)21-26(-29)μ
broadly ellipsoid or broadly obovoid, wall (3-)3.5-4.5(-5)μ
thick, mostly golden brown, verrucose, pores (2?)3(4), equatorial, obscure. Telia adaxial and on sheaths and stems, not
seriate, blackish brown, compact, early exposed; spores (29-)
32-44(-50;-60) x (19-)22-28(-32)μ, mostly broadly ellipsoid or
broadly obovoid, wall (2.5-)3-4.5(-5)μ thick at sides, (4-)
5-8(-10)μ at apex, uniformly chestnut-brown, smooth; pedicels
colorless, persistent, to 165μ long; 1-celled teliospores
sometimes common.

Hosts and distribution: species of Aristida, Hilaria, Tridens:
Africa, India, North and South America.

Type: Cummins 61-230, on Aristida adscensionis, Arizona,
U.S.A. (PUR 59150).

In the United States this fungus has long been confused with
Puccinia subnitens Diet.

Singh (Current Sci. 31:521-522. 1962) proved the life cycle
by inoculating A. adscensionis with spores of Aecidium pupaliae.
Cummins and Husain (Bull. Torrey Bot. Club 93:56-67. 1966)
published a photograph of teliospores of the type.

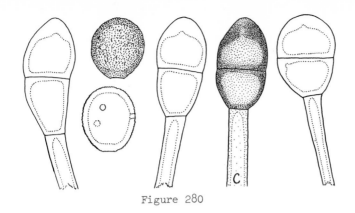

Figure 280

264. PUCCINIA AELUROPODIS Ricker J. Mycol. 11:114. 1905. Fig. 280.

Puccinia tangkuensis Liou & Wang Contr. Inst. Bot. Acad. Peiping 3:448. 1935.

Aecia (Aecidium nitrariae Pat.) occur on species of Nitraria; spores 14-20 x 12-16μ, angularly globoid, wall 1-1.5μ thick, finely verrucose, hyaline. Uredinia mostly amphigenous, yellowish brown, often confluent; spores (22-)24-30(-32) x (20-)22-26(-28)μ, broadly ellipsoid or globoid, wall 2.5-3.5(-4)μ thick, golden to cinnamon-brown, closely and finely verrucose, germ pores 3 or 4, equatorial. Telia usually amphigenous, early exposed, blackish brown, compact; spores (32-)38-48(-52) x (18-)22-28(-32)μ mostly oblong-ellipsoid or elongately obovoid, tending to be dimorphic with the shorter broader spores darker colored, wall (1.5-)2-3.5(-4.5)μ thick at sides, (5-)6-10μ apically, chestnut-brown, pedicels thin- or thick-walled, collapsing or not, to 125μ long but usually less than 100μ, hyaline; brown sporogenous basal cells conspicuous.

Hosts and distribution: Aeluropus lagopoides (L.) Trin., A. littoralis (Willd.) Parl., A. macrostachyus Hack.: the Mediterranean area to India and China.

Type: Frick, on Aeluropus littoralis, Caucasus (WIS).

P. tanghuensis is probably not distinct but no material has been avaiable. Uromyces aeluropodis-repentis Nattrass has similar spores and sporogenous cells.

265. PUCCINIA ABRAMOVIANA Lavrov Trud. Tomsk. gos. Univ. Kuibysheva. Ser. Biol. 110:156. 1951.

Aecia unknown. Uredinia amphigenous, yellowish brown, aparaphysate; spores 18-24 x 18-21μ, globoid, subgloboid or ovate, minutely "verruculosis", germ pores indistinct (probably scattered). Telia amphigenous, blackish brown, covered with the epidermis (paraphyses not mentioned); spores 36-48 x 12-20μ, clavate, apex truncate or rounded, wall pale brown, the apex darker and thickened 3μ, smooth; pedicels short, colorless.

Type: Lavrov?, on Melica nutans L., Okeanskaja USSR (TK?; not seen).

The description is adapted from the original.

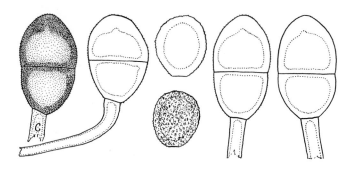

Figure 281

266. PUCCINIA PAZENSIS H. C. Greene & Cumm. Mycologia 50:27.
1958. Fig. 281.

Aecia unknown. Uredinia on adaxial leaf surface, yellow;
spores 23-28 x (16-)20-24(-26)μ, wall (3-)4-7(-7)μ thick,
hyaline or yellowish, labyrinthiformly rugose, pores obscure,
scattered. Telia on adaxial surface, blackish, early exposed,
pulvinate; spores broadly ellipsoid, (36-)42-48(-54) x (24-)
27-30(-32)μ, wall 2.5-5μ thick at sides, 6-8(-10)μ apically,
golden or clear chestnut-brown, minutely verrucose or appearing
smooth; pedicels hyaline, thin-walled, attaining 135μ in length.

Hosts and distribution: Nassella pubiflora (Trin. & Rupr.)
Desv.: Bolivia.

Type specimen: Holway No. 479, LaPaz, Bolivia (PUR; isotype
MIN).

Greene and Cummins (loc. cit.) published a photograph of
teliospores of the type.

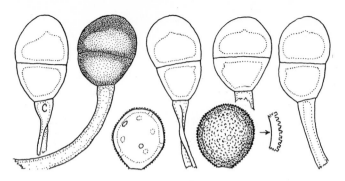

Figure 282

267. PUCCINIA POLLINIAE-QUADRINERVIS Diet. Ann. Mycol. 7:355.
1909. Fig. 282.

Aecia unknown. Uredinia on abaxial leaf surface, cinnamon-
brown; spores (24-)26-30(-33)μ diam, globoid, rarely varying
slightly, wall 2.5μ thick, cinnamon-brown, moderately verrucose
with rod-like papillae, germ pores 7-9, scattered. Telia on
abaxial surface, pulvinate, early exposed, blackish brown;
spores (28-)30-36(-38) x 23-26(-28)μ, mostly broadly ellipsoid,
wall 3-3.5μ thick at sides, 5-7(-9)μ apically, chestnut-brown
or paler basally, smooth; pedicels thick-walled, collapsing
partially or not, colorless, to 60μ long.

Hosts and distribution: Eulalia quadrinervis (Hack.) O.
Ktze.: Japan and the Philippine Islands.

Type: Yoshinaga, on Pollinia quadrinervis (=E. quadrinervis),
Mt. Kiyotaki, Tosa, Japan (S).

A photograph of teliospores of the type was published by
Cummins (Uredineana 4: Pl. XI, Fig. 63. 1953).

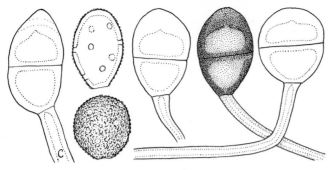

Figure 283

268. PUCCINIA SETARIAE Diet. & Holw. in Holway Bot. Gaz. 24:28.
1897. Fig. 283.

Aecia unknown. Uredinia mostly on abaxial leaf surface,
cinnamon-brown; spores (27-)29-34(-36) x (23-)25-28μ, mostly
broadly ellipsoid or obovoid, wall 2.5-3.5μ thick, golden,
closely verrucose, germ pores 7 or 8, scattered. Telia amphigenous, early exposed, pulvinate, blackish brown; spores
(35-)37-45(-48) x (24-)26-30(-32)μ, mostly ellipsoid, wall
3-5μ thick at sides, 8-11μ apically, chestnut-brown, smooth;
pedicels yellowish, thick-walled, mostly not collapsing, to
100μ long.

Hosts and distribution: Setaria geniculata (Lam.) Beauv.:
the southern United States, southward to Guatemala, Chile and
Argentina.

Type: E. W. D. Holway No. 34a, City of Mexico, Mexico (S;
isotype PUR).

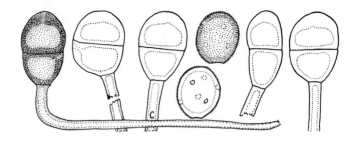

Figure 284

269. PUCCINIA LEPTOCHLOAE Arth. & Fromme Torreya 15:263. 1915.
Fig. 284.

The aecial stage (<u>Aecidium</u> <u>talini</u> Speg.) occurs on species
of <u>Calandrinia</u> and <u>Talinum</u>; spores 17-21(-24) x 14-16(-18)μ,
globoid to ellipsoid, wall 1μ thick, finely verrucose. Uredinia
on abaxial leaf surface, cinnamon-brown; spores 19-26 x (16-)
18-24μ, globoid or obovoid, wall 1.5-2.5μ, golden or cinnamon-
brown, verrucose, pores 4-6, scattered. Telia mostly on abaxial
surface, blackish, early exposed, pulvinate; spores broadly
ellipsoid, 25-34 x 17-24μ, wall 2.5-4μ at sides, 4-7μ apically,
dark chestnut, smooth; pedicels thick-walled, usually not
collapsing, golden, attaining a length of 95μ; 1-celled spores
sometimes are common.

Hosts and distribution: <u>Leptochloa</u> <u>filiformis</u> (Lam.) Beauv.:
southern U.S.A., to Guatemala, Puerto Rico and southward to
Argentina.

Type: Palmer, on <u>Leptochloa</u> <u>filiformis</u>, Guaymas, Sonora,
Mexico (PUR).

Cummins (Mycologia 55:73-78. 1963) produced aecia on <u>Talinum</u>
<u>paniculatum</u> by inoculation. A photograph of teliospores of
the type was published by Hennen and Cummins (Mycologia
48:126-162. 1956).

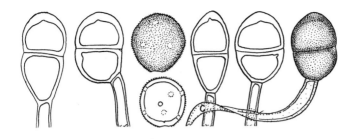

Figure 285

270. PUCCINIA CHIHUAHUANA Cumm. Southw. Nat. 12:75. 1967.
Fig. 285.

Aecia unknown. Uredinia not seen; spores 21-25 x 20-24μ, globoid, wall (1.5-)2-2.5(-3)μ thick, cinnamon-brown, finely verrucose, pores scattered, 6-8. Telia on abaxial surface and on stems, early exposed, pulvinate, blackish brown; spores (26-)30-36(-40) x (17-)19-24(-26)μ, mostly broadly ellipsoid, wall (1.5-)2-3(-4)μ thick at sides, 4-6(-7)μ at apex, chestnut-brown, smooth; pedicel hyaline, collapsing or not, to 90μ long.

Hosts and distribution: _Muhlenbergia fragilis_ Swallen: Mexico.

Type: Cummins 63-412, Chihuahua (State), Mexico (PUR).

This species is similar to P. leptochloae but it has 4-6 germ pores and shorter teliospores. A photograph of teliospores of the type was published with the original description.

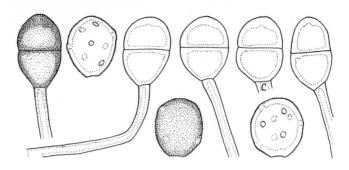

Figure 286

271. PUCCINIA PSEUDOATRA Cumm. Mycologia 34:688. 1942.
Fig. 286.

Aecia unknown. Uredinia mostly on abaxial leaf surface,
pale cinnamon-brown; spores (23-)24-27(-28) x (21-)23-25(-26)μ,
broadly ellipsoid or globoid, wall 2.5-3μ thick, golden or
cinnamon-brown, closely and finely verrucose, the wartlets
often uniting in labyrinthiform patterns, germ pores (5)6-8,
scattered. Telia mostly on abaxial surface, blackish brown,
early exposed; spores (28-)31-37(-39) x (20-)22-25(-26)μ,
wall (2-)2.5-3.5μ thick at sides, 5-8μ apically, chestnut-
brown, smooth; pedicels hyaline, thick-walled or sometimes
thin-walled and collapsing, to about 90μ long.

Hosts and distribution: Digitaria insularis (L.) Mez,
Paspalum pallidum H.B.K., P. penicillatum Hook. f., P. prostratum
Scribn. & Merr.: Argentina, Bolivia, Ecuador, and Peru.

Type: Holway No. 954, on Paspalum pallidum, Quito Ecuador
(PUR; isotypes Reliq. Holw. 100 as Puccinia macra Arth. & Holw.).

A photograph of teliospores of the type was published with
the original description.

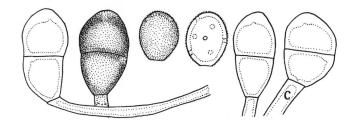

Figure 287

272. PUCCINIA MORIGERA Cumm. Mycologia 43:91. 1951. Fig. 287.

Aecia unknown. Uredinia on abaxial leaf surface, cinnamon-brown, to 1 mm long; spores 19-26 x 18-23μ, globoid or broadly ellipsoid, wall 2-3μ thick, pale cinnamon-brown or golden, verrucose, germ pores 6 or 7, scattered. Telia like the uredinia but pulvinate and blackish brown, early exposed; spores 30-46(-52) x (19-)21-24(-26)μ, broadly ellipsoid or clavate-ellipsoid, wall 2-3.5μ thick at sides, 6-9μ apically, chestnut-brown, smooth; pedicels brownish, thick-walled, not collapsing, to 90μ long.

Hosts and distribution: _Eragrostis_ sp.: China.

Type: S. Y. Cheo No. 385, Fan Ching Shan, Chiang K'ou Hsien, Kweichow Prov., China (PUR).

Cummins (loc. cit.) published a photograph of teliospores of the type.

413

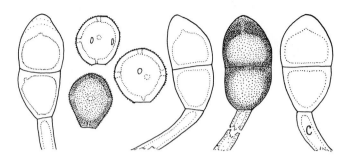

Figure 288

273. PUCCINIA SUBNITENS Diet. Erythea 3:81. 1895. Fig. 288.

Puccinia thalassica Speg. An. Mus. Nac. Buenos Aires 6:225. 1899.

Aecia (Aecidium biforme Peck) occur on genera of the Allioniaceae, Amaranthaceae, Boraginaceae, Capparidaceae, Caryophyllaceae, Chenopodiaceae, Cruciferae, Fumariaceae, Gentianaceae, Hydrophyllaceae, Loasaceae, Lobeliaceae, Onagraceae, Plantaginaceae, Polemoniaceae, Polygonaceae, Primulaceae, Scheuchzeriaceae, Solanaceae, Tetragoniaceae, Tropaeolaceae, and Verbenaceae; spores 15-23 x 13-21, mostly globoid, wall 1-3μ thick, colorless or yellowish, verrucose. Uredinia mostly on adaxial leaf surface, yellowish brown, rather compact; spores (19-)20-24(-26) x 19-24(-25)μ, mostly globoid or broadly ellipsoid, wall (1.5-)2-3(-4)μ thick, golden brown or sometimes darker, finely verrucose with wartlets tending to be in a striate or reticulate pattern, germ pores 4-7, mostly with 3, 4 or 5 equatorial and one apical, less commonly with 2 pores near apex, rarely only equatorial or randomly scattered. Telia mostly on adaxial surface, early exposed, blackish brown, compact; spores often dimorphic with the shorter broader spores more deeply pigmented than the longer spores, (30-)36-46(-55;-64) x (17-)19-24(-27)μ, wall 1.5-2(-4)μ thick at sides, 5-9(-12)μ apically, chestnut-brown, smooth; pedicels mostly thick-walled, collapsing or not, colorless, to 160μ long. Sporogenous basal cells often conspicuous.

Hosts and distribution: species of Distichlis, Monanthochloë littoralis Engelm.: sparingly along the Atlantic Coast of the U.S., from Manitoba to Mexico and west to the Pacific, and in western South America.

Type: Anderson, on Distichlys spicata (now considered to be Distichis stricta), Montana (S; isotype PUR).

Arthur (Bot. Gaz. 35:19. 1903) first proved the aecial stage by inoculation using Chenopodium album as the aecial host. In his "Manual" (1934) he summarized other "cultures" and listed the many proved or suspected aecial hosts.

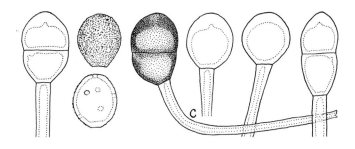

Figure 289

274. PUCCINIA OPUNTIAE Arth. & Holw. in Arthur Proc. Amer. Phil.
Soc. 64:189. 1925. Fig. 289.

The aecial stage (Aecidium opuntiae Magn.) is believed to be
on species of Opuntia. Uredinia on adaxial leaf surface, light
cinnamon-brown; spores 20-26 x 19-26μ, broadly ellipsoid or
globoid, wall 2-2.5μ thick, golden, verrucose, pores 6-8, scattered. Telia amphigenous, blackish, early exposed, pulvinate;
spores 26-45 x 19-26μ, broadly ellipsoid, wall 2-2.5μ thick at
sides, 4-9μ apically, dark chestnut-brown, smooth; pedicels
thick-walled, not collapsing, golden, attaining a length of 130μ;
1-celled spores often are common.

Hosts and distribution: Bouteloua simplex Lag: Bolivia and
Peru.

Type: Holway No. 359, on B. simplex, Cochabamba, Bolivia
(PUR; isotypes Reliq. Holw. No. 56).

Field observation by Holway at Cochabamba, Bolivia indicates
that A. opuntiae is the aecial stage. Hennen and Cummins (Mycologia 48:126-162. 1956) published a photograph of teliospores of
the type.

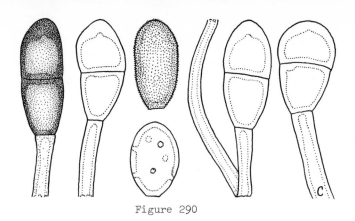

Figure 290

275. PUCCINIA TARRI Cumm. & Husain Bull. Torrey Bot. Club 93:66.
1966. Fig. 290.

Aecia unknown. Uredinia on adaxial surface of leaves,
yellowish brown; spores (28-)32-45(-60) x (16-)18-23(-25)μ,
ellipsoid or oblong-ellipsoid, wall (2-)2.5-3.5μ thick, yellow
or pale golden brown, verrucose, often striately so, pores 6-8,
scattered or bizonate, rarely tending equatorial, obscure.
Telia on adaxial surface and on the sheaths; spores (35-)40-55
(-60) x (18-)21-27(-32)μ, usually ellipsoid or oblong-ellipsoid,
sometimes obovoid, wall (2-)2.5-3.5(-4)μ thick at sides, (4-)5-
7(-9)μ at apex, uniformly chestnut-brown, smooth; pedicels colorless or yellowish, thick-walled, persistent, to 165μ long.

Hosts and distribution: Aristida stipoides Lam., Anglo-
Egyptian Sudan, Tarr No. 971 (Type, PUR F14947: isotype IMI
44854).

P. tarri is distinctive especially because of the long
urediniospores. The teliospores are near the size of those of
P. aristidae var. aristidae but are uniformly and more deeply
pigmented. A photograph of spores of the type was published
with the original description.

276. PUCCINIA MISCANTHICOLA Tai & Cheo Chinese Bot. Soc. Bull. 3:67. 1937.

Aecia, uredia unknown. Telia amphigenous but mostly hypophyllous, roundish, elliptical, or oblong, to 1 mm long, pulvinate, blackish brown; teliospores 32-55 x 15-24μ, sometimes 3- or 4-celled, ellipsoid, oblong-ellipsoid, or nearly fusiform, wall 2-3μ thick at sides, 2-3(-4)μ thick apically, chestnut, smooth; pedicels yellowish, thick-walled and not collapsing, sometimes inserted laterally, to 190μ long, persistent.

Hosts and distribution: Miscanthus sacchariflorus (Maxim.) Hack.: China.

Type: Tai, Nanwutaishan, Shensi, China (Natl. Tsing Hua, Univ. Path. Herb. No. 1283. Not seen.).

This fungus may only be a variant of P. erythropus Diet., which also parasitizes M. sacchariflorus. The 2-celled spores illustrated by Tai and Cheo (Pl. IV, Fig. 24) resemble those of P. erythropus.

277. PUCCINIA LAVROVIANA Cumm. nom. nov.

Puccinia avenastri Lavrov Trud. Tomsk. gos. Univ. Kuibysheva. Ser. Biol. 110:133. Sept. 1951, not Guyot June 1951.

Aecia and uredinia unknown. Telia amphigenous, mostly hypophyllous and on sheaths, blackish, covered by the epidermis; spores 37-62 x 12-19µ, oblong-clavate, truncate or rounded apically, wall brown, thin, thickened to 4µ at apex, smooth; pedicels hyaline, very short.

Type: Lavrov (?), on Avenochloa pubescens (Huds.) Holub (as Avenastrum pubescens), northern Altai, U.S.S.R. (TK?; not seen).

The description is adapted from the original. The species probably belongs in Group VI.

278. PUCCINIA ACHNATHERI-SIBIRICI Wang Acta Phytotax.   Sinica
10:291. 1965.

Aecia and uredinia unknown. Telia amphigenous, mostly
hypophyllous, sometimes on sheaths, becoming exposed, paraphyses capitate, 10-20μ wide apically, the wall 2.5-5μ thick
but 5-10μ in the apex; spores 35-58 x 15-20μ, oblong or oblong-
clavate, wall 1-1.5μ thick at sides, 2.5-5μ apically, yellowish
brown, smooth; pedicels brownish, short.

Type: Wang ?, on Stipa sibirica (L.) Lam. (as Achnatherum
sibiricum), Ning-an, Heilungkiang, China (Inst. Microbiol.,
Peking No. 20577; not seen). One other collection was reported
from Honan.

The teliospores are generally similar but narrower than
those of P. mexicensis and P. mexicensis lacks paraphyses. The
species probably belongs in Group II.

The description is adapted from the original.

Wang (loc. cit.) published a photograph of the teliospores,
presumably of the type.

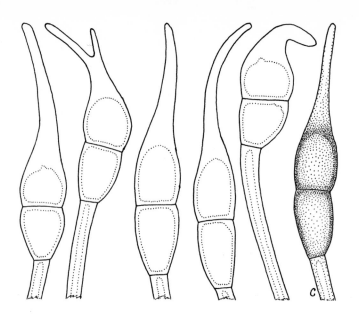

Figure 291

279. PUCCINIA LONGIROSTROIDES Joerst. Ark. Bot. Ser. 2. 4:349-350. 1959. Fig. 291.

Aecia unknown. Uredinia unknown, perhaps not produced. Telia on adaxial leaf surface, early exposed, blackish brown, compact, deeply pulvinate, to 5 mm long and as wide as the (narrow) leaves; spores 60-110(-130) x (14-)16-24(-28)μ, mostly fusiform, wall 1-2μ thick at sides, 20-60(-76)μ apically, the apex extended as a narrow, tapering rostrum, golden brown to clear chestnut-brown, except the rostrum becoming colorless apically, smooth; pedicels colorless, thick-walled, not collapsing, to 150μ long.

Type: Smith No. 1336 on Stipa mongholica (Turcz.) Griseb., Chili Prov. China (UPS). Not otherwise known.

Jørstad (loc. cit.) suggests, because of the similarly rostroid teliospores of the microcyclic Puccinia longirostris Kom., that the aecia may occur on Lonicera. The species probably is an opsis-form.

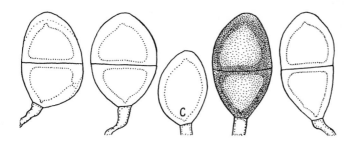

Figure 292

280. PUCCINIA AVOCENSIS Cumm. & H. C. Greene in Greene, Trans. Wis. Acad. Sci. Arts, Letters 43:177. 1954. Fig. 292.

The aecial stage probably is Aecidium avocense Cumm. & H. C. Greene on Callirhoë triangulata (Leavenw.) Gray. Uredinia unknown. Telia epiphyllous, deeply pulvinate, attaining a length of 2 cm, brown; teliospores (32-)37-44(-50) x (19-)25-28(-32)μ, broadly ellipsoid, wall uniformly (2-)3-4(-5)μ thick or only slightly thicker apically, golden or clear chestnut-brown, smooth; pedicels hyaline, thin-walled, collapsing, exceeding 100μ in length but breaking near spore.

Hosts and distribution: Stipa spartea Trin.: U.S.A., one locality in Wisconsin.

Type: H. C. Greene, Avoca, Iowa Co., Wisconsin (PUR; isotype WIS).

A photograph of teliospores of the type was published by Greene and Cummins (Mycologia 50:6-36. 1958).

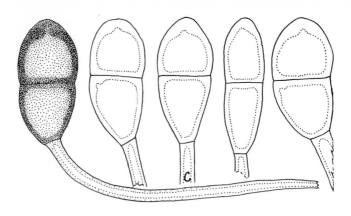

Figure 293

281. PUCCINIA GRAMINELLA Diet. & Holw. in Dietel Erythea
3:80. 1895. Fig. 293.

Aecidium graminellum Speg. An. Soc. Cient. Argentina 12:77.
1881.

Aecia epiphyllous, cylindrical or tongue-like, whitish or
yellowish; spores (18-)22-25(-33) x (16-)20-23(-28)μ, mostly
globoid, wall (2-)3-4.5(-6)μ thick, labyrinthiformly rugose.
Uredinia wanting. Telia on adaxial leaf surface, to 3 mm
long, deeply pulvinate, dark brown; teliospores tending to be
dimorphic, resting type (31-)37-43(-51) x (22-)26-30(-33)μ,
mostly broadly ellipsoid, wall 2-3μ thick at sides, 4-10μ
apically, chestnut-brown, germinating type (40-)50-56(-66) x
(18-)24-28(-32)μ, mostly oblong-ellipsoid, wall 2-2.5μ thick
at sides, 6-18μ apically, golden, smooth; pedicels hyaline or
yellowish, thick-walled, not collapsing, to 200μ long.

Hosts and distribution: Nassella chilensis (Trin. & Rupr.)
Desv., Piptochaetium panicoides (Lam.) Desv., species of Stipa:
western and southern South America and in California, U.S.A.

Type: Blasdale and Holway, on Stipa lepida Hitchc., Berkeley,
California (S; isotype PUR).

A photograph of teliospores of the type was published by
Greene and Cummins (Mycologia 50:6-36. 1958).

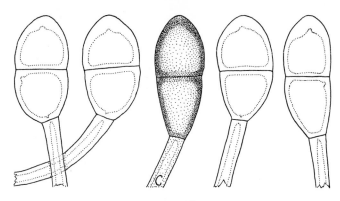

Figure 294

282. PUCCINIA INTERVENIENS Bethel in Blasdale Univ. Calif. Publ.
Bot. 7:119. 1919. Fig. 294.

Aecia, Aecidium modiolae Thuem., on genera of the Malvaceae;
spores 18-27 x 16-24µ, globoid or ellipsoid, wall 5-7µ thick,
colorless, striolate-verrucose. Uredinia wanting. Telia on
adaxial leaf surface, to 5 cm long, deeply pulvinate, dark brown;
spores tending to be dimorphic, resting type (33-)42-47(-56) x
(23-)26-30(-36)µ, broadly ellipsoid, wall 2-3µ thick at sides,
3-12µ apically, chestnut-brown, germinating type (43-)53-60(-80)
x (15-)23-26(-33)µ, mostly ellipsoid, wall 2-2.5µ thick at sides,
4-20µ apically, golden, smooth; pedicels colorless or yellowish,
thick-walled, not collapsing, to at least 200µ long.

Hosts and distribution: species of Nassella and Stipa:
western U.S.A., Mexico, and western South America.

Lectotype: Bethel, on Stipa pulchra Hitchc., Mill Valley,
California (PUR 46787). Lectotype designated by Greene and
Cummins (Mycologia 50:6-36. 1958) who also published a photo-
graph of teliospores.

Mains (Mycologia 25:407-417. 1933) proved the life cycle by
inoculation, using Sidalcea candida as the aecial host.

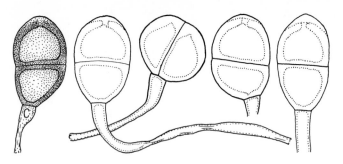

Figure 295

283. PUCCINIA BEWSIAE Cumm. Torrey Bot. Club Bull. 83:226.
1956. Fig. 295.

Aecia and uredinia unknown. Telia on the stems and inflorescence, often confluent, pulvinate, chocolate-brown; teliospores
(32-)34-39(-41) x (22-)24-29(-32)μ, mostly ellipsoid or broadly
ellipsoid, wall 2.5-3.5(-4)μ thick at sides, 5-7(-9)μ apically,
golden or clear chestnut, smooth; pedicels thin-walled and
collapsing, to 90μ or perhaps more but usually broken shorter.

Hosts and distribution: Bewsia biflora (Hack.) Goosens:
Nyasaland.

Type: G. Jackson (comm. P. O. Wiehe as No. 888), Dezda,
Nyasaland (PUR; isotype IMI).

A photograph of spores of the type was published (loc. cit.)
with the diagnosis.

284. PUCCINIA PHAEOPODA H. Syd. in Sydow & Petrak Ann. Mycol. 29:155. 1931.

Aecia and uredinia unknown. Telia on abaxial surface, tardily exposed, blackish brown, small but often confluent, compact; spores 26-34 x 21-27μ, ellipsoid or ovate, usually rounded at both ends or narrowed basally, frequently diorchidioid, wall 1.5-2μ thick at sides, 2.5-5μ thick at apex, smooth; pedicels brown, persistent, to 40μ long.

Type: Clemens 91, on Eulalia cumingii (Nees) A. Camus (as Pollinia cumingii), Bangued, Prov. Abra, Philippines, Feb. 1923. Not seen; not extant?

This is the only reported collection. The description is adapted from the original.

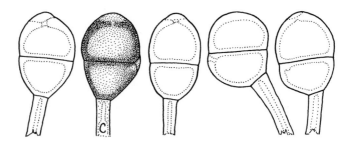

Figure 296

285. PUCCINIA FUSHUNENSIS Hara Fungi Eastern Asia (Japanese)
p. 25. 1928 and in Miura Flora Manchuria & East. Mongolia
3:305. 1928. Fig. 296.

Aecia and uredinia unknown. Telia mostly on abaxial leaf
surface, early exposed, compact, blackish brown; spores 30-34
(-36) x (19-)21-25(-28)μ, mostly broadly obovoid or broadly
ellipsoid, occasionally diorchidioid, wall (1.5-)2-3(-3.5)μ at
sides, 4-6(-7)μ apically, chestnut-brown but with a pale area
apically over the germ pore, smooth; pedicels colorless or
yellowish, thick-walled, not collapsing, to 100μ long.

Type: Hara, on Leersia oryzoides (L.) Swartz var. japonica
Hack., Bujun, South Manchuria, Sept. 1926 (holotype?; isotypes
PUR & Herb. Hiratsuka). Not otherwise reported.

Three urediniospores were seen among the teliospores and
presumably they belong to the species. They were 20 x 18μ and
with a pale yellowish or nearly colorless, echinulate wall.
No paraphyses were seen.

286. PUCCINIA FESTUCAE-OVINAE Tai Farlowia 3:116-117. 1947.

Aecia and uredinia unknown. Telia amphigenous or mostly epiphyllous, to 0.8 mm long, pulvinate, brownish black; spores 28-43 x 12-20μ, ellipsoid or oblong-ellipsoid, wall 1-1.5μ thick at sides, 3-4μ apically, chestnut-brown, smooth; pedicels hyaline, to 57μ long, occasionally inserted laterally.

Type: T. F. Yu and S. T. Chao, on Festuca ovina L., Tali, Yunnan, China, 21 May 1940 (Pl. Pathol. Herb. No. 7834, Tsing Hua Univ., Kunming - not seen). Not otherwise known.

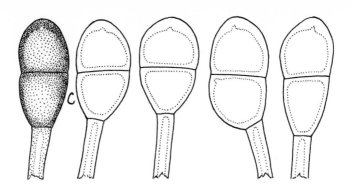

Figure 297

287. PUCCINIA ORYZOPSIDIS H. Syd., P. Syd. & Butler Ann. Mycol.
5:498. 1907. Fig. 297.

Aecia unknown. Uredinia unknown; a few spores presumably
of this species among the teliospores 25-29 x 22-24µ (original
description: 20-25µ diam), wall 1-1.5µ, yellowish, echinulate,
germ pores not visible. Telia amphigenous and on sheaths, early
exposed, chocolate-brown, pulvinate; spores (34-)40-48(-55) x
(20-)22-27(-30)µ, mostly ellipsoid or oblong-ellipsoid, wall
(2-)2.5-3.5(-4)µ thick at sides, (4-)5-7(-8)µ apically, clear
chestnut-brown, smooth; pedicels yellowish or colorless, thick-
walled, not collapsing, to 160µ long.

Type: Butler No. 760, on Oryzopsis molinioides (Boiss.)
Hack., Panikhet, Kumaon, Himalaya (S). Not otherwise reported.

Because few rust fungi on the Stipeae have equatorial germ
pores, it is probable that this species will prove to have
scattered pores and belong in Group VI.

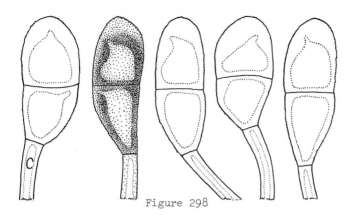

Figure 298

288. PUCCINIA TENELLA Hino & Katumoto Bull. Fac. Agr. Yamaguti Univ. 11:34. 1960. Fig. 298.

Aecia and uredinia unknown. Telia on the abaxial surface of leaves, early exposed, chocolate-brown; spores (40-)42-68 (-80) x 16-25(-34)µ, wall tending to be unilaterally thickened, 2-4µ on the thin side, 4-10µ on the thick side, (6-)8-14(-17)µ apically, bilaminate, the outer layer progressively paler, inner layer golden brown or clear chestnut-brown, finely punctate-verrucose; pedicels colorless, not collapsing, to 250µ long.

Type: On Bambusaceae, collected in Plant Quarantine, Boston, 21 Jan. 1953 as from Hong Kong, China (PUR F15120; isotype BPI). Only this collection is known.

It is impossible to reconcile the original diagnosis and illustration with the type. Apparently, Hino and Katumoto measured and illustrated germinated spores. During germination, most of the pale exterior layer of the wall dissolves. Thus, the germinated spores are quite different from intact spores.

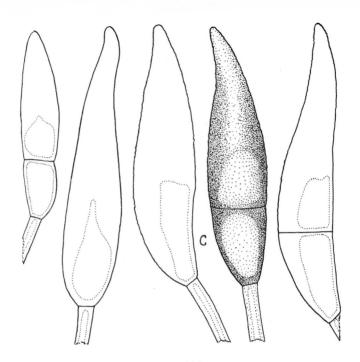

Figure 299

289. PUCCINIA FLAMMULIFORMIS Hino & Katumoto Bull. Fac. Agr.
Yamaguti Univ. 11:31. 1960. Fig. 299.

Aecia and uredinia unknown. Telia on abaxial leaf surface,
early exposed, blackish brown; spores (50-)65-120(-130) x
(16-)18-24(-26)μ, mostly fusiform-ellipsoid or elongate-ovoid,
wall unilaterally thickened, 2-3μ thick on the thin side, somewhat to much thicker on opposite side, 25-75μ apically, yellowish
brown, finely rugose; pedicels colorless, not collapsing, to
270μ long; 1-celled spores common.

Type: W. H. Wheeler, on unidentified Bambuseae from China
(collected in Plant Quarantine, San Francisco No. 9357) (PUR
F3744; isotype BPI).

One other collection (PUR F14858) on _Sasa tesselata_ (Munro)
Makino & Shibata is known. It was collected by Plant Quarantine
officials in Philadelphia as from China.

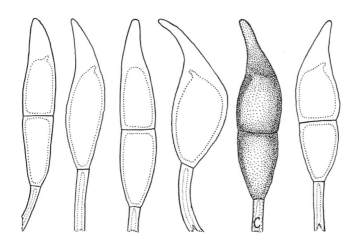

Figure 300

290. PUCCINIA NIGROCONOIDEA Hino & Katumoto Bull. Fac. Agr.
Yamaguti Univ. 11:32. 1960. Fig. 300.

Aecia and uredinia unknown. Telia on abaxial leaf surface,
early exposed, blackish brown; spores (60-)70-85(-92) x 15-22
(-23)μ, mostly ellipsoid or fusiform-ellipsoid, wall 2-3μ thick
at sides or thicker in 1-celled spores, 17-34μ apically, golden
or clear chestnut-brown, or darker in 1-celled spores, minutely
punctate verrucose; pedicels hyaline, non-collapsing, to 150μ
long; 1-celled teliospores common, shorter than the above
measurements.

Type: S. Y. Cheo No. 1584, on Phyllostachys sp., Anhwei
Prov., China (PUR F14381; isotype FH). Known only from the
type.

This species may prove to be synonymous with P. longicornis.

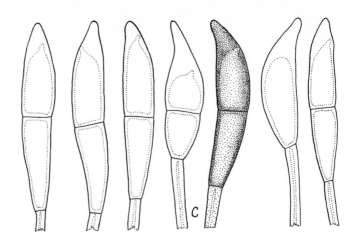

Figure 301

291. PUCCINIA BRACHYSTACHYICOLA Hino & Katumoto Bull. Fac. Agr. Yamaguti Univ. 11:30. 1960. Fig. 301.

Aecia and uredinia unknown. Telia on abaxial surface of leaf, early exposed, blackish brown; spores 60-90(-97) x (14-)16-22(-24)μ, ellipsoid or fusiform-ellipsoid, wall (1-) 1.5-2.5(-3)μ thick at sides, (6-)14-22(-25)μ apically, mostly golden brown, paler in thin-walled narrow spores, chestnut-brown in short robust spores, minutely punctate-verrucose, especially obvious in more robust spores; pedicels colorless, not collapsing, tapering, to 200μ long.

Type: C. Y. Chiao, on Brachystachyum densiflorum (Rendle) Keng, Hanghow, Chekiang Prov., China (PUR F15492; isotype BPI). Only this collection is known.

6. UROMYCES Unger

Exantheme Pflanzen p. 277. 1833

Type species: Uromyces appendiculatus (Pers.) Unger

Key to species

GROUP I: uredinia paraphysate, urediniospores echinulate,
       germ pores equatorial
1. Telia exposed; teliospores with apical digitations (2)
1. Telia covered; teliospore apex without
                  digitations...............1. niteroyensis
2. Urediniospores mostly 26-36µ long...............2. coronatus
2. Urediniospores mostly 22-26µ long...............3. halstedii

GROUP II: uredinia paraphysate, urediniospores echinulate,
       germ pores scattered
1. Urediniospore wall brown; teliospore wall
                  thickened apically.........4. aristidae
1. Urediniospore pale yellowish; teliospore
      wall uniformly 1.5-2µ.....................5. turcomanicum

GROUP III: uredinia paraphysate, urediniospores verrucose,
       germ pores equatorial: no species

GROUP IV: uredinia paraphysate, urediniospores verrucose,
       germ pores scattered: no species

GROUP V: uredinia aparaphysate, urediniospores echinulate,
       germ pores equatorial
1. Telia covered or only tardily exposed, not erumpent (2)
1. Telia early exposed, erumpent (7)
2. Telia with paraphyses, tending to be loculate (3)
2. Telia aparaphysate (4)
3. Urediniospore wall yellowish, germ
                  pores 3...........6. phalaridicola
3. Urediniospore wall about cinnamon-brown,
       germ pores mostly 4......................7. tenuicutis
4. Telia slowly exposed by a narrow slit........8. trichoneurae
4. Telia remaining covered (5)
5. Urediniospores mostly 24-28µ long; apical
      wall of teliospore thickened to 3-5µ.....9. dactyloctenii
5. Urediniospores commonly exceeding 30µ long (6)
6. Germ pores 2, rarely 3; apical wall of telio-
      spore 3-6µ thick.......................10. sporobolicola
6. Germ pores 3; teliospore wall uniformly
                  1-1.5µ.........11. setariae-italicae
7. Teliospores nearly globoid (15)
7. Teliospores oval, ellipsoid, or obovoid (8)
8. Urediniospore wall mostly 1µ thick; teliospore wall
      0.5-1µ at sides, 2-3 apically...........12. costaricensis
8. Urediniospore and teliospore wall thicker (9)

9. Teliospores mostly obovoid and somewhat angular (10)
9. Teliospores mostly oval or ellipsoid, not angular (12)
10. Germ pores 4-7, equatorial or with 1 or 2 extra-
                  equatorial.....................13. tripogonicola
10. Germ pores 3 or mostly 3 (11)
11. Urediniospores mostly 26-32µ long, wall
                          2.5-3µ............14. argutus
11. Urediniospores mostly 20-24µ long, wall
                          1.5µ..........15. schoenanthi
12. Urediniospores mostly 36-40µ long; telio-
            spores mostly 33-40µ long.............16. sporoboli
12. Urediniospores and teliospores less than 33µ long (13)
13. Apical wall of teliospores 5-13µ thick....17. muehlenbergiae
13. Apical wall 5-9µ thick (14)
14. Urediniospores mostly 20-25µ long; teliospores
                  mostly 23-28µ long...........18. graminicola
14. Urediniospores mostly 25-30µ long; teliospores
                  mostly 25-32µ long.............19. penniseti
15. Teliospore pedicels broad, thick-walled, mostly not
                      collapsing (16)
15. Teliospore pedicels slender, thin-walled, mostly
                      collapsing................20. major
16. Teliospores mostly 25-34 x 19-24µ................21. blandus
16. Teliospores mostly 24-29 x 19-24µ...............22. linearis

GROUP VI: uredinia aparaphysate, spores echinulate,
            germ pores scattered
1. Telia covered or only tardily exposed, not erumpent (2)
1. Telia early exposed, erumpent (16)
2. Teliospores loose and powdery beneath the epidermis (3)
2. Teliospores not loose, firmly attached (6)
3. Teliospore with a papilla over the pore........23. brominus
3. Teliospores without a papilla (4)
4. Urediniospores mostly 24-32 x 22-28µ..........24. fragilipes
4. Urediniospores smaller (5)
5. Urediniospores mostly 22-26 x 20-22µ.........25. paspalicola
5. Urediniospores mostly 19-21 x 16-19µ.........26. microchloae
6. Telia without paraphyses (7)
6. Telia paraphysate and usually loculate (10)
7. Urediniospores mostly 25-30 x 22-27µ; teliospores
       tending to be cuboidal, see...............Puccinia cryptica
                                      var. bromicola
7. Urediniospores mostly 23-27 x 18-23µ; teliospores
                      oval or obovoid (8)
8. Urediniospore wall 2-2.5µ thick; teliospores
                  31-37µ long................27. airae-flexuosae
8. Urediniospore wall thinner; teliospores shorter (9)
9. Teliospore pedicel to 60µ long, usually broken
                  shorter...............28. pegleriae
9. Teliospore pedicel to 25µ long, usually broken
                  shorter..................29. tragi
10. Teliospores mostly 29-38 x 20-26µ, wall commonly
              with fine ridges..................30. beckmanniae
10. Teliospores smaller, rarely or not ridged (11)

11. Urediniospore wall 2.5-3.5μ thick, germ pores
                  3-6...............31. koeleriae
11. Urediniospore wall thinner, germ pores more (12)
12. Urediniospore wall golden or near cinnamon-brown (13)
12. Urediniospore wall colorless or pale yellowish (14)
13. Urediniospores mostly 25-30 x 20-24μ..........32. dactylidis
                                        var. dactylidis
13. Urediniospores mostly 21-27 x 17-21μ.....33. calamagrostidis
14. Telia weakly loculate; urediniospores mostly
                  26-30 x 21-25μ...................34. hordeinus
14. Telia strongly loculate (15)
15. Telial paraphyses brown.......................32. dactylidis
                                        var. poae
15. Telial paraphyses colorless...................32. dactylidis
                                        var. poae-alpinae
16. Aecia usually associated with uredinia or telia,
                  autoecious.................35. pencanus
16. Aecia not associated, heteroecious or presumably so (17)
17. Teliospore pedicels thick-walled, terete, not
                              collapsing (18)
17. Teliospore pedicels thin-walled, usually collapsing (19)
18. Teliospores chestnut-brown, mostly 30-38 x
                  21-24μ..............36. nassellae
18. Teliospores golden brown, mostly 32-48 x
                  23-28μ..............37. cuspidatus
19. Teliospore wall uniformly 3-5μ thick,
                  nearly opaque..................38. clignyi
19. Teliospore wall always thickened apically (20)
20. Teliospores ellipsoid, tending to be acuminate apically (21)
20. Teliospore globoid, broadly ellipsoid or obovoid,
        broadly rounded or obtuse apically (28)
21. Urediniospores mostly 35-42 x 30-35μ............39. mcnabbii
21. Urediniospores less than 35μ long (22)
22. Teliospores rarely as much as 30μ long (26)
22. Teliospores commonly exceeding 30μ long (23)
23. Urediniospore wall mostly yellowish, spores mostly
                  27-32 x 23-28μ..................40. acuminatus
23. Urediniospore wall mostly cinnamon-brown, spores
                  smaller (25)
25. Teliospores mostly 27-34 x 19-24μ.............41. danthoniae
25. Teliospores mostly 21-34 x 16-20μ.............42. amphidymus
26. Urediniospores mostly 17-21 x 16-18μ, wall
                  cinnamon-brown...............45. minimus
26. Urediniospores mostly more than 23μ long (27)
27. Teliospores mostly 25-30 x 17-22μ.................43. otakou
27. Teliospores mostly 20-27 x 13-16μ.............44. ehrhartiae
28. Germ pores mostly 5 or 6, equatorial but often with
     1 or 2 extra-equatorial...................13. tripogonicola
28. Germ pore typically scattered (29)
29. Urediniospore wall 1-1.5μ thick (35)
29. Urediniospore wall 2μ or thicker (30)
30. Urediniospore echinulae rather low, broad cones,
        spaced 3.5-4μ...........................46. graminis
30. Urediniospore echinulae not thus (31)

435

31. Urediniospores dark brown, mostly 28-32 x
                    24-30μ................47. epicampis
31. Urediniospores pale cinnamon-brown or paler (32)
32. Teliospore mostly 27-30 x 21-26μ.............48. ferganensis
32. Teliospores smaller (33)
33. Urediospore wall golden to cinnamon-brown, telio-
        spores nearly globoid, not angular (34)
33. Urediniospore wall pale yellowish; teliospores
                    angularly obovoid................51. holci
34. Urediniospores mostly 24-28 x 20-26μ........49. leptochloae
34. Urediniospores mostly 21-24 x 18-21μ...........50. kenyensis
35. Urediniospores mostly 17-20 x 15-18μ..........52. snowdeniae
35. Urediniospores larger (36)
36. Urediniospores mostly 19-24 x 18-22; teliospores
                    nearly globoid............53. aegopogonis
36. Urediniospores mostly 21-29 x 18-23μ........54. eragrostidis

GROUP VII: uredinia aparaphysate, urediniospores verrucose,
            germ pores equatorial
1. Teliospore pedicels thick-walled, not
                    collapsing............55. archerianus
1. Teliospore pedicels thin-walled, usually collapsing (2)
2. Teliospores often punctate-verrucose apically;
        germ pores 3-5, mostly 4....................56. vossiae
2. Teliospores never punctate verrucose; germ pores 2-4,
                    often 3 (3)
3. Teliospore mostly 30-40μ long (4)
3. Teliospores seldom as much as 30μ long (5)
4. Teliospores dimorphic, both slender pale spores and
        robust chestnut-brown spores formed........57. seditiosus
4. Teliospore all similar, robust and
                    chestnut-brown......58. aeluropodis-repentis
5. Teliospores mostly ellipsoid, mostly 20-30 x 13-17μ;
        urediniospores mostly 16-19 x 14-17μ.......59. andropogonis
5. Teliospores mostly broadly ellipsoid or globoid;
        urediniospores mostly 19-24 x 18-21μ (6)
6. Teliospores mostly 23-27 x 18-23μ..........60. mussooriensis
6. Teliospores mostly 20-24 x 18-21μ...............61. inayati

GROUP VIII: uredinia aparaphysate, urediniospores verrucose,
            germ pores scattered
1. Germ pores 4-6, mostly with 1 or 2 apical, the others
        equatorial; teliospores mostly 24-36 x 17-23, tending
                    to be dimorphic.......................62. peckianus

GROUP IX: uredinia either unknown or lacking from the
            life cycle
1. Teliospores with a pale differentiated umbo and
                    fragile pedicels..................63. stipinus
1. Teliospores without such an umbo, pedicels thick-
                    walled, persistent (2)
2. Teliospores mostly 26-36 x 19-24μ, the apex
                    rounded......64. ehrhartiae-giganteae
2. Teliospores mostly 42-65 x 15-19μ, the apex
                    acuminate....................65. procerus

436

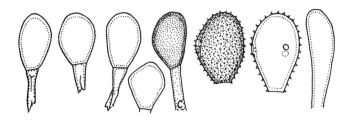

Figure 302

1. UROMYCES NITEROYENSIS Rangel Arch. Mus. Nac. Rio de Janeiro
18:160. 1916. Fig. 302.

Uromyces puttemansii Rangel Arch. Mus. Nac. Rio de Janeiro
18:159. 1916.

Uromyces sepultus Mains Carnegie Inst. Washington Publ.
461:99. 1935.

Aecia unknown. Uredinia amphigenous or mostly on abaxial
leaf surface, yellowish-brown to cinnamon-brown, with inconspicuous, yellowish, thin-walled paraphyses; spores (26-)29-38
(-42) x 20-27μ, mostly broadly ellipsoid or obovoid, wall
1.5-2μ thick, cinnamon-brown, echinulate, pores 3(4), equatorial.
Telia blackish brown, long covered by epidermis, without paraphyses; spores (19-)22-27(-30) x 14-20μ, variable but mostly
angularly obovoid, wall 0-1μ thick at sides, 1.5-2.5μ at apex,
golden to chestnut-brown, smooth; pedicels persistent, yellowish, thin-walled and collapsing, to 25μ long.

Hosts and distribution: Panicum antidotale Retz., species
of Setaria: Cuba and Mexico to Brazil and Argentina.

Lectotype: Rangel No. 1212, on Setaria sp., Cubango-Niteroy,
Brazil (R; isotype PUR).

A photograph of teliospores of the type was published by
Ramachar and Cummins (Mycopathol. Mycol. Appl. 19:49-61. 1963).

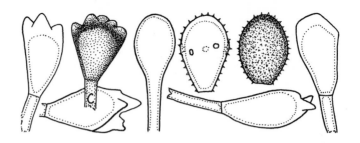

Figure 303

2. UROMYCES CORONATUS Miy. & Nish. ex Dietel in Bot. Centralbl.
105:495. 1907. Fig. 303.

Uromyces coronatus Yoshinaga (error for Miy. & Nish.) ex Dietel
in Ann. Mycol. 5:70. 1907.

Uromyces zizaniae-latifoliae Saw. Descr. Cat. Formosan Fungi
2:93. 1922.

Aecia unknown. Uredinia amphigenous, cinnamon-brown, with
thin-walled capitate paraphyses; spores (23-)26-36(-40) x
(16-)19-22(-25)μ, mostly narrowly obovoid or ellipsoid, wall
1.5-2μ thick at sides, 2-4μ thick at apex, golden below,
cinnamon- to chestnut-brown at apex, echinulate, pores 4 or 5,
equatorial. Telia blackish brown, early exposed, spores (22-)
25-36(-42) x (12-)16-23(-29)μ, including projections, mostly
cuneate or oblong, wall 1-1.5μ thick at sides, 4-13μ at apex
including projections, chestnut-brown, the apex usually coronate
with a few projections from 3-10μ in length, pedicels persistent, brownish, thin-walled, collapsing or not, to 50μ long.

Hosts and distribution: Zizania aquatica L., Z. latifolia
(Griseb.) Turcz.: China, Formosa, and Japan.

Type: Yoshinaga, on Z. aquatica, Tosa, Japan (S).

U. coronatus differs from U. halstedii in having larger
urediniospores whose apical wall is thickened and teliospores
with fewer and inconstant digitations.

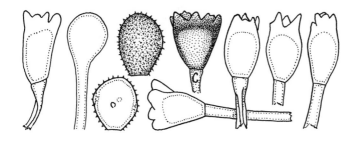

Figure 304

3. UROMYCES HALSTEDII De T. in Saccardo Syll. Fung. 7:557. 1888. Fig. 304.

Uromyces digitatus Halst. J. Mycol. 3:138. 1887, non Winter, 1886.

Uromyces halstedii F. Ludwig Bot. Centrlbl. 37:120. 1889.

Uromyces ovalis Diet. Bot. Jahrb. 37:97. 1905.

Aecia (Aecidium trillii Burr.) occur on species of Trillium; spores 20-24 x 19-22μ, wall 1μ thick, verrucose, colorless. Uredinia amphigenous, yellowish brown, with thin-walled capitate paraphyses; spores (20-)22-26(-28) x (14-)16-21μ, mostly obovoid or broadly ellipsoid, wall (1-)1.5(-2)μ thick, yellowish to cinnamon-brown, echinulate, pores 3 or 4, equatorial. Telia amphigenous, blackish brown, early exposed, pulvinate, compact; spores (20-)24-30(-38) x (12-)15-24(-28)μ including projections, variable but mostly cuneate, the coronate apex often much wider than the body of the spore, wall 1-1.5μ at sides, 5-15μ at apex including the projections which vary from 3-12μ in length, smooth, deep golden to chestnut-brown; pedicels persistent, brown, usually thin-walled and collapsing, to 50μ long but usually shorter.

Hosts and distribution: Brachyelytrum erectum (Schreb.) Beauv., Leersia oryzoides (L.) Swartz, L. sayanuka Ohwi, L. virginica Willd.: Japan, U.S.A.

Neotype: Halsted, on Leersia virginica, Ames, Iowa. (PUR 11952; isotypes, Ellis & Ev. N. Amer. Fungi No. 2227). Neotype designated here.

The life cycle was proved by Barrus (Mycologia 20:117-126. 1928).

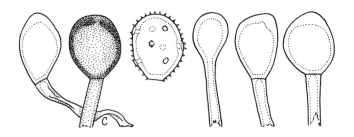

Figure 305

4. UROMYCES ARISTIDAE Ellis & Ev. J. Mycol. 3:56. 1887.
Fig. 305.

Aecia unknown. Uredinia adaxial, cinnamon-brown, paraphyses
capitate or clavate, to 90μ long, 22μ diam; spores (25-)27-33
(-36) x (18-)20-24(-26)μ, broadly ellipsoid or ellipsoid, wall
2.5-3.5μ thick, cinnamon-brown, echinulate, pores 6-9, scattered.
Telia not seen; teliospores in uredinia (23-)25-32 x (19-)
21-27μ, wall (1.5-)2-3(-3.5)μ thick at sides, (4-)5-7(-9)μ at
apex, smooth but tesselately cracked with age, chestnut-brown;
pedicels hyaline, to 100μ long, mostly collapsing.

Hosts and distribution: Aristida arizonica Vasey, New Mexico,
U.S.A.

Type: Vasey, Santa Fe, New Mexico (NY; isotype PUR 11937).

Except for 1-celled teliospores, the species is similar to
Puccinia unica var. unica and probably is derived from it.

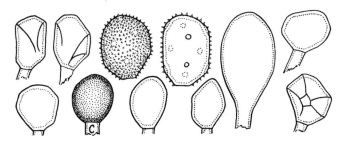

Figure 306

5. UROMYCES TURCOMANICUM Katajev Akad. Nauk Bot. Odt. Sporov.
Rast. Bot. Mater. 8:111. 1952. Fig. 306.

Uromyces iranensis V.-Bourgin Compte Rendu Acad. Sci.
242:412. 1952.

Uromyces boissierae V.-Bourgin Ann. Epiphyt. 1958:151-152.
1958.

Uromyces prismaticus V.-Bourgin Ann. Epiphyt. 1958:161.
1958.

Uromyces viennot-bourginii Wahl & Anikster in Anikster &
Wahl Bull. Soc. Mycol. France 82:554-555. 1966.

Uromyces christensenii Anikster & Wahl Israel J. Bot. 15:98.
1966 (issued 1967).

Aecia occur on species of Bellevalia and Muscari, in groups;
spores (17.5-)19-25(-30) x (12.5-)15-20(-22)μ, wall colorless,
1-1.5μ thick, verrucose. Uredinia amphigenous, with colorless,
mostly saccate, collapsing paraphyses, to 60μ long and to 30μ
wide, wall uniformly 0.5μ thick; spores (22-)24-32(-36) x (17-)
19-25(-28)μ, mostly ellipsoid or broadly ellipsoid, wall
(1-)1.5-2(-2.5)μ thick, pale yellowish to nearly colorless,
echinulate, germ pores 7-11(-13), scattered, difficult to
count. Telia amphigenous, loosely covered by the epidermis or
exposed, always pulverulent, chocolate-brown; spores (16-)
18-24(-26) x (13-)14-20(-22)μ, variable and often angular,
usually more or less obovoid, tending to be dimorphic with the
paler spores more angular than deeply pigmented spores, wall
uniformly 1.5-2μ thick, often with surface ridges, sometimes
appearing to be punctate, otherwise smooth; pedicels thin-
walled, usually collapsing, to 40μ long, usually broken shorter.

Hosts and distribution: Boissiera pumilo (Trin.) Hack.,
Festuca ovina L., Hordeum bulbosum L., H. spontaneum Koch,
H. violaceum Regel, H. vulgare L., Secale montanum Guss.:
southern Russia to Israel, Iraq, and Iran.

Type: Katajev, on Hordeum bulbosum, Kopet-Dagh, Firusa,
Turcomen SSR (LE?; not seen).

The first inoculations, using aeciospores from Bellevalia flexuosa to infect Hordeum bulbosum were by Shabi in 1963 (see Anikster and Wahl, Israel J. Bot. 15:91-105. 1966, issued 1967, who assigned the aecia to Uromyces hordeastri = U. fragilipes). They also report successful inoculations using Muscari parviflorum and H. bulbosum. In 1966 (loc. cit.) they reported aecia on Bellevalia eigii, and inoculations, when they described U. viennot-bourginii.

This species differs from U. fragilipes because of smaller teliospores and the peculiar saccate paraphyses. Both species have pulverulent telia quite unlike most covered telia of the grass rust fungi.

6. UROMYCES PHALARIDICOLA Katajev Akad. Nauk Bot. Otd. Sporov. Rast. Bot. Mater. 7:173. 1951.

Aecia unknown. Uredinia hypophyllous, scattered or linear. "Paraphysibus coalitis, linearibus, pallido-brunneis." Spores 20-30.5 x 20-22.5, subglobose or broadly ellipsoid, wall 1.5-2μ thick, yellowish, finely echinulate, germ pores 3. Telia amphigenous, seriate, covered by the epidermis, blackish, paraphyses linear, pale brown; spores 21-27.5 x 15-22.5μ, ovate, ellipsoid, or pyriform, the apex rounded or truncate, wall at apex 3-5μ thick, yellowish brown but darker apically, smooth; pedicels brownish, persistent, as long as the spore or shorter.

Type: Medvedeva, on Phalaris minor Retz., Kopet-Dag, Chodja-Dere, Turkmen SSR, V 1943 (LE?; not seen).

The description of uredinial paraphyses seems to apply to the telial type. Except that the urediniospores are described as having 3 pores, the species is similar to Uromyces dactylidis. Katajev's drawing (Fig. 4) of teliospores is nearly identical to his Fig. 3 of Uromyces triseti, which I consider to be a synonym of U. dactylidis var. poae.

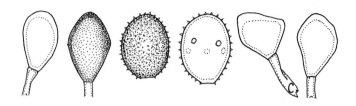

Figure 307

7. UROMYCES TENUICUTIS McAlp. Rusts of Australia, p. 87. 1906. Fig. 307.

Uredo ignobilis Syd. Ann. Mycol. 4:444. 1906.

Uromyces ignobilis (Syd.) Arth. Mycologia 7:181. 1915.

Uromyces wellingtonica T. S. Ramak. & K. Ramak. Indian Acad. Sci. Proc. B. 28:66-67. 1948.

Uromyces sporoboloides Cumm. Bull. Torrey Bot. Club. 83:232. 1956.

Aecia unknown. Uredinia amphigenous or mostly on adaxial surface; yellowish brown; spores (20-)24-30(-35) x (16-)19-23 (-27)μ, wall 1-1.5(-2.5)μ thick, cinnamon-brown or paler, finely echinulate, pores (3-)4(-5), equatorial. Telia adaxial, inconspicuous, covered, paraphyses present and often abundant, greyish; spores (19-)22-28(-35) x (14-)16-23(-25)μ, variable, triangular or angularly obovoid or oblong, wall (1-)1.5-2μ thick at sides, 2-4(-5)μ apically, chestnut-brown or golden, smooth; pedicels brownish, thin-walled and collapsing, to 50μ long but commonly shorter.

Hosts and distribution: On species of Sporobolus: circumglobal in the warmer regions.

Type: G. H. Robinson, on S. indicus, Caulfield (suburb of Melbourne), Australia (MEL).

A photograph of teliospores of the type was published by Cummins and Greene (Brittonia 13:271-285. 1961).

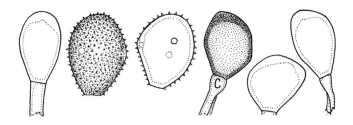

Figure 308

8. UROMYCES TRICHONEURAE Doidge Bothalia 3:512. 1939. Fig.
308.

Aecia unknown. Uredinia amphigenous, about cinnamon-brown;
spores (23-)26-32(-35) x (19-)21-25(-28)μ, mostly obovoid or
ellipsoid, wall 1.5-2μ thick, golden to cinnamon-brown, echinulate, germ pores 3 (rarely 4?), equatorial. Telia amphigenous,
exposed by a slit in the epidermis, blackish brown, compact;
spores (20-)22-29(-32) x (16-)18-20(-24)μ, mostly angularly
obovoid, wall (1-)1.5-2μ thick at sides, 3-5μ apically, chestnut-
brown, smooth; pedicels yellowish, thin-walled and collapsing,
to 45μ long but usually broken short.

Type: Doidge and Bottomley, on Trichoneura grandiglumis
(Nees) Ekman, Donkerpoort, Pretoria, South Africa (PRE 29762;
isotype PUR). Known from one other collection on the same host
in Pretoria. The type was designated by Doidge in 1950
(Bothalia 5:450).

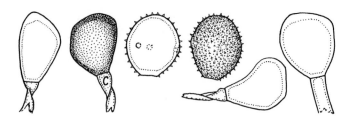

Figure 309

9. UROMYCES DACTYLOCTENII Wakef. & Hansf. Proc. Linn. Soc. Lond. 161:170. 1949. Fig. 309.

Uredo dactylocteniicola Speg. Anal. Mus. Nac. B. Aires 31:392. 1922.

Uromyces dactylocteniicola Lindquist Notas Mus. La Plata 8:136. 1943. (nom. dubium).

Aecia unknown. Uredinia mostly on abaxial leaf surface, cinnamon-brown; spores (21-)23-27(-29) x (18-)20-23(-24)μ, mostly broadly ellipsoid, wall 1.5-2(-2.5)μ thick, golden to cinnamon-brown, echinulate, pores 3(4), equatorial. Telia covered by the epidermis, blackish, without paraphyses; spores (22-)24-28(-30) x (16-)18-22(-26)μ, mostly obovoid or broadly ellipsoid, wall 1-2μ thick at sides, 3-5(-6)μ apically, chestnut-brown, smooth, brittle and easily broken; pedicels hyaline or yellowish, thin-walled, collapsing, and usually broken short, to 25μ long.

Hosts and distribution: Dactyloctenium aegyptium (L.) Beauv., Microchloa indica (L.) Beauv.: Central Africa, the Philippines, and South America.

Type: Hansford No. 1653, Katakwi, Teso (K; isotype PUR).

There is confusion as to the identity of the South American rust. Lindquist described and illustrated the teliospores as verrucose and, in an examination of fragment of the type, a few such spores were found but also present were spores like those in U. dactyloctenii. I believe the verrucose spores to be strays.

A photograph of teliospores of the type was published by Hennen and Cummins (Mycologia 48:126-162. 1956).

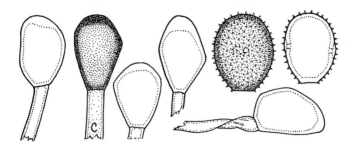

Figure 310

10. UROMYCES SPOROBOLICOLA Lindq. Rev. Fac. Agron. 28:89.
1962. Fig. 310.

Uredo egenula Arth. Bull. Torrey Bot. Club 45:155. 1918.

Uromyces bravensis Cumm. Southw. Nat. 8:193. 1964.

Aecia unknown. Uredinia adaxial, cinnamon-brown; spores
(28-)30-35(-42) x (22-)25-29(-33)µ, wall (1.5-)2-3(-4)µ thick,
golden or cinnamon-brown, echinulate, germ pores 2, rarely 3,
equatorial. Telia amphigenous, blackish, covered by the epidermis, or tardily exposed; spores (24-)26-34(-37) x (15-)17-23
(-25)µ, variable but usually angularly obovoid, wall 1.5-2µ
thick at sides, 3-6(-9)µ thick at apex, chestnut-brown, smooth;
pedicels yellowish, persistent, thin-walled, to 35µ long.

Hosts and distribution: Sporobolus pyramidatus (Lam.)
Hitchc.: Texas, U.S.A. and northeastern Mexico east to the
Dominican Republic, and in Argentina.

Type: Ragonese, on Sporobolus pyramidatus, Argentina (ex
LPM 7763 = LPS 30.926; isotype PUR).

This is one of the few grass rust fungi that have 2 germ
pores in the urediniospores. The telia and teliospores are
generally like those of U. tenuicutis.

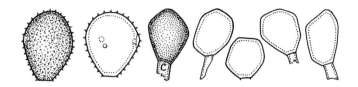

Figure 311

11. UROMYCES SETARIAE-ITALICAE Yosh. Bot. Mag. Tokyo 20:247. 1906 (20 Oct.). Fig. 311.

Uredo setariae-italicae Diet. Bot. Jahrb. 32:632. 1903.

Uredo panici P. Henn. Hedwigia 43:165. 1904.

Uromyces leptodermus H. Syd. & P. Syd. Ann. Mycol. 4:430. 1906 (31 Oct.).

Uredo eriochloae H. Syd. & P. Syd. ibid. 4:444. 1906.

Uredo isachnes H. Syd. & P. Syd. ibid. 4:444. 1906.

Uredo panici-prostrati H. Syd. & P. Syd. ibid. 4:444. 1906.

Uromyces eriochloae H. Syd. P. Syd. & Butl. Ann. Mycol. 5:492. 1907.

Puccinia panicicola Arth. Bull. Torrey Bot. Club 34:586. 1907. Based on uredinia.

Uredo eriochloae Speg. An. Mus. Nac. B. Aires 19:319. 1909.

Uredo henningsii Sacc. & D. Sacc. Syll. Fung. 17:456. 1905.

Uredo eriochloana Sacc. & Trott. in Saccardo Syll. Fung. 21:810. 1912.

Uredo panici-maximi Rangel Arc. Mus. Rio de Janeiro 18:160. 1916.

Uredo panici-villosi Petch Ann. Roy. Bot. Gard. Peradeniya 7:295. 1922.

Uredo melinidis Kern Mycologia 30:550. 1938.

Uredo nampoinae Boriq. & Bassino Rev. Mycol. 31:325. 1966.

Aecia (Aecidium brasiliense Diet.) occur on species of Cordia; spores 20-27 x 18-23μ, globoid or ellipsoid, wall 1μ

thick, verrucose. Uredinia amphigenous, cinnamon-brown; spores
(25-)27-33(-35) x (20-)23-28(-30)μ, broadly obovoid or ellipsoid, wall (1-)1.5(-2)μ thick, cinnamon-brown, echinulate, germ
pores 3, equatorial. Telia amphigenous, covered by the epidermis, blackish, small and inconspicuous; spores (16-)18-25(-28)
x (14-)16-20μ, variable, mostly angularly globoid or obovoid,
wall uniformly 1-1.5μ thick, clear chestnut-brown, smooth;
pedicels colorless, thin-walled and collapsing, to 20μ long but
usually broken near the spore.

Hosts and distribution: species of Brachiaria, Cyrtococcum,
Eriochloa, Melinis, Ottochloa, Panicum, Paspalidium, Pennisetum,
Setaria, Stenotaphrum, and Urochloa: circumglobal in warm
regions.

Neotype: Yoshino, on Setaria italica (L.) Beauv., Kumamoto,
Pref. Kumamoto, Japan, 30 Oct. 1906 (PUR F16520). Neotype
designated by Ramachar and Cummins (Mycopathol. Mycol. Appl.
19:49-61. 1963).

Narasimhan and Thirumalachar (Mycologia 56:555-560. 1964)
proved the life cycle with reciprocal inoculations, using
Cordia rothii Roem. & Schult. and Setaria italica (L.) Beauv.
and S. verticillata (L.) Beauv. as host plants.

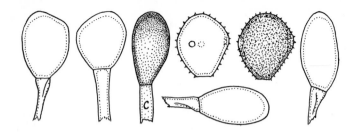

Figure 312

12. UROMYCES COSTARICENSIS H. Syd. Ann. Mycol. 23:312. 1925.
Fig. 312.

Aecia unknown. Uredinia amphigenous, yellowish brown;
spores (21-)24-29(-32) x (16-)20-23(-25)μ, mostly obovoid,
wall 1(-1.5)μ thick, golden to near cinnamon-brown, echinulate,
germ pores 3 or 4, equatorial. Telia amphigenous, chocolate-
brown, exposed; spores (22-)24-30(-34) x (14-)16-18(-20)μ,
mostly ellipsoid or narrowly obovoid, wall 0.5-1μ thick at
sides 2-3(-4)μ apically, golden or clear chestnut-brown, smooth;
pedicels yellowish, thin-walled and collapsing, to 45μ long,
usually broken short.

Hosts and distribution: Lasiacis divaricata (L.) Hitchc.,
L. ruscifolia (H.B.K.) Hitchc., L. sloanei (Griseb.) Hitchc.,
L. sorghoides (Desv.) Hitchc. & Chase: southernmost United
States to Mexico, Venezuela, Brazil, and Trinidad.

Type: Sydow No. 178, on Panicum altissimum (=Lasiacis sorghoides, Grecia, Costa Rica (holotype apparently lost; isotype
BPI).

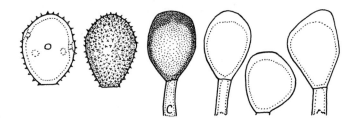

Figure 313

13. UROMYCES TRIPOGONICOLA Payak & Thirum. Sydowia 10:39.
1956 (issued 1957). Fig. 313.

Uromyces tripogonis-sinensis Wang Acta Phytotax. Sinica
10:297. 1965.

Aecia unknown. Uredinia amphigenous and on sheaths, often
conspicuously seriate, about cinnamon-brown; spores (24-)
27-32(-36) x (19-)21-24(-26)μ, mostly obovoid or ellipsoid,
wall (1-)1.5-2(-2.5)μ thick, yellowish to golden, echinulate,
germ pores 4-7, mostly 5 or 6, equatorial or occasionally 1
or 2 are extra-equatorial. Telia amphigenous, narrowly
exposed, blackish brown, compact; spores (22-)26-33(-38) x
(18-)20-25(-27;-30)μ, mostly obovoid or ellipsoid, often
angular, wall (1-)1.5-2(-2.5)μ thick at sides, (3-)4-5(-7)μ
apically, chestnut-brown, smooth; pedicels colorless or yellowish, thin-walled and collapsing, to 35μ long but usually
broken short.

Hosts and distribution: Astrebla elymoides Bailey & F. Muell.,
A. lappacea (Lind.) Domin, A. squarrosa C. E. Hubb., Tripogon
filiformis Nees, T. lisboae Stapf, T. chinensis Hack.: Australia,
China, and India.

Type: Payak, on Tripogon lisboae, Purandhar Hill Fort, Poona,
India, 22 Oct. 1950 (HCIO; isotype PUR).

Wang ascribes smaller spore sizes to his species than are
typical of specimens available to me. He published a photograph
of the teliospores.

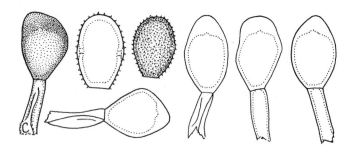

Figure 314

14. UROMYCES ARGUTUS Kern Torreya 11:214. 1911. Fig. 314.

Uredo spartinae-strictae Pat. & Har. Bull. Soc. Mycol. France 21:84. 1905.

Aecia unknown. Uredinia amphigenous, yellowish brown; spores (24-)26-32(-34) x (17-)20-23(-25)μ, mostly ellipsoid or oblong-ellipsoid, wall (2-)2.5-3(-3.5)μ thick, yellowish to dull brown, echinulate, germ pores (2)3(4), equatorial. Telia amphigenous, rather tardily exposed, blackish brown; spores (24-)27-35(-39) x (15-)18-21μ, ellipsoid or obovoid, wall 1.5-2μ thick at sides (5-)6-8(-10)μ apically, golden or clear chestnut-brown, smooth; pedicels yellowish, thin-walled and collapsing, usually broad, to 70μ long.

Hosts and distribution: Spartina alterniflora Loisel: France and U.S.A. (Florida).

Type: Holway, Miami, Florida, 25 Mar. 1903 (PUR).

A photograph of teliospores of the type was published by Hennen and Cummins (Mycologia 48:126-162. 1956).

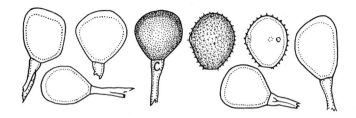

Figure 315

15. UROMYCES SCHOENANTHI H. Syd. & P. Syd. in Sydow & Butler
Ann. Mycol. 4:429. 1906. Fig. 315.

Uromyces apludae H. Syd., P. Syd. & Butl. in Sydow & Butler
Ann. Mycol. 5:493. 1907.

Uromyces polytriadicola Arth. & Cumm. Philippine J. Sci.
59:442. 1936.

Aecia unknown. Uredinia mostly on abaxial leaf surface,
about cinnamon-brown; spores (18-)20-24(-26) x (15-)17-21(-23)μ
mostly broadly ellipsoid or obovoid, wall 1.5(-2)μ thick,
yellowish to cinnamon-brown, echinulate, germ pores 3, equatorial. Telia on abaxial surface, exposed, blackish brown,
compact; spores (18-)22-26(-30) x (14-)17-22(-24)μ, mostly
obovoid, often angular, wall 1.5-2μ thick at sides, 2-4μ
apically, chestnut-brown, smooth; pedicels thin-walled, collapsing, yellowish, to 40μ long.

Hosts and distribution: Apluda mutica L., Cymbopogon
schoenanthus (L.) Spreng., Polytrias amaura (Buse) O. Kuntze:
India, Ceylon, New Guinea, and the Philippine Islands.

Type: Butler No. 485, on Andropogon schoenanthus (=Cymbopogon
schoenanthus), Poona, India, 23 Oct. 1905 (S).

Hennen (Mycologia 57:104-113. 1965) published a photograph
of teliospores of the type.

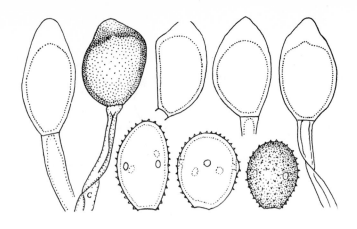

Figure 316

16. UROMYCES SPOROBOLI Ell. & Ev. Proc. Acad. Nat. Sci. Phila.
1893:155. 1893. Fig. 316.

Aecia (Aecidium alliicola Wint.) occur on Allium; spores
(21-)24-28(-35) x (17-)21-24(-28)μ, globoid, broadly ellipsoid,
or oblong, wall 1-1.5(-2)μ thick, finely verrucose. Uredinia
amphigenous, cinnamon-brown; spores (31-)36-40(-49) x (23-)
26-32(-36)μ, ellipsoid or broadly so, wall (1-)1.5-2(-3)μ thick,
cinnamon-brown or golden, echinulate, pores 4 or 5 equatorial.
Telia amphigenous, early exposed, pulvinate, compact, blackish;
spores variable but mostly obovoid or oblong, (28-)35-40(-50)
x (19-)24-28(-35)μ, wall (1-)1.5-2(-3)μ thick at sides, (3-)
5-8(-10)μ apically, chestnut-brown, smooth; pedicels colorless
or yellowish, thick-walled, mostly not collapsing, to 100μ
long.

Hosts and distribution: Sporobolus asper (Michx.) Kunth,
S. cryptandrus (Torr.) A. Gray, S. neglectus Nash, S. vaginiflorus (Torr.) Wood: U.S.A. from Indiana and Wisconsin west
to South Dakota and south to Missouri and Kansas.

Type: E. Bartholomew No. 733, on Sporobolus asper,
Rockport, Kansas, 24 Sept., 1892 (NY; isotype PUR).

Arthur (Mycologia 9:294-312. 1917) first completed the life
cycle using teliospores from S. vaginiflorus and aeciospores
from A. stellatum.

Cummins and Greene (Brittonia 13:271-285. 1961) published
a photograph of teliospores of the type.

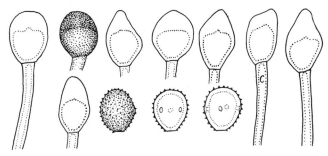

Figure 317

17. UROMYCES MUEHLENBERGIAE Ito J. Tohoku Imp. Univ. Coll. Agr. 3:186. 1909. Fig. 317.

Aecia unknown. Uredinia not seen; urediniospores in the telia globoid or broadly ellipsoid, 18-23(-26) x (16-)17-20(-23)µ, wall (2-)2.5-3.5(-4)µ thick, yellowish, or dull brown, echinulate, pores 3 or 4, equatorial. Telia mostly on abaxial surface, early exposed, pulvinate, blackish; spores mostly globoid or obovoid, (19-)22-27(-35) x (14-)16-18(-22)µ, wall (1-)1.5-2µ thick at sides, 5-13µ apically, chestnut-brown, smooth; pedicels yellow to brownish, thick-walled, mostly not collapsing, to 45µ long.

Hosts and distribution: Muhlenbergia japonica Steud., M. longistolon Ohwi (M. huegelii Auth. not Trin.): northern Japan.

Type: K. Miyabe, on M. japonica, Sapporo, Hokkaido, Japan, Oct. 1890 (SAPA; isotype PUR).

The species differs from U. minimus J. J. Davis in the number and arrangement of the germ pores.

A photograph of teliospores of the type was published by Cummins and Greene (Brittonia 13:271-285. 1961).

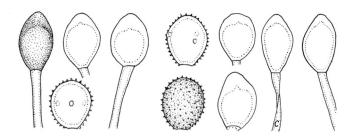

Figure 318

18. UROMYCES GRAMINICOLA Burr. Bot. Gaz. 9:188. 1884. Fig. 318.

Uromyces panici Tracy J. Mycol. 7:281. 1893.

Uredo panici Arth. Bull. Torrey Bot. Club 29:231. 1902.

Aecia (Aecidium crotonopsidis Burr.) occur on Euphorbiaceae; spores 20-32 x 16-23µ, wall 1.5-2µ thick, verrucose, colorless. Uredinia amphigenous or mostly on adaxial surface, cinnamon-brown, spores (18-)20-25(-28) x (17-)19-23(-25)µ, mostly broadly ellipsoid or globoid, wall 1.5-2.5µ thick, golden or cinnamon-brown, echinulate, pores 3 or 4 equatorial. Telia blackish brown, early exposed, pulvinate; spores (20-)23-28(-32) x (12-)17-20(-22)µ, variable but mostly ellipsoid, oval, obovate, often angular, wall 1.5-2.5µ thick at sides, 5-9µ at apex, deep golden or usually chestnut-brown, smooth; pedicels persistent, hyaline to golden, moderately thin-walled, collapsing or not, to 90µ long.

Hosts and distribution: species of Panicum: U.S.A. (New York and South Dakota) to Honduras.

Lectotype: Burrill No. 2347, on P. virgatum, Hudson, Illinois (Ill; isotype PUR).

The species is variable as to urediniospore size, especially.

A photograph of teliospores of the lectotype was published by Ramachar and Cummins (Mycopathol. Mycol. Appl.19:49-61. 1963) who designated the lectotype.

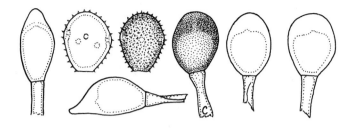

Figure 319

19. UROMYCES PENNISETI S. Ahmad Biologia 6:125. 1960. Fig. 319.

Aecia unknown. Uredinia mostly on abaxial leaf surface, cinnamon-brown; spores (22-)25-30(-32) x (17-)19-22(-24)μ, mostly obovoid or ellipsoid, wall (1.5-)2-3(-3.5)μ thick, echinulate, golden to pale cinnamon-brown, germ pores 4 or 5, equatorial. Telia mostly on abaxial surface, exposed, blackish brown, compact; spores (22-)25-32(-37) x (13-)17-24(-26)μ, ellipsoid, obovoid, or rarely globoid, wall (1.5-)2-3(-4)μ thick at sides, 6-9(-11) apically, clear chestnut-brown, smooth, the spores tend to be dimorphic with the elongate spores paler than the robust spores; pedicels brownish, thin-walled, and collapsing, to 90μ long but usually shorter.

Type: Ahmad No. 14434, on Pennisetum lanatum Klotz., Kagan Valley, Naran, West Pakistan 29 Aug. 1959 (LAH; isotype PUR). Not otherwise reported.

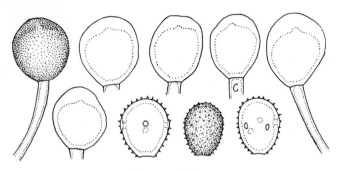

Figure 320

20. UROMYCES MAJOR Arth. Bull. Torrey Bot. Club 38:377. 1911.
Fig. 320.

Aecia unknown. Uredinia amphigenous, cinnamon-brown; spores
(22-)23-26 x (19-)21-23μ, mostly broadly ellipsoid or obovoid,
wall (2-)2.5-3.5μ thick, cinnamon-brown, echinulate, pores (3
or)4(or 5), equatorial but sometimes more or less scattered.
Telia amphigenous, early exposed, pulvinate, blackish; spores
(22-)23-28 x (19-)22-26μ, mostly globoid or broadly ellipsoid,
wall (1.5-)2-2.5(-3)μ thick at sides, (5-)6-7(-9)μ apically,
chestnut-brown, smooth; pedicels yellowish, mostly collapsing,
to 75μ long but often broken short.

Hosts and distribution: Muhlenbergia reverschonii Vasey &
Scribn, M. sp.: southern Texas and central Mexico.

Type: E. W. D. Holway, on M. sp., near Mexico City, 2 Oct.
1896 (PUR).

This is a poorly known species much in need of additional
specimens and study. The Texas specimen is assigned provision-
ally to U. major.

A photograph of teliospores of the type was published by
Cummins and Greene (Brittonia 13:271-285. 1961).

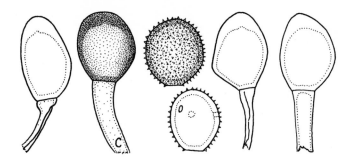

Figure 321

21. UROMYCES BLANDUS H. Syd. Ann. Mycol. 29:145. 1931. Fig.
321.

Aecia unknown. Sori in leaves, mostly on adaxial surface.
Uredinia cinnamon-brown, without paraphyses; spores (21-)23-27
(-29) x 20-25µ, mostly globoid or broadly ellipsoid, wall 2.5
(-3)µ thick, deep golden to cinnamon-brown, closely echinulate,
pores 3 or 4, equatorial or slightly above. Telia blackish
brown, early exposed, pulvinate; spores (23-)25-34(-37) x
19-24(-26)µ, ellipsoid, obovoid, or globoid, wall 2-2.5(-3.5)µ
thick at sides, 4-7(-9)µ at apex, deep golden to chestnut-
brown, smooth; pedicels persistent, hyaline or yellowish, thick-
walled and collapsing or not, to 80µ long.

Hosts and distribution: <u>Phragmites communis</u> Trin.: Philippine Islands.

Type: Clemens No. 6844, Bani, Prov. Pangasinan, March 1925
(isotype PUR). Sydow incorrectly described the urediospores
as being verrucose and as having scattered pores. In globoid
spores such as these it is essential that the spore be oriented
with the hilum in view if the position of the pores is to be
determined.

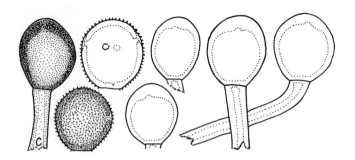

Figure 322

22. UROMYCES LINEARIS Berk. & Br. J. Linn. Soc. London 14:92.
1875. Fig. 322.

Aecia unknown. Uredinia mostly on adaxial leaf surface,
about cinnamon-brown; spores (22-)24-28(-32) x (20-)22-26(-28)μ,
mostly globoid, wall (2-)2.5-3(-3.5)μ thick, finely and closely
echinulate, dull golden or cinnamon-brown, germ pores 3 or 4,
approximately equatorial. Telia mostly on adaxial surface,
blackish brown, early exposed, compact; spores (20-)24-29(-33)
x (17-)19-24(-28)μ, mostly broadly ellipsoid or globoid, wall
(2-)2.5-3(-4)μ thick at sides, (4-)5-7(-8)μ apically, chestnut-
brown, smooth; pedicels yellowish, thick-walled, not collapsing,
to 90μ long.

Hosts and distribution: Panicum repens L.: Morocco and
Mallorca to Uganda east to the Philippines and Japan.

Type: Thwaites No. 597, Peradeniya, Ceylon, Mar. 1868 (K).

A photograph of teliospores of the type was published by
Ramachar and Cummins (Mycopathol. Mycol. Appl. 19:49-61. 1963).

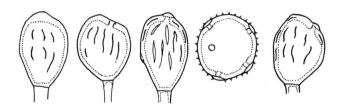

Figure 323

23. UROMYCES BROMINUS Gutsevich Survey of the rust fungi of Crimea. p. 35. 1952. Fig. 323.

Aecia unknown. Uredinia amphigenous, ferruginous, not pulverulent (sic); spores 23-30 x 21-28μ, globoid, wall thickness not given, apparently 1.5-2μ, dull brown, echinulate, germ pores 5, apparently scattered but arrangement not stated. Telia hypophyllous, immersed, shining; spores 19-28.5(-33) x 13.5-22.5μ, mostly obovoid, thickness of side wall not stated, apparently 1.5-2μ, apical wall 3μ, sometimes, at least, with a small papilla over the pore, smooth or undulate-ridged (?); pedicels thin-walled and collapsing, to 23μ, deciduous.

Hosts and distribution: Bromus benekeni (Syme) Beck, B. riparia Rehm., B. scoparius L.: southern U.S.S.R.

Type: On Bromus riparius, Crimea, 23 July 1937 (LE?). Not seen.

The record of B. scoparius is from Uljanischev (Mycoflora Azerbaidzhana 2:273. 1957) but his drawing (Fig. 64) bears little resemblance to that of Gutsevich.

The above description and illustration are adapted from the original.

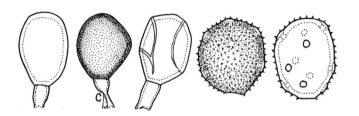

Figure 324

24. UROMYCES FRAGILIPES Tranz. Ann. Mycol. 5:549. 1907. Fig. 324.

Uromyces mysticus Arth. Bull. Torrey Bot. Club 38:377. 1911.

Uromyces jacksonii Arth. & Fromme Torreya 15:260. 1915.

Uromyces hordeastri Guyot Uredineana 1:64. 1938 (issued 1939).

Aecia doubtless on Liliaceae but not yet recognized. Uredinia mostly on the adaxial leaf surface; pale yellowish (dry): spores (20-)24-32(-38) x (20-)22-28(-30)μ; mostly broadly ellipsoid or obovoid, wall (1.5-)2-2.5μ thick, nearly colorless to pale golden, echinulate, germ pores 8-10(-12), scattered, often difficult to count. Telia sometimes mostly abaxial, sometimes mostly adaxial, usually amphigenous, loosely covered by the epidermis or exposed, always pulverulent, chocolate brown; spores (20-)24-30(-34;-40) x (18-)20-25(-28)μ, variable and often angular, usually more or less ellipsoid or obovoid, tending to be dimorphic with the paler spores generally with thinner wall and more angular shape than the deeply pigmented spores, wall uniformly (1.5-)2-2.5μ thick or occasionally to 3μ, rarely to 5μ apically, often with surface ridges and sometimes seemingly punctate, otherwise smooth; pedicels colorless, or brownish next the spore, thin-walled, usually collapsing, to 50μ long.

Hosts and distribution: Agropyron squarrosum Link, Agrostis diegoensis Vasey, A. exarata Trin., A. hallii Vasey, A. pallens Trin., A. palustris Huds., Deschampsia danthonioides (Trin.) Munro, D. caespitosa (L.) Beauv., D. elongata (Hook.) Munro, D. holciformis Presl, Hordeum brachyantherum Nevski, H. bulbosum L., H. jubatum L., H. marinum Huds., H. spontaneum Koch, Secale cereale L., Vulpia dertonensis (All.) Gola, V. pacifica (Piper) Rydb.: the western United States and from southern France to southern Russia, Iran, and Iraq.

Type: Korzinskij, on Agropyron squarrosum As'chabad, Transcaspian region (LE; not seen).

Tranzschel (loc. cit.) suggested that the aecial stage might

occur on Leontice. In 1938, Guyot (Encycl. Mycol. 8:118)
suggested Liliaceae as probably aecial hosts. On the Pacific
Coast of the United States occur the demicyclic Uromyces aureus
Diet. & Holw. on Allium and Chlorogalum and U. brodiaeae Ell.
& Hark. on Brodiaea. Their teliospores are probably not
distinguishable from those of U. jacksonii and U. mysticus.

Until Tranzschel's type is studied, the status of this
complex must remain uncertain. Larger spores and the absence
of uredinial paraphyses separate U. fragilipes from U. turcomanicum.

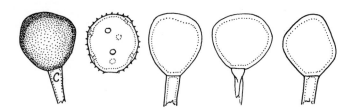

Figure 325

25. UROMYCES PASPALICOLA Arth. & Holw. in Arthur Proc. Amer. Phil. Soc. 64:206. 1925. Fig. 325.

Aecia unknown. Uredinia amphigenous or mostly on abaxial leaf surface, yellow; spores (20-)22-26(-29) x (18-)20-22(-24)μ, mostly broadly ellipsoid, wall 1-1.5μ thick, hyaline or very pale yellowish, echinulate, pores obscure, 6-9, scattered. Telia greyish black, covered by the epidermis but spores loose in sorus; spores (20-)23-28(-33) x (18-)20-26(-28)μ, variable and angular, mostly obovoid or globoid, wall uniformly 2-3μ thick or thickened apically to 3.5μ, chestnut-brown, smooth; pedicels semi-persistent, hyaline, thin-walled and collapsing, to 30μ long.

Type: Holway No. 823, on Paspalum racemosum Lam., Huigra, Chimborazo, Ecuador (PUR F2431; isotypes Reliq. Holw. No. 96).

A photograph of teliospores of the type was published by Ramachar and Cummins (Mycopathol. Mycol. Appl. 19:49-61. 1963).

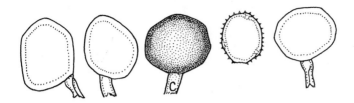

Figure 326

26. UROMYCES MICROCHLOAE H. Syd. & P. Syd. Ann. Mycol. 1:15.
1903. Fig. 326.

Aecia unknown. Uredinia not seen; spores few and mostly
collapsed, 19-21 x 16-19μ, broadly ellipsoid or obovoid, wall
1-1.5μ thick, yellowish, echinulate, germ pores probably
scattered and few. Telia hypophyllous, tardily exposed, blackish brown; spores (21-)23-25(-32) x (12-)23-27(-30)μ, angularly
globoid, depressed globoid, or oblong, wall uniformly 2.5-3.5μ
thick, deep golden brown or clear chestnut-brown, smooth; pedicels colorless, thin-walled and collapsing, to 30μ long, usually
broken shorter.

Type: Schweinfurth, on <u>Microchloa</u> <u>setacea</u> R. Br. (=<u>M. indica</u>
(L.) Beauv.), Seriba Ghattas, Central Africa, 12 Sept. 1869 (S).
Not otherwise known.

A photograph of teliospores of the type was published by
Hennen and Cummins (Mycologia 48:126-162. 1956). South
American specimens, referred by them to this species, are now
considered to be <u>U</u>. <u>dactyloctenii</u>.

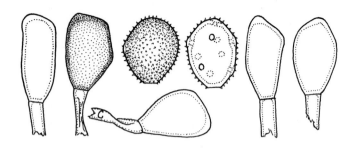

Figure 327

27. UROMYCES AIRAE-FLEXUOSAE Ferd. & Winge Bull. Soc. Mycol.
France 36:164. 1920. Fig. 327.

Uredo airae-flexuosae Liro Bidr. Kaenned. Finl. Nat. Folk
65:573. 1908.

Aecia unknown. Uredinia on the adaxial leaf surface, orange-
yellow; spores (21-)23-26(-30) x (18-)21-23(-24)µ, mostly
broadly ellipsoid, wall 2-2.5(-3)µ thick, colorless to yellowish,
echinulate, germ pores scattered, 7-9. Telia amphigenous and
on sheaths, covered by the epidermis, blackish, with few or no
paraphyses, the sori not loculate; spores (25-)31-37(-41) x
(14-)18-20(-22)µ, ellipsoid, oblong, or mostly obovoid, wall
1-1.5(-2)µ thick at sides, (2-)2.5-3.5(-4)µ apically, uniformly
golden or sometimes chestnut-brown apically, smooth; pedicels
yellowish, thin-walled and collapsing, to 40µ long but usually
broken near the spore.

Hosts and distribution: Deschampsia flexuosa (L.) Trin.,
D. discolor (Thuill.) Roem. & Schult.: Europe from the British
Isles to Bulgaria and Russia.

Type: Ferdinandsen and Winge, on Aira flexuosa (=Deschampsia
flexuosa), Hareskoven near Copenhagen, Denmark (CP).

466

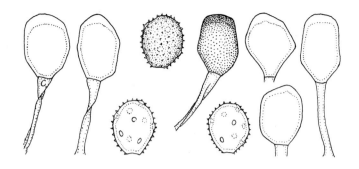

Figure 328

28. UROMYCES PEGLERIAE Pole Evans ex Sydow Ann. Mycol. 12:263.
1914 var. pegleriae. Fig. 328.

Uredo paspali-longiflorae Petch Ann. Roy. Bot. Gard.
Peradeniya 6:216. 1917.

Uredo tacita Arth. Bull. Torrey Bot. Club 60:476. 1933.

Uromyces digitariae-adscendentis Wang Acta Phytotax. Sinica
10:296-297. 1965.

Aecia unknown. Uredinia mostly on adaxial leaf surface,
yellowish brown; spores (21-)23-27(-30) x (16-)18-22(-24)μ,
mostly broadly ellipsoid, wall 1.5-2μ thick, yellowish to
golden, echinulate, germ pores 7-9, scattered or tending to be
bizonate. Telia amphigenous, blackish brown, covered by epidermis, only tardily or not exposed, sometimes with a few
peripheral, pale golden paraphyses, the sori not loculate;
spores (22-)25-30(-34) x (15-)17-20(-24)μ, variable, mostly
angularly obovoid, wall (1-)1.5-2(-2.5)μ thick at sides, 3-5μ
apically, chestnut-brown, smooth; pedicels colorless or brownish, thin-walled and collapsing, to 60μ long, usually broken
shorter.

Hosts and distribution: species of Digitaria: Africa to
New Guinea, the Philippines, and Brazil.

Type: Pegler No. 7755, on D. ternata, Kentani, Cape Prov.,
South Africa (PRE).

Ramachar and Cummins published a photograph of teliospores
of the type (Mycopathol. Mycol. Appl. 19:49-61. 1963), as did
Wang (loc. cit.) of U. digitariae-adscendentis. The latter is
described as having longer teliospore than typical.

UROMYCES PEGLERIAE Pole Evans var. beckeropsidis (E.
Castellani) Ramachar in Ramachar & Cummins Mycopathol. Mycol.
Appl. 19:57. 1963.

Uromyces beckeropsidis E. Castellani Nuovo G. Bot. Ital.
53:224. 1946.

Urediniospores (20-)22-26 x (17-)19-22μ, wall 1.5-2μ thick, yellowish, germ pores 8-10, scattered; teliospores (22-)23-27 (-29) x (17-)19-23(-25)μ, wall 2μ thick at sides, (2.5-)3-4 (-5)μ apically, chestnut-brown; pedicels yellowish, collapsing, to 60μ long.

Type: Castellani, on _Beckeropsis_ _nubica_ Fig. & de Not. (=_Pennisetum_ _nubicum_ (Fig. & de Not.) Chiov.), Enda Cioa, pr. Adua, Erytraea (Herb. Castellani; isotype PUR). Not otherwise known.

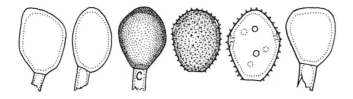

Figure 329

29. UROMYCES TRAGI Wakef. & Hansf. Proc. Linn. Soc. Lond.
161:175. 1949. Fig. 329.

Aecia unknown. Uredinia amphigenous or mostly on adaxial
surface, cinnamon-brown or yellowish brown; spores (21-)23-27
(-30) x (17-)19-22(-23)µ, ellipsoid or broadly ellipsoid,
wall 1-1.5µ thick, golden to cinnamon-brown, echinulate, germ
pores (5)6 or 7(8), scattered. Telia blackish, covered by
the epidermis, or developing in old uredinia, without paraphyses; spores (21-)23-30 x (14-)17-22(-26)µ, mostly oval or
obovate, commonly angular and sometimes with fine surface
ridges along the angles, wall 1.5µ thick at sides, 2-4µ at
apex, chestnut-brown, smooth; pedicels persistent, hyaline,
thin-walled and collapsing, to 25µ long but usually broken
shorter.

Hosts and distribution: Tragus berteroanus Schult.: Kenya
and Uganda to South Africa.

Type: Maitland No. 976, Ruwenzori, Uganda (K).

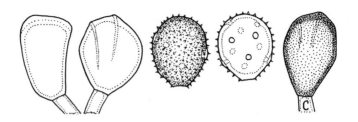

Figure 330

30. UROMYCES BECKMANNIAE Jacks. Brooklyn Bot. Gard. Mem. 1:274. 1918. Fig. 330.

Aecia unknown. Uredinia amphigenous, yellowish brown; spores (23-)25-29(-32) x (19-)21-24(-26)μ, mostly broadly ellipsoid, wall 2-3μ thick, yellowish to pale golden, echinulate, pores difficult to count, 8-11, scattered. Telia amphigenous and on sheaths, blackish, covered by the epidermis or tardily dehiscent, weakly loculate with brownish paraphyses; spores (25-)29-38(-42) x (18-)20-26(-29)μ, variable, mostly oblong-ellipsoid or obovoid, angular, commonly with fine surface ridges on the angles, wall 1.5-2(-2.5)μ thick at sides, 3-5(-6)μ at apex, chestnut-brown, or golden below, smooth; pedicels hyaline or yellowish, thin-walled and collapsing, to 40μ long but usually broken near the spore.

Hosts and distribution: Beckmannia syzigachne (Steud.) Fernald: U.S.A. (Oregon).

Type: Jackson No. 3145, on B. erucaeformis (=B. syzigachne), Corvallis, Oregon (PUR).

The species has spores considerable like U. fragilipes but differs especially in having long-covered and paraphysate telia.

31. UROMYCES KOELERIAE Uljan. Mycoflora Azerbaidzhana 2:263. 1959.

Aecia unknown. Uredinia amphigenous or mostly on abaxial leaf surface; spores 23-27 x 16-19μ, ovoid or ellipsoid, wall 2.5-3.5μ thick, light olivaceous, echinulate, germ pores 3-6, scattered, obscure. Telia epiphyllous, covered by the epidermis, blackish; spores 18-29 x 14-21μ, mostly obovoid, wall 1-1.5μ thick at sides, to 3.5μ apically, dark brown apically, light brown below, smooth; pedicels brownish, apparently thin-walled, to 19μ long, easily deciduous.

Type: Uljanischev, on Koeleria caucasica (Trin.) Dom., Dastafjur district, Azerbaijan, U.S.S.R., 28 Aug. 1937 (BAK). Not seen.

The description is adapted from the original. Uljanischev did not describe telial paraphyses but states that the species is near to U. dactylidis. He illustrated urediniospores with "cuticular caps" but the spores were apparently randomly oriented and there are almost certainly more than the 3-6 pores described.

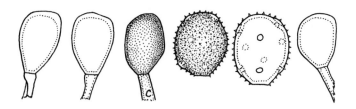

Figure 331

32. UROMYCES DACTYLIDIS Otth Mitt. Nat. Ges. Bern 1861:85.
1861 var. dactylidis. Fig. 331.

Uromyces festucae H. Syd. & P. Syd. Hedwigia 39:117. 1900.

Uromyces phyllachoroides P. Henn. Hedwigia Beibl. 40:129.
1901.

Uromyces ranunculi-festucae Jaap Verh. Bot. Vereins Prov.
Brandenb. 47:90. 1905.

Uromyces festucae-nigricantis Gz. Frag. Trab. Mus. Nac.
Cienc. Nat. Madrid Ser. Bot. 3:33. 1914.

Aecia (on species of Ranunculus) in groups; spores 18-22 x
15-20µ, wall hyaline, 1µ thick, verrucose. Uredinia amphigenous
or on the adaxial leaf surface of fescues, yellowish brown to
cinnamon-brown; spores (22-)25-30(-34) x (18-)20-24(-26)µ, mostly
broadly ellipsoid, wall 1.5-2µ thick, about golden brown, echinulate, germ pores 7-9(10), scattered, relatively obvious because
of the "cuticular caps." Telia amphigenous, or epiphyllous on
fescues, blackish, covered by the epidermis, loculate with
conspicuous, brown, mostly abundant paraphyses; spores (18-)
22-30(-34) x (12-)16-20(-24)µ, variable but mostly oblong-ellipsoid or angularly obovoid, wall 1-1.5µ thick at sides, 2-4(-6)µ
apically, chestnut-brown, smooth; pedicels colorless to brownish,
thin-walled and collapsing, to 40µ long but usually less than
25µ.

Hosts and distribution: species of Cynosurus, Dactylis, and
Festuca: Europe and northern Africa to Russia, New Zealand,
and the United States.

Lectotype: Otth (?), on Dactylis glomerata, Bern, date, and
collector not given (BERN). Lectotype designated here.

The two following varieties are more or less recognizeable;
both have smaller urediniospores and, in addition, var. poae-
alpinae has colorless paraphyses in the telia. Teliospores are
not distinctive within the complex.

The life cycle was demonstrated first by Schroeter (Beitr. Biol. Pfl. 3:51-93. 1879) using spores from _Dactylis_ to produce aecia on _Ranunculus_ _bulbosus_ and _R_. _repens_.

UROMYCES DACTYLIDIS Otth var. poae (Rabenh.) Cumm. comb.
nov.

Uromyces poae Rabenh. in Marcucci Unio itin. Crypt. No. 38.
1866.

Uromyces alopecuri Seym. Proc. Bost. Soc. Nat. Hist. 24:186.
1889.

Uromyces sclerochloae Tranz. Ann. Mycol. 5:550. 1907.

Uromyces alopecuri Seym. var. japonica Ito J. Coll. Agr.
Tohoku Imp. Univ. 3:184. 1909.

Uromyces atropodis Tranz. Ann. Mycol. 5:550. 1907.

Uromyces lygei P. Syd. & H. Syd. Monogr. Ured. 2:331. 1910.

Uromyces ranunculi-distichophylli Semad. Centralbl. Bakt.
II. 46:463. 1916.

Uromyces poae Rabenh. f. agrostidis Gz. Frag. Trab. Mus. Nac.
Cienc. Nat. Madrid Ser. Bot. 15:134. 1918.

Uromyces adelphicus H. Syd. Svensk. Bot. Tidsk. 29:71. 1935.

Uromyces agrostidis (Gz. Frag.) Guyot Uredineana 1:69. 1938.

Uromyces vulpiae Losa Espana An. Jard. Bot. Madrid 6:422.
1946.

Uromyces vulpiae Camara Agron. Lusit. 11:166. 1949.

Uromyces triseti Katajev Akad. Nauk Bot. Otd. Spor. Rast.
Bot. Mater. 7:172. 1951.

Uromyces volkartii Gaeum. & Terrier Ber. Schweiz Bot. Ges.
62:299. 1952.

Uromyces brizae Gaeum., Mueller & Terrier Sydowia Beih.
1:187-188. 1957.

Aecia (Aecidium ficariae Pers.) occur on species of Ficaria
and Ranunculus; in groups; spores 18-24 x 10-18µ, wall 1µ thick,
hyaline, verrucose. Uredinia amphigenous, orange color (almost
colorless dry); spores (17-)20-25(-27) x (16-)17-20(-23)µ,

474

mostly ellipsoid or broadly ellipsoid, wall (1-)1.5-2μ thick,
pale yellowish to colorless, echinulate, germ pores (5-)7-9(-10),
scattered, obscure, detectable mostly because of slight "cuticular caps." Telia amphigenous, blackish, covered by the epidermis, variously loculate with brown paraphyses; spores (18-)
22-30(-36;-40) x (12-)16-20(-22)μ, mostly obovoid or oblong-
ellipsoid, wall 1-2μ thick at sides, 2.5-4(-5)μ apically,
golden brown to chestnut-brown, smooth; pedicels mostly yellowish, thin-walled and collapsing, to 35μ long, usually shorter.

Hosts and distribution: Agrostis, Alopecurus, Briza, Lygeum,
Milium, Poa, Puccinellia, Sclerochloa, Scleropoa, Trisetum,
Vulpia: Europe and North Africa to Russia, Iran, China, Japan,
Canada and the United States.

Type: Marcucci, Macomer, Giungo Sardinia (isotypes, Marcucci
Unio Itin. Crypt. No. 38; probable isotypes Rabenhorst-Winter
F. Europaei No. 2705).

The first inoculations proving the life cycle were made by
Schroeter (Beitr. Biol. Pfl. 3:51-93. 1879) using Poa nemoralis
and Ficaria verna as hosts.

UROMYCES DACTYLIDIS Otth var. poae-alpinae (Rytz) Cumm. comb. nov.

Uromyces poae-alpinae Rytz Mitt. Naturf. Ges. Bern 1910:70. 1910.

Uromyces phlei-michelii P. Cruchet Bull. Soc. Vaud. Sci. Nat. 51:75. 1916.

Aecia (on Ranunculus montanus Willd.) grouped; spores 17-24 x 15-20µ, wall thin (1µ?), hyaline, verrucose. Uredinia amphigenous, yellowish brown; spores (20-)23-27(-30) x (17-)19-23 (-24)µ, wall 1.5(-2)µ thick, pale yellowish, echinulate, germ pores 7-9(10), scattered, obscure. Telia amphigenous, covered by the epidermis, blackish, loculate with abundant, thick-walled, colorless paraphyses; spores (18-)20-28(-32) x (14-)18-22(-24)µ, globoid, ellipsoid, or obovoid, often somewhat angular, sometimes with surface ridges, wall 1.5-2µ thick at sides, 2-4µ apically, chestnut-brown, smooth; pedicels thin-walled and collapsing, yellowish to 30µ long but usually broken near spore.

Hosts and distribution: Phleum alpinum L., P. michelii All., Poa alpina L.: alpine regions of France and Switzerland.

Type: Rytz, on Poa alpina, Fuss des Telli, Kientales, Switzerland (BERN).

The colorless telial paraphyses separate the variety from U. dactylidis vars. dactylidis and poae.

Cruchet (loc. cit.) and Semadini (Centralbl. Bakt. 46:451-468. 1916) demonstrated the life cycle of the Poa and Phleum rusts, respectively.

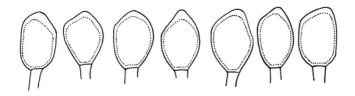

Figure 332

33. UROMYCES CALAMAGROSTIDIS Uljan. Mykoflora Azerbaidzhana
2:259. 1959. Fig. 332.

Aecia unknown. Uredinia amphigenous or mostly on abaxial
leaf surface, light brown; spores 21-27 x 17-21μ, globoid,
ellipsoid, or oblong, wall 1.5-2.5μ thick, cinnamon-brown,
densely echinulate, germ pores obscure (but doubtless scattered!).
Telia epiphyllis, covered by the epidermis, blackish; spores
18-26 x 13-18, mostly obovoid, or oblong-ellipsoid, wall 1-1.5μ
thick at sides, 2.5-3.5(-5)μ apically, cinnamon-brown, smooth;
pedicels brownish, thin-walled, fragile, to 11μ long.

Type: Uljanischev, on Calamagrostis arundinacea (L.) Roth,
Kusarski district, Azerbaijan, U.S.S.R., 9 Aug. 1951 (BAK).
Not seen.

The description and illustration are adapted from the
original. It is probably that the urediniospores have several
scattered pores with inconspicuous or no "cuticular caps."

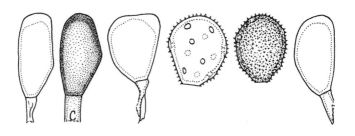

Figure 333

34. UROMYCES HORDEINUS (Arth.) Barth. Handb. N. Amer. Ured.
Ed. 1. p. 63. 1928. Fig. 333.

Uromyces hordei Tracy J. Mycol. 7:281. 1893. Not Nielsen,
1875.

Nigredo hordeina Arth. N. Amer. Fl. 7:749. 1926.

Aecia occur on Nothoscordium bivalve (L.) Britt.; in groups,
cupulate; spores 23-27 x 19-23µ, ellipsoid to globoid, wall
1.5µ thick, colorless, verrucose. Uredinia amphigenous, yellow-
orange when fresh, nearly colorless when dry; spores mostly
broadly ellipsoid (24-)26-30(-34) x (19-)21-25µ, wall 1.5-2µ
thick, yellowish or pale golden, echinulate, germ pores 9-12,
scattered, indistinct. Telia amphigenous and on sheaths,
blackish, covered by the epidermis, weakly loculate with brownish paraphyses; spores (23-)26-34(-38) x (15-)17-23(-25)µ,
variable, ellipsoid, oblong-ellipsoid, obovoid, or rarely
globoid, wall 1-1.5(-3)µ thick at sides, 3-5(-6)µ apically,
smooth; pedicel yellowish, thin-walled and collapsing, 25µ or
less long.

Hosts and distribution: Festuca octoflora Walt., Hordeum
brachyantherum Nevski, H. pusillum Nutt., Scribnera bolanderi
(Thurb.) Hack.: Virginia to Nebraska, Texas, and Colorado.

Type: Tracy, on Hordeum pratense (=H. pusillum) New Orleans,
Louisiana, May 1891 (BPI).

Arthur (Mycologia 8:139. 1916) first reported inoculations
that proved the life cycle.

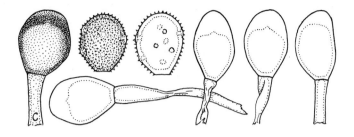

Figure 334

35. UROMYCES PENCANUS Arth. & Holw. in Arthur Proc. Amer. Phil.
Soc. 64:211. 1925. Fig. 334.

Uredo pencana Diet. & Neger Bot. Jahrb. 27:15. 1899.

Sori in adaxial side of leaves. Spermogonia unknown. Aecia
cylindrical, the peridium becoming variously lacerated; spores
(23-)24-28(-31) x (20-)22-26(-27)μ, wall 3.5-5(-6)μ thick,
hyaline or pale yellowish, verrucose. Uredinia cinnamon-brown;
spores (23-)26-30(-36) x (21-)23-27(-30)μ, mostly broadly
ellipsoid, wall (2-)2.5-3.5(-4)μ thick, pale cinnamon-brown,
echinulate, pores obvious, 5-7(-9) scattered. Telia erumpent,
pulvinate, blackish brown; spores (25-)27-34(-40) x (18-)
21-25(-28)μ, mostly oval or obovate, wall 2-3μ thick at sides,
(3-)6-10(-13)μ at apex, chestnut-brown, smooth; pedicels persistent, hyaline to brownish, moderately thin-walled and usually
collapsing, to 70μ long.

Hosts and distribution: Nassella chilensis (Trin. & Rupr.)
Desv., Stipa manicata Desv., S. mucronata H.B.K., S. neesiana
Trin. & Rupr., S. setigera Presl: Chile and Argentina.

Lectotype: Holway No. 307 on Stipa manicata, Zapallar, Chile
(PUR; isotypes Reliq. Holw. No. 47).

This is one of the few autoecious grass rusts.

Arthur (loc. cit.) published a photograph of teliospores of
the lectotype as did Greene and Cummins (Mycologia 50:6-36.
1958).

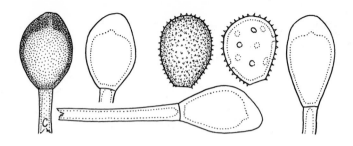

Figure 335

36. UROMYCES NASSELLAE Cumm. Torrey Bot. Club Bull. 83:231.
1956. Fig. 335.

Aecia unknown. Uredinia yellowish, in adaxial side of leaves,
(28-)30-35(-38) x (21-)23-26(-28)μ, mostly ellipsoid or broadly
ellipsoid, wall (2-)2.5(-3)μ thick, hyaline or very pale yellowish, echinulate, pores very obscure, scattered, probably 10-13.
Telia erumpent, pulvinate, blackish brown; spores (27-)30-38(-43)
x (19-)21-24(-28)μ, obovoid, oval, or oblong-ellipsoid, wall
2-3μ thick at sides, 5-7(-9)μ at apex, chestnut-brown, smooth;
pedicels persistent, brownish, thick-walled and not collapsing,
to 70μ long.

Hosts and distribution: <u>Nassella</u> <u>pubiflora</u> (Trin. & Rupr.)
Desv.: Bolivia.

Type: Holway No. 464, La Paz, Bolivia, (PUR; isotypes Reliq.
Holw. No. 72 as <u>Uromyces</u> <u>pencanus</u>).

Photographs of teliospores of the type were published with
the diagnosis and by Greene and Cummins (Mycologia 50:6-36.
1958).

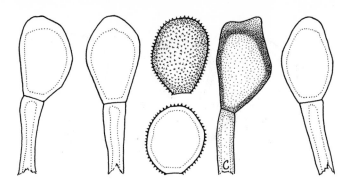

Figure 336

37. UROMYCES CUSPIDATUS Wint. Hedwigia 26:15. 1887. Fig. 336.

Uromyces fuegianus Speg. Bol. Acad. Nac. Cienc. Cordoba 11:181. 1888.

Uromyces chubutensis Speg. An. Mus. Nac. Buenos Aires 3:60. 1902.

Aecia unknown. Uredinia mostly in adaxial side, yellowish, spores 30-40 x 25-33μ, ellipsoid to nearly globoid, wall 2-3μ thick, hyaline or very pale yellowish, echinulate, pores obscure, 10-13, scattered. Telia blackish brown, erumpent, pulvinate, without paraphyses; spores (28-)32-48(-53) x (16-)23-28(-32)μ, mostly obovoid, wall 1.5-2.5μ thick at sides, 4-8μ at apex, deep golden to clear chestnut-brown, smooth; pedicels persistent, rather thick-walled and usually not collapsing, yellowish, to 90μ long.

Hosts and distribution: *Festuca commersonii* Spreng., *F. dissitiflora* Steud., *F. hieronymi* Hack., *F. lasiorachis* Pilger, *F. purpurascens* Banks & Sol., *F. rigescens* (Presl) Kunth, *Poa chubutensis* Speg.: Bolivia and Argentina.

Type: Hariot No. 7, on *Festuca commersonii*, Cape Horn, Argentina (PC).

Guyot (Les Uredinee. I Genre Uromyces. p. 438, Lechevalier, Paris) published drawings of the type, suggested that U. *fuegianus* is synonymous and that a rust of *Festuca procera* was probably undescribed. For this species see *Uromyces procerus* Lindq.

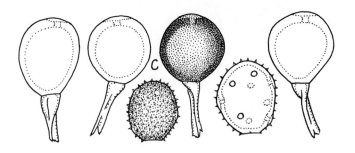

Figure 337

38. UROMYCES CLIGNYI Pat. & Har. J. Bot. 14:237. 1900. Fig. 337.

Uromyces andropogonis-annulati H. Syd., P. Syd. & Butl. Ann. Mycol. 5:492. 1907.

Uredo anthistiriae Petch. Ann. R. Bot. Gard. Peradeniya 5:254-255. 1912.

Uredo themedicola Cumm. Mycologia 33:151. 1941.

Uromyces triandrae T. S. Ramak. & Srin. Current Sci. 19:26. 1950.

Uromyces amphilophidis-insculptae T. S. Ramak. & Srin. Proc. Indian Acad. Sci. B. 36:92. 1952. Nom. confusum.

The aecia (Aecidium hartwegiae Thuem.) occur on species of Chlorophytum; spores 20-25 x 18-20µ, wall 1-1.5µ thick, verrucose. Uredinia mostly on abaxial leaf surface, yellowish brown; spores (20-)22-28(-32) x (17-)19-25(-27)µ, mostly broadly ellipsoid, wall 1.5-2.5µ thick, yellow to golden brown, echinulate, germ pores 7-10, scattered. Telia amphigenous or mostly on abaxial surface, exposed, pulverulent, blackish brown; spores (23-)25-30(-34) x (23-)25-30(-32)µ, mostly globoid, wall uniformly 3-5µ thick or 3-6µ apically, chestnut-brown, usually nearly opaquely so, smooth; pedicels colorless, thin-walled and collapsing, to 110µ long but usually broken near the spore.

Hosts and distribution: species of Andropogon (incl. Schizachyrium), Bothriochloa, Cymbopogon, Dichanthium, Eremopogon, Exotheca, Hemarthria, Heteropogon, Hyparrhenia, Monocymbium, Sorghastrum, Themeda: Africa to India, New Guinea, the Philippines, Mexico, Central America, and the British West Indies.

Type: Chevalier, on Andropogoneae, between Segou and Bammako, Moyen Niger, Oct. (1899?) (FH).

Narasimhan and Thirumalachar (Mycologia 58:456-459. 1966) demonstrated the life cycle with reciprocal inoculations using Chlorophytum laxum R. Br. and Heteropogon contortus (L) P. Beauv.

482

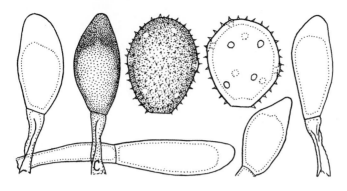

Figure 338

39. UROMYCES MCNABBII Cumm. sp. nov. Fig. 338.

Aeciis ignotis. Urediniis epiphyllis (adaxialibus), cinnamomeo-brunneis, pulverulentis; sporae (30-)35-42(-46) x (27-)30-35(-39)μ, late ellipsoideae vel obovoideae, membrana 2.5-3(-3.5)μ crassa, echinulata, cinnamomeo-brunnea, poris germinationis 9-12, sparsis. Teliis epiphyllis, atro-brunneis, pulvinatis, compactis; sporae (26-)30-42(-46;-52) x (14-)18-21 (-24)μ, plerumque ellipsoideae, membrana ad latere (1-)1.5-2 (-2.5)μ, ad apicem 8-14μ, castaneo-brunnea vel pallidiore; pedicello flavido, tenui tunicato, usque ad 125μ longo, persistenti.

Type: McNabb, on Danthonia raoulii Steud. var. rubra Ckne., Boyle River, Canterbury, New Zealand, 23 Feb. 1961 (PUR F16460; isotype PDD). Known otherwise from Rangipo Desert, Wellington.

The species differs from U. danthoniae because of longer teliospores and larger urediniospores with echinulations spaced 3-4μ versus about 2μ.

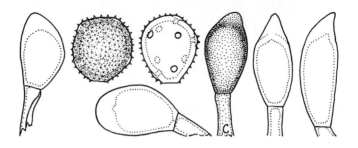

Figure 339

40. UROMYCES ACUMINATUS Arth. Bull. Minn. Acad. 2:35. 1883.
Fig. 339.

Uromyces spartinae Farl. Proc. Am. Acad. 18:77. 1883.

Uromyces polemonii Barth. N. Amer. Ured. No. 597. 1913.

Uromyces steironematis Arth. Mycologia 9:311. 1917.

Uromyces magnatus Arth. Mycologia 9:311. 1917.

Uromyces acuminatus Arth. var. steironematis (Arth.) J. J. Davis Trans. Wis. Acad. 7:410. 1922.

Uromyces acuminatus Arth. var. polemonii (Arth.) J.J. Davis Trans. Wis. Acad. 7:411. 1922.

Uromyces acuminatus Arth. var. magnatus (Arth.) J. J. Davis Trans. Wis. Acad. 7:410. 1922.

Uromyces acuminatus Arth. var. spartinae (Farl.) Arth. Man. Rusts, p. 168. 1934.

The aecia (Aecidium polemonii Peck) occur on hosts in the Caryophyllaceae, Liliaceae, Polemoniaceae, and Primulaceae, spores 17-28 x 15-24μ, wall 1.5-2μ thick, verrucose, colorless. Uredinia on adaxial leaf surface, yellowish brown, spores (24-) 27-32(-36) x (20-)23-28(-30)μ, mostly broadly ellipsoid, wall 2-3.5μ thick, yellow to golden, echinulate, pores 7-10, scattered. Telia blackish brown, erumpent, pulvinate; spores (23-)26-36 (-42) x (13-)15-20(-24)μ, ellipsoid, oblong or obovoid, the apex truncate, rounded, acuminate, or occasionally semicoronate, wall 1-2μ thick at sides, 5-12μ at apex, golden to chestnut-brown, smooth; pedicels persistent, yellowish, moderately thin-walled and collapsing or not, to 70μ long.

Hosts and distribution: species of Spartina: Canada and U.S.A.

Type: Arthur, on S. cynosuroides (=error for S. pectinata) Fort Dodge, Iowa (PUR).

Arthur (Mycologia 8:136. 1916) first proved the life cycle by inoculation.

484

A photograph of teliospores of the type was published by
Hennen and Cummins (Mycologia 48:126-162. 1956).

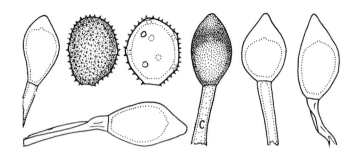

Figure 340

41. UROMYCES DANTHONIAE McAlp. Rusts of Australia. p. 85. 1906.
Fig. 340.

Uredo haumata Cunn. Trans. N. Zeal. Inst. 59:499. 1928.

Grouped, cup-shaped aecia with nearly globoid spores 16 x 12μ
diam were described on the grass by McAlpine. Uredinia on adaxial
leaf surface, brownish (dry), probably orange-brown when fresh;
spores (22-)24-30(-32) x (19-)22-26(-29)μ, mostly broadly elli-
soid, wall (1.5-)2-2.5(-3)μ thick, golden to cinnamon-brown,
echinulate, germ pores 8-10, scattered. Telia on adaxial surface,
early exposed, blackish brown, compact; spores (24-)27-34(-37) x
(16-)19-24(-26)μ, mostly broadly ellipsoid or obovoid, wall
(1.5-)2-2.5(-3)μ thick at sides, (5-)7-11(-14)μ apically, smooth,
chestnut-brown; pedicels yellowish, thin-walled and collapsing,
to 100μ long.

Hosts and distribution: Danthonia gracilis Hook. f., D.
pilosa R. Br., D. semiannularis R. Br., D. unarede Raoul:
Australia, New Zealand, and Tasmania.

Lectotype: Robinson, on D. pennicillata (=D. semiannularis),
Killara, Australia 9 Oct. 1902 (MEL), designated by McNabb
(Trans. Roy. Soc. N. Zealand 1:235-257. 1962).

The species is assumed to be autoecious. McNabb (loc. cit.)
also records as hosts D. cunninghami Hook. f., D. flavescens
Hook. f., and D. setifolia (Hook.) Ckne. but, since he states
that the rust has larger spores on the endemic hosts, it is
possible that they should be assigned to Uromyces mcnabii.

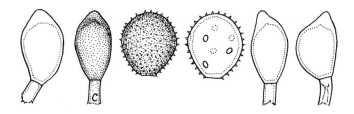

Figure 341

42. UROMYCES AMPHIDYMUS P. Syd. & H. Syd. Ann. Mycol. 4:29.
1906. Fig. 341.

Uromyces glyceriae Arth. Bull. Torrey Bot. Club 37:572.
1910.

Aecia unknown. Uredinia amphigenous, yellowish brown; spores
22-26(-28) x 19-23(-25)μ, mostly broadly ellipsoid or globoid,
wall (1.5-)2-2.5μ thick, golden to cinnamon-brown, echinulate,
germ pores 6-8, scattered. Telia chocolate-brown, early exposed,
compact; spores (20-)21-34(-37) x (14-)16-20(-22)μ, mostly
ellipsoid, wall 1-1.5μ thick at sides, 2.5-5μ at apex, smooth,
deep golden or light chestnut-brown; pedicels brownish, persistent, thin-walled and collapsing, to 40μ long.

Hosts and distribution: Glyceria acutiflora Torr., G.
borealis (Nash) Batchelder, G. septentrionalis Hitchc.: central
and eastern U.S.A.

Type: Waite, or Glyceria fluitans (=G. septentrionalis),
Oregon, Illinois (S; isotype PUR).

Arthur (Manual of Rusts) treated U. amphidymus as correlated
with Puccinia rubigo-vera (=P. recondita) but such a relationship is doubtful. The telia are early erumpent and without
paraphyses. Many, and in some collections most, of the teliospores germinate before winter, possibly indicating an unusual
time of infection of the still unknown aecial hsot.

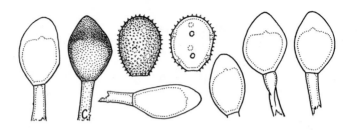

Figure 342

43. UROMYCES OTAKOU Cunn. Trans. N. Zeal. Inst. 54:627. 1923.
Fig. 342.

Aecia unknown. Uredinia on the adaxial leaf surface, orange-
yellow (fresh), pale yellowish brown when dry; spores (20-)
23-30 x 19-22(-23)μ, ellipsoid, broadly ellipsoid, or obovoid,
wall 1.5(-2)μ thick, pale yellowish or nearly colorless, echinulate, germ pores 6-10, scattered. Telia on adaxial surface,
early exposed, blackish brown, compact; spores (23-)25-30(-31)
x (15-)17-22(-24)μ, mostly obovoid, wall 2-2.5μ thick at sides,
(6-)8-10(-12)μ apically, chestnut-brown, smooth; pedicels colorless to yellowish, thin-walled, mostly collapsing, to 60μ long.

Hosts and distribution: _Poa_ _anceps_ Forst. f., _P_. _caespitosa_
Forst. f., _P_. _litorosa_ Cheesem.: New Zealand.

Type: Reid, on _Poa_ _caespitosa_, Otago (PDD 1323; isotype PUR).

488

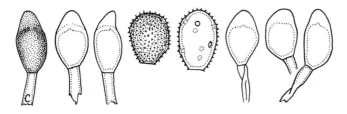

Figure 343

44. UROMYCES EHRHARTAE McAlp. Rusts of Australia p. 86. 1906.
Fig. 343.

Uredo ehrhartae McAlp. Agr. Gaz. New So. Wales 6:855. 1895.

Aecia unknown. Uredinia amphigenous, yellowish brown; spores
(21)23-27(-29) x (16-)18-22(-24)µ, mostly ellipsoid or broadly
ellipsoid, wall (1-)1.5(-2)µ thick, yellowish to pale cinnamon-
brown, echinulate, germ pores (5)6-8, scattered or often bizonate.
Telia mostly on adaxial leaf surface, early exposed, blackish
brown, compact; spores (18-)20-27(-30) x (11-)13-16(-18)µ, mostly
ellipsoid or obovoid, wall 1.5(-2)µ at sides 6-11µ apically,
golden to chestnut-brown, smooth; pedicels colorless or yellowish, thin-walled and collapsing, to 40µ long.

Hosts and distribution: Microlaena stipoides (Labill.) R.
Br., Tetrarrhena acuminata R. Br.: Australia and New Zealand.

Lectotype: Robinson, on Microlaena stipoides, Killara,
Australia 16 Mar. 1903 (MEL); lectotype designated here.

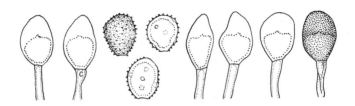

Figure 344

45. UROMYCES MINIMUS J. J. Davis Bot. Gaz. 19:415. 1894.
Fig. 344.

Aecia unknown. Uredinia abaxial, small, golden to cinnamon-
brown; spores (16-)17-21(-24) x (14-)16-18(-22)μ, mostly globoid
or obovoid, wall (1-)1.5-2.5(-3)μ thick, cinnamon-brown, echinulate, pores 4-6, scattered. Telia abaxial, early exposed,
pulvinate, compact, blackish; spores (14-)19-24(-29) x (12-)
14-17(-19)μ, mostly obovoid or narrowly oval, wall (1-)1.5-2(-3)μ
thick at sides, 5-10(-13)μ apically, chestnut-brown, smooth;
pedicels colorless or tinted, thin-walled, collapsing, to 40μ
long but usually broken shorter.

Hosts and distribution: Muhlenbergia andina (Nutt.) Hitchc.,
M. racemosa (Michx.) B. S. P., M. sylvatica Torr.: Canada and
the U.S.A. from Ontario to northern Michigan, Wisconsin, and
Oregon.

Type: J. J. Davis, on M. sylvatica, Somers, Kenosha County,
Wisconsin, 8 Oct., 1893 (WIS; isotype PUR).

A photograph of teliospores of the type was published by
Cummins and Greene (Brittonia 13:271-285. 1961).

490

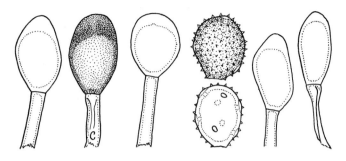

Figure 345

46. UROMYCES GRAMINIS (Niessl) Diet. Mitth. Thuering. Bot. Ver.,
Neue Folge 2:18. 1892. Fig. 345.

Capitularia graminis Niessl in Rabenhorst Fungi eur. No. 1191.
1868.

Uromyces laserpitii-graminis Ed. Fisch. Centralbl. Bakt.
17:204. 1906.

Uromyces seseli-graminis Ed. Fisch. Centralbl. Bakt. 17:204.
1906.

Aecia (Aecidium ferulae Mont.) occur on many members of the
Umbelliferae, the peridium bulliform, opening by a pore; spores
22-32μ diam, globoid, wall 2.5-3(-4)μ thick, yellowish, verrucose, germ pores fairly obvious. Uredinia on adaxial surface
of leaves, about cinnamon-brown; spores 24-29(-33) x 21-24μ,
mostly broadly ellipsoid, wall 2.5-3.5μ thick, golden, echinulate, pores obscure, 6-8, scattered. Telia blackish brown,
early exposed, compact; spores 22-31 x 17-24μ, mostly ellipsoid
or obovoid, wall 1.5-3μ thick at sides, 4-8μ at apex, deep golden
to clear chestnut-brown, smooth; pedicels hyaline to pale brownish, persistent, thin-walled and mostly collapsing, to 50μ long.

Hosts and distribution: species of Melica: southern Europe
and northern Africa.

Type: Niessl (Rab. Fungi eur. No. 1191) on an undetermined
grass, near Brunn, Czechoslovakia.

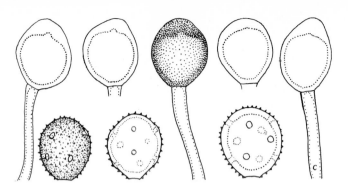

Figure 346

47. UROMYCES EPICAMPIS Diet. & Holw. in Holway Bot. Gaz. 24:23.
1897 var. epicampis. Fig. 346.

Aecia unknown. Uredinia adaxial, seriate, cinnamon-brown;
spores (25-)28-32(-37) x (21-)24-30(-32)μ, mostly globoid or
broadly ellipsoid, wall (2-)2.5-3.5(-4)μ thick, cinnamon-brown
or olivaceous, echinulate, pores (6)7-10, scattered. Telia
adaxial, early erumpent, compact, blackish; spores (23-)28-32
(-40) x (16-)22-25(-29)μ, ovoid, obovoid, or oblong, wall (1-)
1.5-2(-3)μ thick at sides, (3-)4-7(-10)μ apically, chestnut-
brown, smooth; pedicels yellowish, thin-walled, collapsing, to
100μ long but usually broken shorter.

Hosts and distribution: Melica laxiflora Cav., species of
Muhlenbergia: U.S.A. from southern Arizona and California
south to Guatemala, Equador and Chile.

Type: Holway, on Epicampes macroura (=M. macroura), near
Mexico City, Mexico, 30 Sept., 1896 (S; isotype PUR).

The confusion in the assignment of records to this and other
species has been pointed out by Cummins and Greene (Brittonia
13:271-285. 1961), who published a photograph of teliospores
of the type.

Uromyces epicampis is remarkably similar to U. graminis,
differing mainly in having urediniospores that are closely and
finely echinulate rather than sparsely beset with prominent,
spaced cones and less conspicuous "cuticular caps" over the
pores.

UROMYCES EPICAMPIS Diet. & Holw. var. durangensis Cumm.
Southw. Nat. 12:34. 1967.

Urediniospores (20-)22-26(-28) x (18-)20-22(-23)μ, wall 1.5-2μ
thick, cinnamon-brown, echinulate, pores scattered, 6-8. Teliospores (21-)23-26(-29) x (19-)21-24(-25)μ, wall (1.5-)2(-3)μ
thick at sides, (4-)5-7(-8)μ at apex.

Hosts and distribution: Muhlenbergia glauca (Nees) Mez:
Mexico.

492

Type: Cummins 63-547 (=PUR 60269), near Durango, Dgo.,
Mexico.

The variety has smaller urediniospores and teliospores than
the typical.

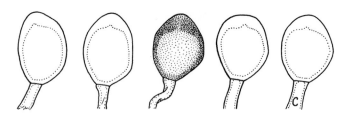

Figure 347

48. UROMYCES FERGANENSIS Tranz. & Eremeeva in Tranzschel
Conspectus Ured. U.S.S.R. p. 100. 1939. Fig. 347.

Sori not described. Urediniospores 21-27 x 21-27μ, globoid,
wall (3.5μ ? thick, golden or cinnamon-brown ?), echinulate,
pores 5 or 6, (scattered?). Teliospores 21-30 x 21-24μ, globoid,
oval, or obovoid, wall (2-3μ ?) thick at sides, 5?-8μ at apex,
golden ? (fuscus), smooth; pedicels persistent, length not
stated.

Hosts and distribution: Stipa barbata Desf., S. lessingiana
Trin. & Rupr.: U.S.S.R. and Morocco.

Type: Eremeeva, on Stipa lessingiana, Alai Mountains,
Kirghiz, U.S.S.R. (LE). Not seen.

The description is adapted from the original text and
illustrations. Tranzschel states that the species is similar
to U. graminis and differs from U. mussooriensis which has
verrucose urediospores.

The sori are doubtless aparaphysate and the telia erumpent.

Greene and Cummins (Mycologia 50:6-36. 1958) reported and
illustrated what they considered might be this species on S.
barbata from Morocco. The single telium was 4 mm long, the
spores were (23-)27-30(-35) x (20-)21-26μ, and the wall was
(1.5-)2-2.5(-3.5)μ thick at the sides and 4-6(-8)μ at the apex.

494

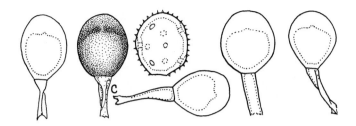

Figure 348

49. UROMYCES LEPTOCHLOAE Wakef. in Wakefield and Hansford Proc. Linn. Soc. London 161:172. 1949. Fig. 348.

Aecia unknown. Uredinia not seen; spores in telia 24-28 x 20-26μ, globoid or broadly ellipsoid, wall 2.5-3(-3.5)μ thick, golden to cinnamon-brown, echinulate, germ pores 8-10, scattered. Telia hypophyllous, exposed, blackish brown, compact; spores 22-27 x 20-24μ, mostly broadly obovoid or globoid, wall 2-3.5μ thick at sides, 7-10μ apically, clear chestnut-brown, smooth; pedicels colorless, thin-walled and collapsing, to 35μ long.

Type: Hansford No. 999, on Leptochloa obtusiflora Hochst., Tororo, Uganda, Jan. 1929 (K). Known otherwise from one other Hansford collection in Uganda.

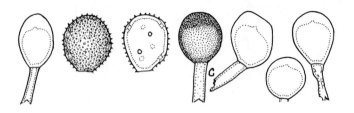

Figure 349

50. UROMYCES KENYENSIS Hennen in Hennen & Cummins Mycologia
48:158. 1956. Fig. 349.

Aecia unknown. Uredinia mostly on adaxial side of leaf,
cinnamon-brown; spores (19-)21-24(-26) x (16-)18-21(-23)μ,
mostly broadly ellipsoid or globoid, wall (1.5-)2-2.5(-3)μ thick,
echinulate, cinnamon-brown, germ pores 7 or 8, scattered. Telia
on adaxial surface, early exposed, blackish brown; spores (18-)
20-24 x 17-20μ, mostly broadly obovoid, wall 2-2.5μ thick at
sides, 5-8μ apically, chestnut-brown, smooth; pedicels colorless
or yellowish, thin-walled and collapsing, to 60μ long.

Type: Nattrass No. 1427, on Chloris roxburghiana Schult.
(C. myriostachya Hochst.), Nairobe, Kenya (PUR; isotype IMI).
Not otherwise known.

496

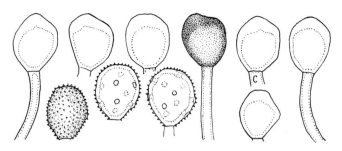

Figure 350

51. UROMYCES HOLCI Joerst. Ark. Bot. Ser. 2. 3:577. 1956.
Fig. 350.

Uromyces schismi Joerst. Ark. Bot. Ser. 2. 3:577. 1956.

Aecia unknown. Uredinia on adaxial leaf surface, orange-yellow; spores (20-)22-26(-28) x (17-)20-23(-24)μ, mostly broadly ellipsoid or broadly obovoid, wall 2-2.5(-3)μ thick, pale yellow-ish, echinulate, germ pores 7-9, scattered. Telia amphigenous, blackish, loosely covered by the epidermis; spores (18-)20-25(-28) x (14-)17-21(-23)μ, mostly obovoid, tending to be dimorphic with the larger more robust spores darker colored and thicker-walled and with thick-walled pedicels, wall (1.5-)2-3.5(-4)μ thick at sides, 4-6(-8)μ apically, golden in the smaller, thinner-walled spores, chestnut-brown in the robust, thick-walled spores, smooth; pedicels thin-walled and collapsing in the golden spores, thick-walled and not collapsing in the chestnut spores, to 60μ long.

Hosts and distribution: Holcus setiger Nees, Schismus scaberrimus Nees: South Africa.

Type: Drege, on Holcus setiger, between Pedroskloff and Leliefontein, Cape Prov., So. Africa (S).

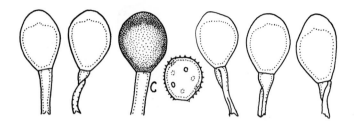

Figure 351

52. UROMYCES SNOWDENIAE Cumm. Torrey Bot. Club Bull. 83:231.
1956. Fig. 351.

Aecia unknown. Uredinia amphigenous, yellow; spores 17-20
x 15-18µ, mostly broadly ellipsoid, wall 1-1.5µ thick, hyaline
to pale yellowish, echinulate, pores 5-7, scattered, obscure.
Telia blackish brown, early exposed, pulvinate; spores (21-)
23-27 x (16-)18-20(-22)µ, obovoid or broadly ellipsoid, sometimes slightly angular, wall 2(-2.5)µ thick at sides, 3.5-5.5µ
at apex, chestnut-brown, smooth; pedicels persistent, hyaline
or yellowish, thin-walled and collapsing, to 45µ long.

Hosts and distribution: Snowdenia polystachya (Fresen.)
Pilger, S. scabra (Pilger) Pilger: Kenya and Ethiopia.

Type: A. Bogdan No. 3272, Bahati Forest, Kenya (PUR; isotype
K).

This is the only species known on the tribe Arthropogoneae.
It is similar in general to U. aegopogonis and U. schoenanthi
but has smaller and paler urediniospores.

A photograph of teliospores of the type was published with
the original diagnosis.

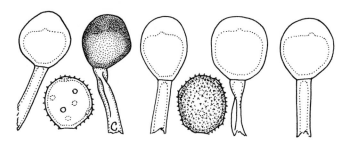

Figure 352

53. UROMYCES AEGOPOGONIS Diet. & Holw. in Holway Bot. Gaz.
24:25. 1897. Fig. 352.

The aecia (Aecidium roseum Diet. & Holw.) occur on species
of Eupatorium and Stevia, and are not clearly distinguishable
from those of Puccinia aegopogonis; spores 19-36 x 15-26μ, wall
1-2μ thick at sides, to 7μ apically, verrucose, colorless.
Uredinia on abaxial side of leaves, yellowish brown; spores
19-24(-26) x (16-)18-22μ, broadly ellipsoid or obovoid, wall
1.5μ thick, yellowish to golden, echinulate, pores 6-8, scattered.
Telia on abaxial surface, blackish brown, early exposed, compact;
spores (22-)24-28(-30) x (19-)21-27(-30)μ, mostly globoid or
broadly obovoid, wall 2-2.5(-3.5)μ thick at sides, 5-9μ at apex,
chestnut-brown, smooth; pedicels persistent, hyaline to brownish,
thin-walled, usually collapsing, to 60μ long.

Hosts and distribution: Aegopogon cenchroides Humb. &
Bonpl., A. geminiflorus H.B.K., A. gracilis Vasey: Mexico.

Type: Holway, on A. cenchroides, Mexico City (S; isotype
PUR).

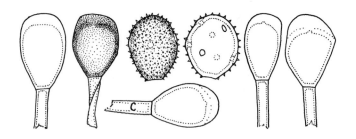

Figure 353

54. UROMYCES ERAGROSTIDIS Tracy J. Mycol. 7:281. 1893. Fig. 353.

Uromyces pedicellata P. Evans Bull. Misc. Inf. Kew 1918:228. 1918.

Aecia (Aecidium anthericicola Arth.; A. antherici P. Henn. & Pole Evans?) occur on species of Anthericum; spores 19-23 x 16-20µ, wall 1.5µ thick, colorless, verrucose. Uredinia in leaves and sheaths, amphigenous, yellowish brown; spores (20-) 21-29(-31) x (16-)18-23(-26)µ, mostly broadly ellipsoid or nearly globoid, wall 1.5µ thick, golden to pale cinnamon-brown, echinulate, pores variable (3)4-8(-10), equatorial or tending equatorial when 3-5, scattered when 5-8, or bizonate when 7-10. Telia blackish brown, early exposed, compact; spores (22-) 23-31(-34) x (16-)18-23(-25)µ, mostly obovoid, wall 1.5-2.5(-3)µ thick at sides, 4-6(-8)µ at apex, chestnut-brown, smooth; pedicels yellowish to brownish, thin-walled and usually collapsing, to 75µ long.

Hosts and distribution: Cypholepis yemenica (Schweinf.) Chiov., Desmostachya bipinnata (L.) Stapf, species of Eragrostis: U.S.A. to Argentina, Africa, Palestine, India and Australia.

Type: Tracy, on Eragrostis pectinacea, Starkville, Miss., U.S.A. (BPI; isotype PUR).

Cummins (Mycologia 55:73-78. 1963) proved the life cycle by inoculation.

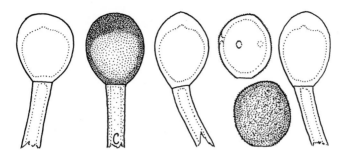

Figure 354

55. UROMYCES ARCHERIANUS Arth. & Fromme Torreya 15:261. 1915.
Fig. 354.

Uromyces chloridis Doidge Bothalia 2:207. 1927.

Aecia unknown. Uredinia on the abaxial leaf surface,
cinnamon-brown; spores (21-)23-27(-29) x (20-)22-26(-27)μ,
broadly ellipsoid or globoid, wall (2-)2.5-3.5(-4)μ thick,
golden to cinnamon-brown, rugose-verrucose, pores 2 or 3,
equatorial. Telia early exposed, blackish brown, compact;
spores (20-)24-29(-32) x (17-)20-24(-26)μ, mostly obovate or
globoid, wall 2-3(3.5)μ thick at sides, 6-8μ at apex, chestnut-
brown, smooth; pedicels persistent, yellowish, usually thick-
walled and non-collapsing, to 120μ long; brown basal cells often
obvious.

Hosts and distribution: Chloris breviseta Benth., C. virgata
Swartz, Enteropogon monostachya (Vahl) K. Schum.: South Africa,
Tanganyika, Uganda, Mexico, and U.S.A. (New Mexico).

Type: Archer, on Chloris virgata, Mesilla Park, New Mexico,
12 Nov. 1914. (PUR).

Hennen and Cummins (Mycologia 48:126-162. 1956) published
a photograph of teliospores of the type.

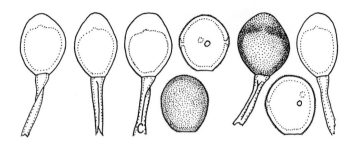

Figure 355

56. UROMYCES VOSSIAE Barclay J. Asiat. Soc. Bengal 59:76. 1890.
Fig. 355.

Uromyces rottboelliae Arth. Bull. Torrey Bot. Club 29:228.
1902.

Uromyces superfluus P. Syd. & H. Syd. Monogr. Ured. 2:337.
1910.

Aecia unknown. Uredinia mostly on abaxial leaf surface,
yellowish brown; spores (16-)18-24(-26) x (16-)18-22µ, mostly
broadly ellipsoid or globoid, wall 1.5-2µ thick, golden, finely
verrucose or striolate-verrucose, germ pores 3-5, mostly 4,
equatorial; amphispores often associated with telia, 25-30 x
18-25µ, mostly ellipsoid or obovoid, wall 3µ thick, golden to
near cinnamon-brown, striolate verrucose. Telia amphigenous,
exposed, blackish brown, more or less compact; spores (20-)
24-29(-32) x (18-)20-24(-26)µ, mostly obovoid, wall (1.5-)2-2.5
(-3)µ thick at sides, 5-8(-10)µ apically, chestnut-brown, smooth
or minutely punctate-verrucose, especially apically; pedicels
colorless or yellowish, mostly thin-walled and collapsing, to
115µ long, usually less than 85µ.

Hosts and distribution: Panicum antidotale Retz., Phacelurus
speciosus (Steud.) C.E. Hubb.: northwestern India and Kashmir.

Neotype: Butler, on Rottboellia speciosa (=Phacelurus
speciosus), Machobra, Simla, India, 11 Aug. 1904 (PUR F2487;
isotypes Sydow Ured. No. 2108 as Uromyces rottboelliae).
Neotype designated by Hennen (Mycologia 57:104-113. 1965).

Hennen (loc. cit.) published a photograph of teliospores of
the neotype. Ramachar & Cummins (Mycopathol. Mycol. Appl.
19:49-61. 1963) published a photograph of the type of U.
superfluus. The two fungi are indistinguishable.

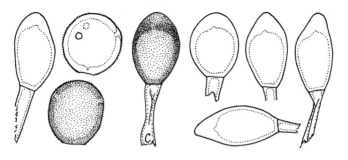

Figure 356

57. UROMYCES SEDITIOSUS Kern Torreya 11:212. 1911 var. seditiosus.
Fig. 356.

Aecia (A. oldenlandianum Ell. & Tracy) occur on species of
Plantago and probably Houstonia; spores (16-)18-24(-26) x
(13-)16-19(-20)μ, wall 1.5μ thick, hyaline, verrucose. Uredinia
adaxial, cinnamon-brown; spores (18-)22-26(-29) x 18-26μ,
globoid or depressed globoid, wall 2-3.5μ thick, mostly golden
brown, verrucose, pores 2 or 3 (4?), equatorial or slightly
superequatorial, difficult to count. Telia adaxial, blackish
brown, compact, early exposed; spores (23-)26-40(-44) x
(13-)16-25(-27)μ, usually dimorphic with the shorter more robust
spores deep chestnut-brown, the elongate spores golden or clear
chestnut-brown, wall (1.5-)2.5-3.5(-4)μ thick at sides with the
robust spores in the thicker range, (5-)7-10(-12)μ at apex,
smooth; pedicels yellowish, persistent, to 100μ long but
usually shorter.

Hosts and distribution: species of Aristida: in the United
States from New York and Virginia southwestward to Oklahoma
and Texas.

Type: Bartholomew (PUR 11913 = Barth. F. Columb. No. 2390),
on Aristida oligantha, Kansas.

The species has been confused with U. peckianus Farl. Aecial
records on Plantago from Wyoming, Montana, Alberta, and Washington
doubtless belong to Puccinia subnitens.

U. seditiosus is obviously closely related to Puccinia
aristidae.

UROMYCES SEDITIOSUS Kern var. mexicensis Cumm. & Husain
Bull. Torrey Bot. Club 93:66. 1966.

Aecia unknown. Urediniospores in telia 26-29 x 23-26μ,
broadly obovoid or nearly globoid, wall 3.5-4.5(-5)μ thick, golden brown or pale golden, verrucose, pores 3 or 4, equatorial.
Telia adaxial, blackish brown, compact, early exposed; spores
(20-)24-30(-34) x (18-)20-25(-28)μ, broadly ellipsoid, broadly
obovoid, or globoid, wall (2-)2.5-3.5(-4)μ thick at sides,
(4-)5-7(-8)μ at apex, chestnut-brown, smooth; pedicels colorless
or yellowish, persistent 80-150μ long.

Hosts and distribution: Aristida adscensionis L., A.
orizabensis Fourn.: the northern half of Mexico.

Type: Cummins No. 63-607 (PUR 59560), on Aristida adscensionis,
Durango, Mexico.

This variety differs from var. seditiosus in having larger
urediniospores with thicker walls and uniformly robust deeply
pigmented teliospores.

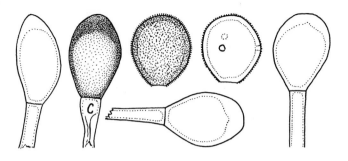

Figure 357

58. UROMYCES AELUROPODIS-REPENTIS Nattrass A First List of
Cyperus Fungi. p. 21. 1937. Fig. 357.

Uredo aeluropodina Maire Bull. Soc. Hist. Nat. Afrique
Nord 18:152-153. 1917.

Uromyces aeluropodinus Tranz. Conspectus Uredinalium
U.R.S.S. p. 101. 1939.

Aecia unknown. Uredinia amphigenous, about cinnamon-brown;
spores (24-)26-30 x (20-)22-26(-29)μ, mostly globoid or nearly
so, wall (2-)2.5-3(-3.5)μ thick, about golden brown, finely
and closely verrucose or the wartlets merging in rugose patterns,
germ pores 3 or 4 (5), equatorial. Telia amphigenous and on
sheaths and culms, early exposed, blackish, compact; spores
(26-)30-40 x (18-)20-26(-28)μ, mostly ellipsoid or obovoid, wall
(1.5-)2-3(-4)μ thick at sides, 4-8μ apically, chestnut-brown,
smooth; pedicels colorless or yellowish, thin-walled, collapsing
or not, to 115μ long.

Hosts and distribution: Aeluropus littoralis (Willd.) Parl,
A. repens (Willd.) Parl : the Mediterranian region and southern
Russia.

Type: Nattrass No. 650, near Nicosia, Cyprus (IMI; isotype
PUR).

Except for 1-celled teliospores the fungus is like Puccinia
aeluropodis and also has conspicuous basal cells.

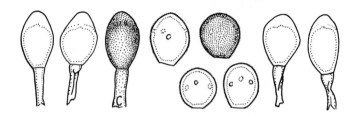

Figure 358

59. UROMYCES ANDROPOGONIS Tracy J. Mycol. 7:281. 1893. Fig. 358.

Uromyces pedatatus Sheldon Torreya 10:90. 1910. Nom. nudum.

Aecia (Caeoma (Aecidium) pedatatum Schw.) occur on species of Viola; spores 12-19μ diam., globoid, wall 1-1.5μ thick, yellowish, verrucose. Uredinia on abaxial leaf surface, yellowish brown; spores (15-)16-19(-21) x (13-)14-17μ, mostly obovoid or globoid, wall (1.5-)2-2.5(-3)μ thick, golden or dull cinnamon-brown, minutely verrucose, usually striately so, germ pores 3(4) approximately equatorial. Telia on abaxial surface, exposed, blackish brown, compact; spores (18-)20-30(-36) x (11-)13-17(-20)μ, mostly ellipsoid or obovoid, wall 1.5(-2)μ thick at sides, 4-8(-10)μ apically, chestnut-brown except progressively paler in the apical thickening, smooth; pedicels yellowish to brownish, thin-walled and mostly collapsing, to 70μ long; basal sporogenous cells usually obvious, golden-brown.

Hosts and distribution: species of Andropogon: U.S.A. from New England States to Florida, the Midwest, and Texas.

Type: Tracy, on Andropogon virginicus, Starkville, Miss., Oct. 1891 (NY; isotype PUR).

Inoculations demonstrating the life cycle were made first by Sheldon (Torreya 9:54-55. 1909). A photograph of teliospores of the type was published by Hennen (Mycologia 57:104-113. 1965).

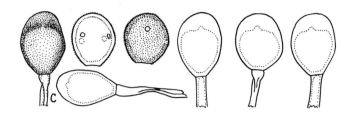

Figure 359

60. UROMYCES MUSSOORIENSIS H. Syd. & P. Syd. in Sydow & Butler
Ann. Mycol. 4:430. 1906. Fig. 359.

Aecia unknown. Uredinia on the adaxial leaf surface, yellowish brown; spores (17-)19-24(-25) x (15-)18-21(-23)μ, mostly
globoid, wall 2-2.5(-3)μ thick, yellowish to golden brown, densely
and finely verrucose, mostly striately so, germ pores 3 or 4(5),
equatorial or slightly above. Telia on adaxial surface, early
exposed, chocolate-brown, compact; spores (19-)23-27(-28) x (16-)
18-23(-25)μ, mostly broadly obovoid or globoid, wall 1.5-2(-3)μ
thick at sides, (3-)5-7(-8)μ apically, deep golden brown or clear
chestnut-brown, smooth; pedicels colorless or yellowish, thin-
walled and mostly collapsing, to 65μ long.

Type: Butler No. 542, on *Stipa sibirica* Lam., Mussoorie,
India (S). Not otherwise known.

Greene & Cummins (Mycologia 50:6-36. 1958) published a photograph of teliospores of the type.

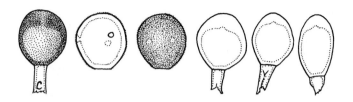

Figure 360

61. UROMYCES INAYATI H. Syd. & P. Syd. in Sydow and Butler Ann. Mycol. 5:493. 1907. Fig. 360.

Uredo apludae Barcl. J. Asiatic Soc. Bengal 59:99. 1890.

Aecia unknown. Uredinia on abaxial leaf surface, yellowish brown; spores (18-)20-24(-26) x (16-)18-21(-23)μ, globoid or broadly ellipsoid, wall (2-)2.5-3μ thick, dull golden or pale cinnamon-brown, verrucose, germ pores 3 or 4, equatorial. Telia on abaxial surface, exposed, blackish brown, more or less compact; spores (18-)20-24(-27) x (16-)18-21(-23)μ, mostly broadly obovoid or globoid, wall 1.5-2.5(-3)μ thick at sides, 4-7μ apically, chestnut-brown, smooth; pedicels colorless or yellowish, thin-walled and collapsing, to 50μ long.

Hosts and distribution: Apluda mutica L.:   India and China.

Type: Inayat (Butler No. 883), on Apluda aristata Hock. (=A. mutica), Kumaon, Himalaya, 15 June 1907 (S; probable isotypes in HC10 Indian Ured. Fasc. 2, No. 95).

Hennen (Mycologia 57:104-113. 1965) published a photograph of teliospores of a syntype (Butler No. 884).

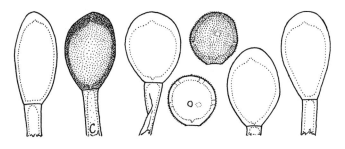

Figure 361

62. UROMYCES PECKIANUS Farl. Proc. Amer. Acad. Arts & Sci. 18:78. 1883. Fig. 361.

Aecia occur on species of Atriplex, Dondia, Chenopodium, and Salicornia, grouped, cupulate or cylindrical; spores (16-)18-20 (-22) x (14-)16-18µ, mostly globoid, wall 1-1.5µ, pale yellowish, verrucose. Uredinia mostly on adaxial leaf surface, yellowish brown; spores (16-)18-22 x 18-22µ in eastern material, (19-) 21-24(-26) x 20-24µ in western specimens, mostly globoid or slightly depressed globoid, wall (2-)2.5-3(-3.5)µ thick, golden or dull brownish, densely and finely verrucose, pores 4-6, mostly with 1 or 2 apical, the others approximately equatorial. Telia amphigenous or often only on adaxial surface, early exposed, blackish, compact; spores (20-)24-36(-45) x (13-)17-23(-26)µ, mostly obovoid or ellipsoid, tending to be dimorphic, the longer spores usually paler and with thinner side wall than the shorter spores, wall (1.5-)2-3(-3.5)µ thick at sides (3-)4-6(-7)µ apically, uniformly chestnut-brown, smooth; pedicels colorless or yellowish, thin-walled and collapsing, to 80µ long.

Hosts and distribution: Distichlis spicata (L.) Greene: east and west coasts of United States and Canada.

Type: Farlow on Brizopyrum spicatum (=Distichlis spicata), Gloucester, Mass. (FH; isotypes Ellis N. Amer. Fungi No. 240).

Fraser (Mycologia 3:67-74. 1911) first proved the life cycle by inoculation, using Atriplex patula and Chenopodium album as aecial hosts.

Except for the 1-celled teliospores, the species is similar to and probably derived from Puccinia subnitens.

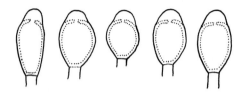

Figure 362

63. UROMYCES STIPINUS Tranz. & Eremeeva Conspectus Uredinalium
URSS. p. 101-102. 1939. Fig. 362.

Aecia and uredinia unknown. Telia not described; spores
24-32 x 16-21µ, ovate to oblong, side wall thickness not given,
apparently 1.5-2µ, pale brown, to 11µ thick apically as a pale
differentiated umbo, smooth; pedicels fragile (apparently thin-
walled and collapsing; length not given).

Type: Collector not given, on Stipa rubens Smirn.?, Karkaralen
mountain, Kazakhstan, U.S.S.R. (LE). Not seen.

The description and illustration are adapted from the original. The hyaline umbo is distinctive among the rust fungi on
Stipa and most other grasses.

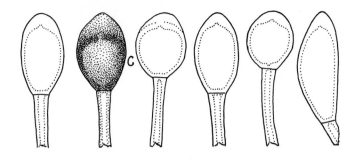

Figure 363

64. UROMYCES EHRHARTAE-GIGANTEAE Doidge Bothalia 2(1a):207. 1927. Fig. 363.

Aecia unknown. Uredinia unknown. Telia mostly on stems, exposed, cushion-like, chocolate-brown, loosely felt-like; spores (23-)26-36(-38;40) x (16-)19-24(-26)μ, tending to be dimorphic with broadly ellipsoid or obovoid spores mostly less than 30μ long, ellipsoid or oblong-ellipsoid spores mostly more than 30μ long, wall (1.5-)2-2.5(-3)μ thick at sides, 5-8(-12)μ apically, chestnut-brown in the short spores, about golden brown in the long spores, smooth; pedicels colorless, thick-walled, not collapsing, to at least 160μ long.

Type: van der Merwe, on _Ehrharta_ _gigantea_ Thunb., Mowbray, Cape Prov. South Africa 10 Feb. 1914 (PRE 7392; isotype PUR). Known only from this locality.

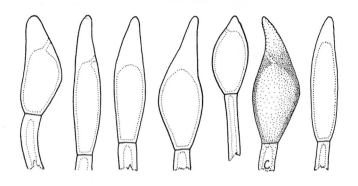

Figure 364

65. UROMYCES PROCERUS Lindq. Rev. Fac. Agron. La Plata 36:106. 1960. Fig. 364.

Aecia and uredia unknown. Telia in adaxial leaf surface, early exposed, dark brown; spores (33-)42-65(-73) x (12-)15-19 (-23)μ, mostly elongate-ellipsoid, wall 1-2μ thick at sides, (9-)15-21(-26)μ apically, yellowish to golden, smooth; pedicels to 130μ, usually shorter, thick-walled, not collapsing; the spores germinate without dormancy.

Hosts and distribution: Festuca procera H.B.K.: Chile.

Type: Holway No. 260, on Festuca procera, Termas de Chilan, Chile (PUR; isotypes, Reliq. Holw. No. 39, issued as Uromyces cuspidatus Wint.).

Holway noted "stage of Uromyces on Festuca procera No. 260" on a packet of aecia on some composite. No test of the relationship has been made. The specimen was issued as Uromyces cuspidatus in Reliq. Holw. No. 38.

Arthur (Proc. Amer. Philos. Soc. 54:131-223. 1925) published a photograph of teliospores of the type.

512

UREDO Pers., a Form Genus

The species are arranged alphabetically, using the same group system as in Puccinia and Uromyces.

GROUP I: uredinia paraphysate, spores echinulate, germ pores equatorial.

1. UREDO ARUNDINELLAE-NEPALENSIS Cumm. Bull. Torrey Bot. Club 72:218. 1945.

Uredinia amphigenous, brownish, paraphyses peripheral, in-curved, 35-45 x 9-12μ, wall 1-1.5μ ventrally and basally, thickened to 6μ apically and dorsally, colorless or yellowish; spores 25-33(-39) x 17-23μ, mostly ellipsoid or obovoid, wall 1-1.5μ thick, yellowish or golden, finely echinulate, germ pores inconspicuous, 4 or 5, equatorial.

Type: Clemens on Arundinella nepalensis Trin.: Australia (PUR F10853).

The type and one other Clemens specimen, both from near Brisbane, are known. The species is doubtless a Phakopsora or a Physopella.

2. UREDO BAMBUSAE-NANAE Yen Rev. Mycol. 34:322. 1970.

Sori hypophyllous, orange-brown or pale brown, paraphyses clavate or cylindrical clavate, incurved, rarely 1-3 septate, thick-walled, to 60μ long; spores 24-30 x 18-24μ, mostly broadly ellipsoid or obovoid, wall 1.5-2μ thick, brown or yellowish brown, echinulate, germ pores 2-5, equatorial.

Hosts and distribution: Bambusa nana Roxb.: Singapore.

3. UREDO CYNODONTIS-DACTYLIS Tai Farlowia 3:133. 194. 1947.

Uredinia amphigenous, chestnut-brown, paraphyses cylindrical or clavate-cylindrical, 37-57 x 13-18μ, yellowish, wall 1.5-2μ thick; spores 29-42 x 23-33μ, globoid or ovoid, rarely ellipsoid, wall 1.5-2μ thick, echinulate, chestnut-brown, germ pores 2, equatorial.

Hosts and distribution: Cynodon dactylon (L.) Pers.: China.

There appears to be great similarity between this fungus and Uredo ophiuri and, in turn, Puccinia cacao. One suspects an error in the identification of the grass.

4. UREDO DENDROCALAMI Petch Ann. Roy. Bot. Gard. Peradeniya 7:296. 1922.

Uredinia on abaxial leaf surface, small, pale brown, seriate, paraphyses incurved, clavate, 7-12µ wide, wall thick, the lumen occupying one-half to two-thirds the length and situated nearer the concave side; spores 26-35 x 19-22µ, oval or pyriform, wall colorless, echinulate, germ pores not described.

Hosts and distribution: Dendrocalamus strictus Nees: Ceylon.

There is a similar fungus (PUR F14921) on D. latiflorus Munro from China. The spores are of the same magnitude and have (5)6 or 7(8) equatorial germ pores. The paraphyses are dorsally thick-walled but the lumen occupies more of the length than described by Petch.

5. UREDO EULALIAE-FULVAE Cumm. Bull. Torrey Bot. Club 70:529. 1943.

Uredinia amphigenous, opening by a slit in the epidermis, brownish; paraphyses inconspicuous, 23-30 x 15-28µ, obovoid, wall colorless, uniformly 1µ thick; spores 29-38(-42) x (17-)19-26(-28)µ, mostly obovoid or oblong-ellipsoid, often angular, wall 1.5µ thick or 2-4µ apically in occasional spores, echinulate, yellowish, germ pores obscure, probably about 5, equatorial.

Hosts and distribution: Eulalia fulva (R. Br.) Kunth: New Guinea.

The paraphyses collapse readily and can easily be overlooked.

6. UREDO GENICULATA Cumm. Ann. Mycol. 35:104. 1937.

Uredinia on abaxial leaf surface, paraphyses abundant, incurved, usually geniculate, clavate-capitate, yellowish, 45-65 x 12-18µ, wall progressively thicker from base upward, to 4µ thick apically; spores 23-29 x 18-24µ, mostly broadly ellipsoid, wall 1.5-2µ thick, cinnamon-brown or darker, echinulate, germ pores 6-8, equatorial or tending to be bizonate.

Hosts and distribution: Sorghum nitidum (Vahl) Pers. (Andropogon serratus Thunb.): New Guinea and the Philippines.

The species is similar to Puccinia andropogonis-hirti.

7. UREDO IGNAVA Arth. Bull. Torrey Bot. Club 46:121. 1919.

Uredinia amphigenous, pale brown, paraphyses abundant, incurved, colorless or brownish, the wall 1-1.5µ thick on ventral side, 3-5µ dorsally and apically; spores (21-)23-28(-31) x (14-)16-19(-21)µ, obovoid or ellipsoid, wall 1-2µ thick, yellowish or pale brownish, echinulate, germ pores probably 4, equatorial, very obscure.

Hosts and distribution: Species of Bambusa, Arthrostylidium, Dendrocalamus, Sinocalamus, Schizostachyum: Central and South America, the West Indies, Africa, Malaya, and China.

This fungus doubtless will prove to be a _Dasturella_ or something similar. Both the identity of the fungus and the hosts leave something to be desired.

8. UREDO OPHIURI H. Syd., P. Syd. & Butl. Ann. Mycol. 4:445. 1906.

Uredinia on adaxial leaf surface or amphigenous, dark brown, paraphyses mostly cylindrical, mostly 12-20μ wide, wall uniformly 1-2μ thick, yellowish; spores (26-)30-38(-42) x (20-)24-30(-33)μ, broadly ellipsoid or ellipsoid, wall 1.5μ thick, dark cinnamon- or chestnut-brown, echinulate, germ pores 2(3), equatorial, in flattened sides.

Hosts and distribution: _Ophiuros exaltatus_ (L.) Kuntze (_O. corymbosus_ Gaertn.): India.

The fungus differs from _Puccinia cacao_ only in typically having 2, rather than 3, pores.

9. UREDO PALMIFOLIAE Cumm. Mycologia 33:151. 1941.

Uredinia amphigenous or mostly on abaxial leaf surface, golden when fresh, pale brownish when dry, paraphyses abundant, incurved, cylindrical, colorless or pale yellow, 30-50 x 8-12μ, wall 1.5-2μ ventrally, 3-6μ dorsally and apically; spores 21-27(-29) x 17-20μ, mostly obovoid or broadly ellipsoid, wall 1-1.5μ thick, yellowish or pale brownish, echinulate, germ pores obscure but apparently 4, equatorial.

Hosts and distribution: _Setaria palmifolia_ (Koen.) Stapf : New Guinea.

This fungus will prove to be a _Phakopsora_ or a _Physopella_.

10. UREDO STIPAE-LAXIFLORAE Wang Acta Phytotax. Sinica 10:298. 1965.

Sori hypophyllous, yellowish brown, with capitate paraphyses to 50μ long, the head to 20μ diam, wall 5μ thick in the apex of head; spores 17-25 x 15-20μ, ovoid, subgloboid, or ellipsoid, wall 1.5μ thick, echinulate, germ pores 6 or 7, equatorial.

Type: Wang Ching-tze No. 620, on _Stipa laxiflora_ Keng, Yunnan Prov., China (Inst. Microbiol. Acad. Sinica 34721). Not seen.

GROUP II: uredinia paraphysate, spores echinulate, germ pores scattered.

11. UREDO ANTHISTIRIAE-TREMULAE Petch Ann. Roy. Bot. Gard. Peradeniya 5:255. 1912.

Uredinia hypophyllous, brownish, paraphyses incurved, clavate, 10-12µ wide, yellowish, wall 2-6µ thick apically; spores (18-) 20-26(-29) x 17-21µ, ovate or ellipsoid, wall 1-1.5µ thick, echinulate, germ pores 6-10, scattered.

Hosts and distribution: Themeda tremula (Nees) Hack.: Ceylon.

Petch described the spores as echinulate, Sydow (Monogr. Ured. 4:540. 1924) as densely verruculose. The incurved paraphyses suggest Phakopsora incompleta.

12. UREDO BROMI-PAUCIFLORAE Ito J. Coll. Agr. Tohoku Imp. Univ. 3:246. 1909.

Uredinia on adaxial leaf surface, yellowish brown, paraphyses abundant, mostly capitate, to 22µ diam in the head, wall uniformly (1-)1.5-2(-3)µ thick, colorless or yellowish; spores (22-)25-30(-32) x (20)22-25(-28)µ, broadly ellipsoid, obovoid, or ellipsoid, wall 1.5-2µ thick, yellowish brown, echinulate, germ pores 7-10, scattered.

Hosts and distribution: Bromus pauciflorus (Thunb.) Hack.: Japan.

The species resembles Puccinia pygmaea.

13. UREDO DITISSIMA Cumm. in Hino & Katumoto Bull. Fac. Agr. Yamaguti Univ. 11:27. 1960.

Puccinia ditissima H. Syd. in Sydow & Petrak Ann. Mycol. 29:152. 1931.

Uredinia on abaxial leaf surface, conspicuously seriate, brown, paraphyses abundant, incurved, 10-18µ wide, wall uniformly 1.5-2µ thick or slightly thicker dorsally and apically, incurved, yellowish or pale brownish; spores (25-)28-38(-42) x (21-)23-27(-30)µ, mostly obovoid, wall 1-1.5(-2)µ thick, about cinnamon-brown, echinulate, germ pores very numerous, about 15-20, scattered.

Hosts and distribution: Dendrocalamus latiflorus Munro, Schizostachyum lumampao (Blco.) Merr.: the Philippines and Taiwan.

Teliospores, but too rare to describe, occur on the type. The fungus almost certainly will prove to belong in the genus Dasturella.

14. UREDO ISCHAEMI-CILIARIS Petch Ann. Roy. Bot. Gard. Peradeniya 5:254. 1912.

Uredo ischaemi-commutati Petch Ann. Roy. Bot. Gard. Peradeniya 5:254. 1912.

Uredinia on abaxial leaf surface, yellowish brown, paraphyses varying from cylindrical to capitate, colorless or yellowish, wall nearly uniformly 1.5-2(-2.5)µ thick; spores (27-)30-36(-38) x

(23-)26-30(-32)μ, mostly broadly ellipsoid or obovoid, sometimes
globoid or oblong-ellipsoid, wall 2-2.5(-3)μ thick, golden brown,
echinulate, germ pores 7-9(-10), scattered.

Hosts and distribution: Ischaemum commutatum Hack., I. indicum
(Hoult.) Merr.: Ceylon.

15. UREDO KARETU Cunn. Trans. N. Zealand Inst. 55:41. 1924.

Uredinia hypophyllous, with a few hyaline capitate paraphyses,
sori orange-yellow; spores (24-)26-35(-37) x (20-)24-28(-31)μ,
obovoid or nearly globoid, wall 2-2.5μ thick, colorless or yellowish, finely and closely echinulate, germ pores 6-10, scattered,
obscure.

Hosts and distribution: Hierochloë redolens (Vahl) Roem. &
Schult.: New Zealand.

In Chile, there is an undescribed species of Puccinia on the
same host. It has abundant colorless, mostly cylindrical paraphyses and urediniospore (29-)32-43(-45) x (24-)26-30(-32)μ whose
wall is colorless and has 12-14 germ pores. Dr. E. Oehrens B.
has found telia and presumably will describe a new species.

16. UREDO OCHLANDRAE Petch Ann. Roy. Bot. Gard. Peradeniya
5:255. 1912.

Uredinia hypophyllous, small, in striiform brownish spots,
paraphyses cylindrical, yellowish, 7-12μ wide, the wall thick
apically; spores 21-25 x 17-20μ, oval or nearly globoid, wall
1μ thick, yellowish or colorless, closely echinulate, germ pores
obscure.

Hosts and distribution: Ochlandra stridula Thwait.: Ceylon.

17. UREDO SETARIAE-EXCURRENS Wang Acta Phytotax. Sinica 10:298.
1965.

Sori amphigenous, yellowish brown, with incurved paraphyses,
ventral wall 1.5μ thick, dorsal wall 3-5μ thick; spores 20-23
x 18-23μ, subglobose, ellipsoid, or subovoid, wall 1.5μ thick,
yellowish brown, densely echinulate, germ pores 4-6, scattered.

Type: Wang Ching-tse No. 605, on Setaria excurrens (Trin.)
Miq., Kweichow Prov., China (Inst. Microbiol. Acad. Sinica
34707).

This fungus will prove to be a Physopella or a Phakopsora.

18. UREDO TRINIOCHLOAE Arth. & Holw. Amer. J. Bot. 5:538.
1918.

Uredinia mostly on adaxial leaf surface, yellowish, paraphyses
clavate or capitate, 10-29μ diam, wall uniformly 1-2μ thick or

517

slightly thicker apically; spores 19-26 x 16-19μ, ellipsoid or obovoid, wall 1μ thick, yellowish or pale brownish, echinulate, germ pores 4-6, scattered or occasionally in the equatorial region.

Hosts and distribution: _Triniochloa stipoides_ (H.B.K.) Hitchc.: Colombia and Guatemala.

19. UREDO TANZANIAE Cumm. sp. nov.

Urediniis plerumque hypophyllis, flavidis, paraphysibus 30-60 x 8-12μ, incurvatis, membrana ventralis 1-1.5μ crassa, dorsalis et apicalis 4-7μ crassa, hyalina; sporae (16-)18-22 (-25) x (13-)14-18μ, ellipsoideae vel late ellipsoideae, membrana 1μ crassa, hyalina, dense echinulata, poris germinationis perobscuris, sparsis, verisimiliter 6-8.

Type: Hitchcock No. 24463, on _Panicum brevifolium_ L., Amani, Tanganyika, 28-30 Aug. 1929 (PUR F14780).

The species undoubtedly is a _Physopella_ or a _Phakopsora_.

GROUP III: uredinia paraphysate, spores verrucose, germ pores equatorial. No species known.

GROUP IV: uredinia paraphysate, spores verrucose, germ pores scattered.

20. UREDO MISCANTHI-SINENSIS Sawada in Hiratsuka Trans. Mycol. Soc. Japan 2:11. 1959.

Uredinia amphigenous, pale yellowish brown, paraphyses abundant, clavate-cylindrical, straight or incurved, 8-12μ wide, wall thin, colorless; spores 18-27 x 15-24μ, ellipsoid or globoid, wall 1.5-2μ thick, yellowish brown, verrucose, germ pores 6-8, scattered.

Hosts and distribution: _Miscanthus sinensis_ Anderss.: Taiwan.

GROUP V: uredinia aparaphysate, spores echinulate, germ pores equatorial.

21. UREDO ARUNDINELLAE Arth. & Holw. in Arthur Mycologia 10:148. 1918.

Puccinia arundinellae Barth. Handb. N. Am. Ured. Ed. 1, p. 88. 1928. Based on uredinia.

Uredinia adaxial, cinnamon-brown; spores (27-)29-37(-42) x (22-)24-29(-31)μ, mostly obovoid, wall 1-1.5μ thick or often slightly (2-2.5μ) thicker apically, cinnamon-brown, echinulate,

518

germ pores (2)3(4), equatorial or usually slightly subequatorial.

Type: Holway No. 431, on _Arundinella deppeana_ Nees: Costa Rica (PUR 18276).

There are no other records. The spores are similar to those of _Puccinia substriata_, a widely distributed rust of the Paniceae.

22. UREDO AVENOCHLOAE Urban Ceska Mycol. 17:23. 1963.

Uredinia on adaxial leaf surface, about cinnamon-brown; spores 25-33 x 21-26µ, broadly ovoid or nearly globoid, wall 3-4µ thick, echinulate, yellowish brown, germ pores 4 or 5(6), equatorial.

Hosts and distribution: _Avenochloa pubescens_ (Huds.) Holub: Czechoslovakia.

The species differs from others on _Avenochloa_ because of equatorial pores.

23. UREDO GAYANAE Lindq. Rev. Fac. Agron. La Plata 39:118. 1963.

Uredinia amphigenous, cinnamon-brown; spores (27-)30-38(-42) x (22-)24-28(-31)µ, ellipsoid, obovoid, or broadly ellipsoid, wall (1-)1.5-2µ thick, about cinnamon-brown, echinulate, germ pores 4 or 5, equatorial.

Hosts and distribution: _Chloris gayana_ Kunth: Brazil.

Lindquist points out that the species is nearest to _Puccinia cacabata_ of any rust fungus on Chlorideae, but it certainly is not synonymous.

24. UREDO MOROBEANA Cumm. Bull. Torrey Bot. Club 70:528-529. 1943.

Uredinia mostly on abaxial leaf surface, about cinnamon-brown; spores 25-32 x 20-27µ, mostly broadly ellipsoid, wall 1.5(-2)µ thick, pale cinnamon-brown, echinulate, germ pores 3 or 4, equatorial.

Hosts and distribution: _Eulalia fulva_ (R. Br.) Kuntze : New Guinea.

25. UREDO NAKANISHIKII P. Henn. Bot. Jahrb. 37:158. 1905.

Uredinia adaxial, in linear series, cinnamon-brown; spores 20-26 x (18-)20-24µ, broadly ellipsoid or globoid, wall 2-3µ thick, cinnamon-brown, closely and finely verrucose-echinulate, germ pores 3 or 4, equatorial.

Type: Nakanishiki, on _Arundinella anomala_ Steud.: Japan (B?, isotype S).

There seem to be no subsequent records, indicating that the species is rare or that the host plant may have been misidentified.

26. UREDO PANICI-MONTANI Petch Ann. Roy. Bot. Gard. Peradeniya 6:215. 1917.

Uredinia on abaxial leaf surface, yellowish brown, very small, seriate; spores 20-25(-27) x (15-)17-20(-21)μ, mostly obovoid, wall 1μ thick, pale brownish, echinulate, germ pores 4 or 5, equatorial, obscure.

Hosts and distribution: Panicum montanum Roxb.: Ceylon.

This fungus will probably prove to be a Phakopsora or a Physopella.

27. UREDO PHRAGMITIS-KARKAE Sawada Coll. Agr. Natl. Univ. Taiwan Spec. Bull. 8:96. 1959.

Uredinia epiphyllous, to 3 mm long, brown; spores 25-40 x 18-26μ, obovoid, ellipsoid, or oblong, wall 3-5μ thick at sides, 4.5-8μ apically, echinulate, pale brown, germ pores not described, doubtless equatorial.

Hosts and distribution: Phragmites karka (Retz.) Trin.: Taiwan.

28. UREDO RAVENNAE Maire Bull. Soc. Nat. Hist. Africa Nord 8:153. 1917.

Uredo fragosoana Cabal. Publ. Secc. Cien. Nat. Univ. Barcelona 1920:99. 1920.

Uredinia hypophyllous, yellowish brown to pale cinnamon-brown, usually linear; spores (28-)30-38(-42) x (22-)24-28(-30)μ, mostly ellipsoid or obovoid, wall uniformly 1.5-2(-2.5)μ thick or 3-5μ apically in some spores, sparsely echinulate, golden to pale cinnamon-brown, germ pores 3-6, mostly 4 or 5, equatorial, or sometimes scattered in short broad spores.

Hosts and distribution: Erianthus ravenna (L.) Beauv.: the Mediterranean region.

29. UREDO SETARIAE Speg. An. Mus. Nac. B. Aires 23:33. 1912.

Uredinia amphigenous, cinnamon-brown; spores 27-30(-33) x 22-28μ, obovoid or broadly ellipsoid, wall 2μ thick, echinulate, germ pores 3, equatorial.

Hosts and distribution: Setaria macrostachya H.B.K.: Argentina.

30. UREDO TRIBULIS Cumm. Ann. Mycol. 35:105. 1937.

Uredinia on abaxial leaf surface, dark brown; spores 24-30
x (15-)18-22(-23)μ, mostly obovoid, wall 2-2.5μ thick at sides,
3.5-5μ apically, chestnut-brown, echinulate, germ pores 3 or 4,
equatorial.

Hosts and distribution: ?Rottboellia ophiuroides Benth.:
the Philippines.

31. UREDO UROMYCOIDES Speg. An. Mus. Nac. B. Aires 6:240. 1899.

Uredinia amphigenous, cinnamon-brown; spores (22-)25-28(-30)
x (20-)21-24(-26)μ, ellipsoid, obovoid, or globoid, wall 2-2.5μ
thick, rather dull brown, approaching cinnamon-brown or chestnut-
brown, echinulate, germ pores 3 or 4, equatorial or often 3
equatorial and 1 apical, the pedicels tend to persist.

Hosts and distribution: Panicum phyllanthum Steud.:   Argentina.

The species has some resemblance to Puccinia levis but the
spores are generally smaller and the pores different.

32. UREDO VICTORIAE Cumm. sp. nov.

Urediniis hypophyllis, obscure brunneis; sporae (22-)25-30
(-32) x (22-)24-28μ, plerumque globoideae, membrana prope basim
(1-)1.5μ crassa, apicem versus leniter crassiore, ad apicem 2-3
(-4)μ crassa, ad apicem castaneo-brunnea, deorsum pallidiore,
echinulata, poris germinationis (3)4 vel 5(6), prope hilum.

Type: Hennen, on unidentified grass (possibly Andropogoneae),
19 miles southwest of Ciudad Victoria, Tamps., Mexico, 17 Oct.
1967 (PUR 63277).

Unfortunately, the grass cannot be identified but, because
of the basal pores, the fungus is readily recognizeable.

33. UREDO ZEUGITIS Arth. & Holw. Amer. J. Bot. 5:538. 1918.

Uredinia mostly on the abaxial leaf surface, cinnamon-brown;
spores 23-26 x 19-21μ, mostly broadly ellipsoid, wall 1.5-2.5μ
thick, about cinnamon-brown, echinulate, germ pores 3(4),
equatorial.

Hosts and distribution: Zeugites hartwegii Fourn.:   Colombia
and Guatemala.

GROUP VI: uredinia aparaphysate, spores echinulate, germ
           pores scattered.

34. UREDO ANDROPOGONIS-LEPIDI P. Henn. in Engler Die Pflanzenwelt
Ost-Afrikas und der Nachbargebiete, C, p. 52. 1895.

Uredinia amphigenous, sometimes seriate, rather long covered by the epidermis, pale ochraceous; spores 17-27 x 16-23μ, globoid, obovoid, or ellipsoid, wall of variable thickness, 1.5-3μ thick, "aculeate", colorless or pale yellowish, germ pores 6-8, scattered.

Hosts and distribution: Andropogon lepidus Nees: Tanzania.

The variable thickness of the wall suggests Puccinia agrophila.

35. UREDO ANDROPOGONIS-ZEYLANICAE Petch Ann. Roy. Bot. Gard. Peradeniya 6:215. 1917.

Uredinia amphigenous, yellowish; spores 22-28 x 20-25μ, globoid, ovoid, or ellipsoid, wall 2μ thick, yellowish or nearly colorless, echinulate, germ pores obscure.

Hosts and distribution: Chrysopogon zeylanicus (Nees) Thwait.: Ceylon.

The description indicates a possible similarity to Puccinia agrophila.

36. UREDO EHRHARTAE-CALYCINAE Doidge Bothalia 4:907. 1948.

Uredinia amphigenous, large and Puccinia graminis-like, cinnamon-brown; spores (20-)22-27(-29) x (17-)19-21μ, mostly broadly ellipsoid or broadly obovoid, wall 3-4μ thick, about golden brown, echinulate, germ pores (4)5-8, scattered.

Hosts and distribution: Ehrharta calycina J. E. Smith: South Africa.

37. UREDO MARTYNII Dale Commonw. Mycol. Inst. Mycol. Papers 60:14. 1955.

Uredinia hypophyllous, pale brown; spores 23-33 x 18-24μ, broadly ellipsoid or obovoid, wall 1.5μ thick, pale yellow or golden, echinulate, germ pores obscure.

Hosts and distribution: Isachne arundinacea Griseb.: Jamaica.

38. UREDO NASSELLAE H. C. Greene & Cumm. Mycologia 50:35. 1958.

Uredinia on adaxial leaf surface, yellow; spores (22-)24-30 (-33) x (18-)19-23(-25)μ, ellipsoid, obovoid, or broadly ellipsoid, wall 1(-1.5)μ thick, colorless or pale yellowish, echinulate, germ pores 7-10, scattered, obscure.

Hosts and distribution: Nassella pubiflora (Trin. & Rupr.) Desv.: Bolivia.

39. UREDO TOETOE Cunn. Trans. N. Zealand Inst. 55:41. 1924.

Uredinia amphigenous and on culms, cinnamon-brown; spores (24-)27-32(-34) x (20-)23-28µ, mostly broadly ellipsoid, wall (2-)2.5-3µ thick, cinnamon-brown or darker, echinulate, germ pores 7-12, scattered.

Hosts and distribution: Arundo conspicua Forst. f.: New Zealand.

GROUP VII: uredinia aparaphysate, spores verrucose, germ
           pores equatorial.

40. UREDO ISCHAEMI Syd. & Butl. Ann. Mycol. 5:509. 1907.

Uredinia amphigenous, in striiform leaf spots, pale yellowish brown; spores 16-22 x 13-17µ, ovoid or globoid, wall 1-1.5µ thick, colorless, "verruculose or verruculose-echinulate", germ pores 8-10, obscure.

Hosts and distribution: Ischaemum timorensis Kunth: India.

GROUP VIII: uredinia aparaphysate, spores verrucose, germ
            pores scattered.

41. UREDO CHASCOLYTRI Diet. & Neger Bot. Jahrb. 27:15. 1899.

Uredinia hypophyllous, ochraceous; spores 23-32 x 20-25µ, ovoid, ellipsoid, or nearly globoid, wall 1.5-2µ thick, densely verruculose, pale yellow, germ pores 6-8, scattered.

Hosts and distribution: Chascolytrum trilobum (Nees) E. Desv.: Chile.

42. UREDO SUSICA Maire in Maire & Werner Mem. Soc. Sci. Nat. Maroc 45:75. 1937.

Uredinia amphigenous, linear, rusty brown; spores 21-26 x 18-25µ, globoid or nearly so, wall 2µ thick, golden, finely and densely verruculose, germ pores 3-5, irregularly scattered, hilum not conspicuous (hence the position of the pores is difficult to determine. The small number might suggest that the pores actually are equatorial.).

Hosts and distribution: Dichanthium annulatum (Forssk.) Stapf: Morocco.

Excluded Species

<u>Chrysomyxa bambusae</u> Teng Sinensis 9:226. 1938.

Not a rust fungus.

<u>Kweilingia bambusae</u> (Teng) Teng Sinensis 11:124. 1940.

Not a rust fungus.

<u>Puccinia campulosi</u> Thuem. Bull. Torrey Bot. Club 6:215. 1878.

No specimen exists nor has a rust fungus on the genus <u>Ctenium</u> been found in searches of the grass collections of the U.S. National Herbarium and the Field Museum.

<u>Puccinia gracilenta</u> Syd. & Butl. Ann. Mycol. 10:263. 1912.

Not a rust fungus.

<u>Puccinia neoporteri</u> Hino & Katum. J. Japan. Bot. 40:89. 1965.

  <u>Puccinia porteri</u> Hino & Katum. Bull. Fac. Agr. Yamaguti Univ. 11:33. 1960, not Peck 1874.

The type (PUR) is a very meager specimen and the teliospores have germinated and collapsed. This, together with the unidentifiable host material, make it relatively certain that the species can never be recognized.

<u>Puccinia poae-aposeridis</u> Gaeum. & Poelt. Z. Phytopathol. 37:346. 1960.

The host is a species of <u>Carex</u> and the fungus <u>Puccinia dioicae</u> Magn. <u>vel</u>. <u>aff</u>.

<u>Puccinia pseudophakopsora</u> Speg. An. Mus. Nac. B. Aires 31:31. 1922.

Not a rust fungus.

<u>Puccinia sasae</u> Kusano Bull. Coll. Agr. Tokyo Imp. Univ. 8:9. 1908.

Repeated efforts have failed to locate the type and hence there is doubt that it exists. Until the type is found, it is impossible to recognize the species, if indeed, it differs from <u>Puccinia longicornis</u>.

<u>Sphaerophragmium sorghi</u> Batista & Bezerra Nova Hedw. 2:347. 1960.

Not a rust fungus.

<u>Uredo danthoniae</u> P. Henn. Hedwigia 41:211. 1902.

Apparently no type exists and it is doubtful if the species can be recognized.

Uredo isachnes Sawada Taiwan Agr. Res. Inst. Rept. 87:45. 1944.

A nomen nudum.

Uromyces agropyri Barcl. J. Asiatic Soc. Bengal 60:212. 1891.

The type has not been found and probably does not exist. No subsequent collections have been reported. A misidentification of host is suspected.

Uromyces scleropoae Baudys & Picb. in Picbauer Bull. Inst. Jard. Bot. Univ. Belgrade 1:62. 1928.

Tye type has no rust fungus present.

531

536

pseudocesatii Cumm., 398, _400_
pseudo-myuri Kleb., 317
pseudophacopsora Speg., 524
pugiensis Tai, _130_
pumilae-coronatae Paul, 141
purpurea Cke., _179_
pusilla Syd., _116_
pustulata Arth., 367
puttemansii P. Henn., _234_
pygmaea Eriks., _154_, _161_, 516
pygmaea var. ammophilina (Mains) Cumm. & Greene, _156_
pygmaea var. angusta Cumm. & Greene, _159_
pygmaea var. chisosana Cumm., 163
pygmaea var. major Cumm. & Greene, _157_
pygmaea var. minor Cumm. & Greene, _158_
rangiferina Ito, 147
recondita Rob., 286, 303, 309, 310, 316, 319, _320_, 487
redfieldiae Tracy, _394_
rottboelliae Syd., _224_
rubigo-vera Wint., 286, 320, 487
rufipes Diet., _115_
sagittata Long, 395
saltensis Cumm., _185_
saltensis Cumm. var. faldensis Greene & Cumm., _186_
sardonensis Gaeum., 322
sanguinea Diet. & Atkinson, 179
sasae Kus., 524
sasicola Hara ex Hino & Katum., _139_
scaber Ell. & Ev., 290
scarlensis Gaeum., 321
schedonnardi Kell. & Swing., 307, _361_
schismi Bub., 317
schmidtiana Diet., 311
schoenanthi Cumm. & Guyot, _401_
schottmuelleri P. Henn., 66
scillae-rubrae Cruchet, 294
scleropogonis Cumm., _345_, 346
secalina Grove, 321
sertata Preuss, 141
sesleriae Reich., _212_
sesleriae-coeruleae Ed. Fisch., 210
sessilis Schneider, _311_
sessilis var. minor Cumm., _313_
setariae Diet. & Holw., _409_
setariae-longisetae Wakef. & Hansf., _272_
setariae-forbesianae Tai, _200_
setariae-viridis Diet., _224_
seymouriana Arth., _215_
sierrensis Cumm., _354_
simillima Arth., 182
simplex (Koern.) Eriks. & Henn., 317
simulans (Peck) Barth., 270
sinica H. Syd., _344_
sinkiangensis Wang, 275
sonorica Cumm. & Husain, _133_

sonorica var. minor Cumm. & Husain, 134
sorghi-halepensis Speg., 179
sorghi Schw., 260
sparganioides Ell. & Barth., 218
spegazziniella Sacc. & Trott., 361
spicae-venti Buch., 168
sporoboli Arth., 202
sporoboli var. robusta Cumm. & Greene, 203
stakmanii Presley, 262
stapfiolae Mundk. & Thirum., 151
stenotaphri Cumm., 86
stichosora Diet., 115
stipae Arth., 281, 368, 370, 380
stipae (Opiz) Hora, 370
stipae var. stipae-sibiricae (Ito) H.C. Greene & Cumm.,   369
stipae var. stipina (Tranz.) H.C. Greene & Cumm.,   370
stipae-sibiricae Ito,   369
stipina Tranz.,   370
straminis Fuckel, 151
straminis Fuckel f. simplex Koern.,   317
striatula Peck,   311
striiformis Westend.,   151, 194
striiformis var. dactylidis Manners,   152
subalpina Lagerh., 321
subandina Speg., 210
subcentripora Arth. & Cumm.,   207
subdigitata Arth. & Holw.,   145
subdiorchidioides P. Henn.,   328
subglobosa Speg.,   361
sublesta Cumm.,   97
subnitens Diet.,   404, 414, 503, 509
substerilis Ell. & Ev.,   290
substerilis var. oryzopsidis H.C. Greene & Cumm., 291
substerilis var. scribneri H.C. Greene & Cumm., 292
substriata Ell. & Barth.,   237, 519
substriata var. imposita (Arth.) Ramachar & Cumm.,   237
substriata var. indica Ramachar & Cumm.,   239
substriata var. insolita (Syd.) Ramachar & Cumm.,   240
substriata var. penicillariae (Speg.) Ramachar & Cumm.,   239
subtilipes Speg.,   343
sydowiana Diet.,   220
symphyti-bromorum F. Muell.,   320
taiwaniana Hirat. f. & Hashioka,   231
takikibicola Y. Morimoto,   373
taminensis Gaeum.,   315
tangkuensis Liou & Wang,   405
tarri Cumm. & Husain,   416
tenella Hino & Katum.,   429
tepperi F. Ludwig,   126
tetuanensis Guyot,   318
thalassica Speg.,   414
thalictri-distichophylli Ed. Fisch. & Mayor,   321
thalictri-koeleriae Gaeum.,   321
thalictri-poarum Ed. Fisch. & Mayor,   166
themedae Hirat.,   332